# SOLID STATE PHYSICS

VOLUME 36

## Contributors to This Volume

A. MacKinnon

A. Miller

P. S. Peercy

G. A. Samara

K. S. Singwi

M. P. Tosi

D. Weaire

# SOLID STATE PHYSICS

## Advances in Research and Applications

*Editors*

**HENRY EHRENREICH**

*Division of Applied Sciences*
*Harvard University, Cambridge, Massachusetts*

**FREDERICK SEITZ**

*The Rockefeller University, New York, New York*

**DAVID TURNBULL**

*Division of Applied Sciences*
*Harvard University, Cambridge, Massachusetts*

VOLUME 36

1981

ACADEMIC PRESS

*A Subsidiary of Harcourt Brace Jovanovich, Publishers*

NEW YORK LONDON TORONTO SYDNEY SAN FRANCISCO

ACADEMIC PRESS, INC.
111 Fifth Avenue, New York, New York 10003

*United Kingdom Edition published by*
ACADEMIC PRESS, INC. (LONDON) LTD.
24/28 Oval Road, London NW1 7DX

Library of Congress Catalog Card Number: 55-12200

ISBN 0-12-607736-3

PRINTED IN THE UNITED STATES OF AMERICA

81 82 83 84 9 8 7 6 5 4 3 2 1

# Contents

## Contributors to Volume 36

Numbers in parentheses indicate the pages on which the authors' contributions begin.

A. MACKINNON, *Physikalisch-Technische Bundesanstalt, Braunschweig, Federal Republic of Germany* (119)

A. MILLER,* *Physics Department, North Texas University, Denton, Texas 76203* (119)

P. S. PEERCY, *Sandia National Laboratories, Department 5130, Albuquerque, New Mexico 87185* (1)

G. A. SAMARA, *Sandia National Laboratories, Department 5130, Albuquerque, New Mexico 87185* (1)

K. S. SINGWI, *Department of Physics and Astronomy, Northwestern University, Evanston, Illinois 60201* (177)

M. P. TOSI, *International Center for Theoretical Physics, Trieste, Italy* (177)

D. WEAIRE,† *Physics Department, Heriot-Watt University, Edinburgh, Scotland* (119)

* Present address: Royal Signals and Radar Establishment, Great Malvern, Worcestershire, United Kingdom.

† Present address: Department of Experimental Physics, University College, Dublin, Ireland.

# Preface

In the first article of this volume Samara and Peercy review the theory of "soft-mode" phase transitions and the research on the effects of hydrostatic pressure of these transitions. Displacive ferroelectric, antiferroelectric, and antidistortive transitions are given special emphasis. This article constitutes the first extensive treatment of such transitions in this publication since that of Kanzig in Volume 4.

The electrical, optical, and structural properties of the elemental and binary compound semiconductors have been discussed extensively in a number of articles here. Interest in the more complex tetrahedrally coordinated semiconducting compounds, e.g., chalcopyrite, often composed of three or more elements, has been developing recently. The atomic and band structures of these compounds are reviewed by Miller, MacKinnon, and Weaire in the second article of this volume.

The final article by Singwi and Tosi summarizes some of the formalisms and approximations which can be used to estimate correlation effects among electrons in simple metals for appropriate ranges of electron densities, when band effects are relatively unimportant. Much of the discussion revolves around calculations of the dielectric and polarization functions. Applications are made to a variety of problems including pseudopotentials, phonon dispersion, surfaces, X-ray scattering, and positron annihilation. There are numerous articles in this serial publication devoted to one or another aspect of the matters discussed here. Among these are the reviews by Hedin and Lundquist (Volume 23), Joshi and Rajagopal (Volume 22), Lang (Volume 28), and Rice (Volume 32).

HENRY EHRENREICH
FREDERICK SEITZ
DAVID TURNBULL

## Supplements

Supplement 1: T. P. Das and E. L. Hahn
Nuclear Quadrupole Resonance Spectroscopy, 1958

Supplement 2: William Low
Paramagnetic Resonance in Solids, 1960

Supplement 3: A. A. Maradudin, E. W. Montroll, G. H. Weiss, and I. P. Ipatova, Theory of Lattice Dynamics in the Harmonic Approximation, 1971 (Second Edition)

Supplement 4: Albert C. Beer
Galvanomagnetic Effects in Semiconductors, 1963

Supplement 5: R. S. Knox
Theory of Excitons, 1963

Supplement 6: S. Amelinckx
The Direct Observation of Dislocations, 1964

Supplement 7: J. W. Corbett
Electron Radiation Damage in Semiconductors and Metals, 1966

Supplement 8: Jordan J. Markham
*F*-Centers in Alkali Halides, 1966

Supplement 9: Esther M. Conwell
High Field Transport in Semiconductors, 1967

Supplement 10: C. B. Duke
Tunneling in Solids, 1969

Supplement 11: Manuel Cardona
Optical Modulation Spectroscopy of Solids, 1969

Supplement 12: A. A. Abrikosov
An Introduction to the Theory of Normal Metals, 1971

Supplement 13: P. M. Platzman and P. A. Wolff
Waves and Interactions in Solid State Plasmas, 1973

Supplement 14: L. Liebert, Guest Editor
Liquid Crystals, 1978

Supplement 15: Robert M. White and Theodore H. Geballe
Long Range Order in Solids, 1979

SOLID STATE PHYSICS, VOLUME 36

# The Study of Soft-Mode Transitions at High Pressure*

G. A. SAMARA
AND
P. S. PEERCY

*Sandia National Laboratories,† Albuquerque, New Mexico*



* This work was supported by the U.S. Department of Energy under Contract DE-AC04-76DP00789.

† A Department of Energy facility.




ISBN 0-12-607736-3

## I. Introduction

High-pressure research has contributed much to our current understanding of the properties of matter. Perhaps there is no one area in which this is truer than that of phase transitions. High-pressure studies have improved our understanding of the properties of materials near phase transitions and the mechanisms of certain types of transitions. In addition, these studies have led to the discovery of a multitude of new phase transitions that can be induced by pressure alone. Examples are many, ranging from subtle electronic transitions to a wide variety of structural transitions. In this review we shall restrict our discussion to the effects of hydrostatic pressure on structural phase transitions and specifically those transitions involving "soft" lattice modes.

The concept of soft modes[1,2] has in recent years played an important role in our understanding of the static and dynamic properties of many types of structural phase transitions in solids. By a soft mode we simply mean a normal mode of vibration whose frequency decreases (and thus the lattice literally softens with respect to this mode) and approaches zero as the transition point is approached. From a lattice dynamical viewpoint, the limit of stability of a crystal lattice is approached as the frequency of any one mode decreases and approaches zero, because once the atoms are displaced in the course of the particular vibration, there is no restoring force to return them to their original positions, and they assume new equilibrium positions determined by the symmetry (eigenvector) of this soft mode.

[1] W. Cochran, *Phys. Rev. Lett.* **3**, 412 (1959).
[2] W. Cochran, *Adv. Phys.* **19**, 387 (1960); **20**, 401 (1961).

The soft mode concept was introduced[1,2] and beautifully confirmed[3,4] for displacive ferroelectric transitions. Since then soft modes have been discovered in a variety of classes of materials.[5,6] In ferroelectrics similar ions in adjacent unit cells undergo identical displacements but with negative and positive ions undergoing opposite displacements and thereby producing a macroscopic polarization. The soft mode is thus necessarily a long wavelength (or zone center, $\mathbf{q} = 0$) transverse optic (TO) phonon. In general, however, the soft mode can be either an optic or acoustic phonon of either long or short wavelength. Furthermore, in more complicated cases the soft mode can be the coupled response of two or more fundamental excitations. Examples of these various possibilities have been examined, and several will be discussed in this review. The reader is also referred to the literature for detailed reviews of the field.[5–8]

It is generally accepted that the decrease in the soft-mode frequency as the transition is approached from the high-symmetry phase results from cancellation between competing forces. It is not difficult to imagine that pressure can significantly influence the balance between such forces and thereby strongly influence, or even induce, soft mode behavior. Indeed there is now a body of literature showing this to be the case. In reviewing the highlights of this literature, we shall emphasize the value of pressure as a complementary parameter to temperature for investigating soft-mode systems. Furthermore, it will be shown that in many cases pressure is an essential variable to the occurrence and/or understanding of these transitions. Examples will be given to illustrate various facets of the general soft-mode problem and the usefulness of pressure in investigating soft-mode systems.

Ferroelectric, antiferroelectric, and other antiferrodistortive structural phase transition have received the most attention and are the best understood soft-mode systems. The literature on high-pressure studies of soft-mode transitions reflects this emphasis. Thus the bulk of the present review will necessarily deal with these types of transitions. Our emphasis will be on experimental results and their interpretation in terms of relevant theory. It is not our intent to give a complete summary of all the

[3] A. S. Barker and M. Tinkham, *Phys. Rev.* [2] **125,** 1527 (1962).

[4] R. A. Cowley, *Phys. Rev. Lett.* **9,** 159 (1962); *Phys. Rev.* [2] **134,** A981 (1964).

[5] E. Samuelsen, E. Anderson, and J. Feder, eds., "Structural Phase Transitions and Soft Modes." Universitetsforlaget, Oslo, Norway, 1971.

[6] J. F. Scott, *Rev. Mod. Phys.* **46,** 83 (1974).

[7] R. Blinc and B. Zeks, "Soft Modes in Ferroelectrics and Antiferroelectrics." Am. Elsevier, New York, 1974.

[8] *Proc. Int. Conf. on Low Lying Vibrational Modes Their Relationship Supercond. and Ferroelectr., Ferroelectrics* **16,** 1 (1977).

available literature. Rather we have selected examples that illustrate important features or principles. Special attention will be given, insofar as is possible, to systematic trends and generalizations of behavior, for it is such trends that lead to greater understanding.

The effects of hydrostatic pressure on ferroelectric transitions were reviewed several years ago by one of the present authors (GAS).[9] Since then a great deal of new work has been done in the field, and new tools (e.g., Raman, Brillouin, and neutron scattering) have been introduced that permit closer examination of the microscopic details. Interesting and important new results have emerged in the study of both ferroelectric and nonferroelectric transitions. In this article we review these new results and assess their contribution to the understanding of the phenomena involved. We deal exclusively with the effects of hydrostatic pressure on transition properties. Uniaxial stress is also important in the study of structural phase transitions and soft modes, especially in connection with stress-induced soft-mode behavior[10,11] and tricritical points.[12] We shall not deal with these effects.

The review is organized as follows. We begin in Section II with a brief theoretical background giving some of the essential results needed for the interpretation of the results presented later in the review. Section III deals with displacive ferroelectric transitions with emphasis on the behavior of the soft ferroelectric mode and the vanishing of the transition at high pressure. Section IV reviews results on displacive antiferroelectric and antiferrodistortive transitions and contrasts their pressure behavior with that of the ferroelectric transitions. This contrast along with results of lattice dynamical calculations suggest that there is a reversal in the roles of the short-range and the Coulomb interactions in the lattice dynamics of the two cases. In Section V we discuss results on hydrogen-bonded ferroelectrics and antiferroelectrics of the $KH_2PO_4$ family. In these materials the important feature is the coupling between proton and phonon modes, the pressure investigations have been crucial to the understanding of the phenomena involved. Among the most important pressure results are (*a*) the resolution of long-standing questions concerning the nature of the soft mode, and (*b*) the vanishing of the ordered state. Section VI summarizes pressure results on a few interesting, but unrelated soft-mode transitions, specifically, transitions in $TeO_2$, the rare earth pentaphosphates, and the A-15 superconductors. In many crystals as the

[9] G. A. Samara, *in* "Advances in High Pressure Research" (R. S. Bradley, ed.), Chapter 3. Academic Press, New York, 1969.

[10] H. Uwe and T. Sakudo, *Phys. Rev. B: Solid State* [3] **15,** 337 (1977), and references therein.

[11] H. Uwe and T. Sakudo, *Phys. Rev. B: Solid State* [3] **13,** 271 (1976).

[12] A. Aharony and A. D. Bruce, *Phys. Rev. Lett.* **33,** 427 (1974).

frequency of a soft mode decreases on approaching the transition, this mode interacts with other degrees of freedom of the crystal. Section VII reviews the effects of pressure on mode couplings in a number of crystals. Finally Section VIII provides concluding remarks and suggestions for future work.

Throughout this review the pressure unit used is the kilobar (kbar). This is a cgs unit: 1 kbar = $10^9$ dynes $cm^{-2}$. It is simply related to the SI pressure unit, the pascal (Pa): 1 Pa = 1 N $m^{-2}$. 1 kbar = $10^8$ Pa = 0.1 gigapascal (GPa). In the text the following abbreviations are used: FE for ferroelectric; AFE, antiferroelectric; PE, paraelectric; $T_c$, transition temperature; $T$, temperature; and $p$, pressure.

## II. Theoretical Background

### 1. Conditions for Lattice Stability and the Soft-Mode Concept

The conditions for lattice stability were discussed from a lattice dynamical viewpoint by Born and co-workers.[13] For a structure to be stable, the lattice must be stable with respect to three distinct types of deformations: long wavelength homogeneous deformations (zone-center acoustic modes), long wavelength inhomogeneous deformations (zone-center optic modes), and short wavelength deformations (e.g., zone-boundary modes). Hence, the general condition for the stability of a particular structure in terms of lattice dynamics is that all normal modes of vibration have real and finite frequencies. The limit of stability is approached when the frequency of any mode decreases and approaches zero; if the frequency of the mode is zero, once atoms are displaced in the course of that particular vibration, there is no restoring force to return them to their original equilibrium position. The atoms then assume new equilibrium positions determined by the symmetry of the mode, and the structure of the crystal changes. Structural transitions accompanied by these three types of deformations have been observed and will be discussed in the present review.

This connection between lattice stability and the behavior of the normal modes of vibration appears not to have received much attention[14] until

[13] See, e.g., M. Born and K. Huang, "Dynamical Theory of Crystal Lattices." Oxford Univ. Press, London and New York, 1954.

[14] Apparently this connection was first made by C. V. Raman and T. M. K. Nedungadi [*Nature* (*London*) **145**, 147 (1940)] in their spectroscopic study of the phase transition in quartz.

1960 when the important work of Cochran[1,2] appeared. This work dealt specifically with displacive ferroelectric (FE) transitions and introduced the concept of "soft" phonon modes. The concept was confirmed around the mid-1960s for several FE crystals. Since then soft mode properties in a variety of crystals exhibiting structural phase transitions have been investigated in detail using a variety of experimental tools (primarily inelastic scattering of light and neutrons).[5,6] These studies have led to detailed descriptions of the static and dynamic properties of several displacive phase transitions. By displacive phase transitions we refer to those transitions accompanied by small, and in principle continuous, spontaneous displacements of some of the ions within the unit cell from their equilibrium positions in the high-symmetry phase to yield the structure of the lower-symmetry phase. These displacements, which can be thought of as the order parameter of the transition (e.g., they are directly related to the spontaneous polarization in ferroelectrics), increase and ultimately saturate as the solid moves away from the transition.

Much of the early theoretical and experimental work on soft modes was aimed at understanding temperature-induced displacive transitions in ferroelectrics. These transitions are associated with long wavelength ($\mathbf{q} = 0$) transverse optic phonons. However, the soft mode concept is more general than that. The soft mode can be an acoustic or optic phonon of long or short wavelength, or it can even be a coupled response of the crystal. Considered from a lattice dynamical viewpoint, the decrease and ultimate vanishing of the frequency of the soft mode as the transition is approached is caused by cancellation between competing forces. This cancellation can be induced by changes in composition, temperature, or other external fields. In particular, the application of pressure can significantly influence the balance between competing forces and thereby strongly influence, or induce, soft-mode behavior. The emphasis of this review is on these pressure effects, and we shall deal with a variety of examples. However, before doing so it is necessary to summarize some theoretical considerations that will be needed for the interpretation of the results. Important among these considerations are results from Landau's theory of phase transitions and from the theories of harmonic and anharmonic lattice dynamics.

## 2. Landau's Theory for Continuous Phase Transitions

A general theory for continuous phase transitions, based on symmetry arguments, has been given by Landau.[15] According to this treatment, a

[15] See, e.g., L. D. Landau and E. M. Lifshiftz, "Statistical Physics." Addison-Wesley, Reading, Massachusetts, 1958.

continuous, or second-order, transition can occur between two states only if certain symmetry conditions are satisfied for the two phases. If the system changes continuously, a density distribution $\rho_0(\mathbf{r})$ invariant under the symmetry operations of the high-symmetry phase can be written as

$$\rho(\mathbf{r}) = \rho_0(\mathbf{r}) + \Delta\rho(\mathbf{r}) \tag{2.1}$$

in the low-symmetry phase. Landau assumed that $\rho(\mathbf{r})$ must be a subgroup of $\rho_0(\mathbf{r})$ and the change in $\rho_0(\mathbf{r})$ occurs for a single irreducible representation. If the symmetry change involves only a one-dimensional representation $g(\mathbf{r})$, then $\Delta\rho(\mathbf{r})$ can be written as

$$\Delta\rho(\mathbf{r}) = \eta g(\mathbf{r}), \tag{2.2}$$

where $\eta$ is the order parameter of the transition.

It is further assumed that near the transition the free energy $\varphi(p, T, \eta)$ can be expanded in a power series in $\eta$,

$$\varphi(p, T, \eta) = \varphi_0(p, T) + a\eta + \tfrac{1}{2}A(p, T)\eta^2 + \tfrac{1}{3}B(p, T)\eta^3 + \tfrac{1}{4}C(p, T)\eta^4 + \cdots, \tag{2.3}$$

where the expansion coefficients are functions of temperature $T$ and pressure $p$. If the transition is continuous, the terms containing odd powers in $\eta$ must vanish by symmetry, whereupon Eq. (2.3) reduces to

$$\varphi(p, T, \eta) = \varphi_0(p, T) + \tfrac{1}{2}A(p, T)\eta^2 + \tfrac{1}{4}C(p, T)\eta^4 + \tfrac{1}{6}D(p, T)\eta^6 + \cdots. \tag{2.4}$$

For $D > 0$, Eq. (2.4) describes a first-order transition if $C < 0$, a second-order transition if $C > 0$, and a critical point (where a line of first-order transitions goes continuously into a line of second-order transitions) at $C = 0$. The equilibrium value $\eta_0$ of $\eta$ must satisfy the conditions $(\partial\varphi/\partial\eta)_{\eta_0} = 0$ and $(\partial^2\varphi/\partial\eta^2)_{\eta_0} > 0$. For $C > 0$, a second-order transition thus occurs when $A(p, T) = 0$, with the state $\eta = 0$ stable for $A > 0$, whereas the state $\eta \neq 0$ is stable for $A < 0$.

For an FE transition the order parameter is the spontaneous polarization $P$. The free energy expansion in terms of the polarization and stress $X$ thus becomes[16]

$$\varphi(X, T, P) = \varphi_0(X, T) + \tfrac{1}{2}s_{ijkl}X_{ij}X_{kl} + a_{ijk}X_{ij}P_k + \tfrac{1}{2}q_{ijkl}X_{ij}P_kP_l + \tfrac{1}{2}\gamma_{ij}P_iP_j + \tfrac{1}{3}\omega_{ijk}P_iP_jP_k + \tfrac{1}{4}\xi_{ijkl}P_iP_jP_kP_l + \cdots, \tag{2.5}$$

where repeated indices are to be summed over the $x$, $y$, and $z$ coordinates. In Eq. (2.5), $s$, $a$, and $q$ are the elastic compliance, piezoelectric, and

[16] F. Jona and G. Shirane, "Ferroelectric Crystals." Macmillan, New York, 1962, and references therein.

electrostrictive tensors, respectively. If we restrict the discussion to crystals that are allowed by symmetry to undergo continuous transitions and are nonpiezoelectric ($a = 0$) in the paraelectric (PE) phase, and if the applied pressure is hydrostatic (i.e., $X_{ij} = \delta_{ij}p$) and the FE axis of the crystal is chosen to be the $z$ axis (i.e., $P_1 = P_2 = 0$ and $P_3 = P$), then $\varphi(X, T, P)$ becomes

$$\varphi(p, T, P) = \varphi_0(p, T) + \tfrac{1}{2}Sp^2 + \tfrac{1}{2}(Qp + \gamma)P^2 + \tfrac{1}{4}\xi P^4 + \tfrac{1}{6}\zeta P^6 + \cdots, \tag{2.6}$$

where

$$S = \sum_{i,j=1}^{3} s_{ij} \quad \text{and} \quad Q = \sum_{i=1}^{3} q_{ii} \tag{2.7}$$

in the contracted notation. For notational convenience, we define

$$\alpha(p, T) = Qp + \gamma. \tag{2.8}$$

The condition $(\partial\varphi/\partial P)_{P_0} = 0$ yields

$$(\alpha + \xi P_0^2 + \zeta P_0^4)P_0 = 0 \tag{2.9}$$

for the equilibrium value $P_0$ of $P$. Equation (2.9) has a solution $P_0 = 0$ and a solution with $P_0 \neq 0$. The stable solution must satisfy the condition $(\partial^2\varphi/\partial P^2)_{P_0} > 0$; thus the solution $P_0 = 0$ is stable for $\alpha > 0$ and corresponds to the PE phase. If there exists a solution of Eq. (2.9) for some $T$ and $p$ with $P_0 \neq 0$ for which $(\partial^2\varphi/\partial P^2)_{P_0} > 0$, then the system will undergo a transition to an FE phase.

This transition occurs when $\alpha(T, p) = 0$ if $\zeta > 0$. If it is assumed that the temperature and pressure dependences of $\zeta$ are negligible in the vicinity of the temperature $T_0$ and pressure $p_0$ where $\alpha(T, p)$ vanishes, and that $\alpha(T, p)$ vanishes, and that $\alpha(T, p)$ may be expanded about this point as

$$\alpha(T, p) = \alpha(T_0, p_0) + \gamma_0(T - T_0) + Q(p - p_0) + \cdots. \tag{2.10}$$

then $\varphi(p, T, P)$ becomes

$$\partial(p, T, P) = \partial_0(p, T) + \tfrac{1}{2}Sp^2 + \tfrac{1}{2}[Q(p - p_0) + \gamma_0(T - T_0)]P^2 + \tfrac{1}{4}\xi P^4 + \tfrac{1}{6}\zeta P^6 + \cdots, \tag{2.11}$$

where only the leading terms in $\alpha(T, p)$ are retained.

When the effect of a driving field $E$ is considered, one has

$$(\partial\varphi/\partial P) = E, \tag{2.12}$$

and the inverse susceptibility $\chi^{-1} \equiv \partial E/\partial P$ is

$$\chi^{-1} = \partial^2\varphi/\partial P^2. \tag{2.13}$$

Near $T = T_0$ and $p = p_0$, Eqs. (2.11) and (2.13) thus yield

$$\chi^{-1} = \gamma_0(T - T_0) + Q(p - p_0) \tag{2.14}$$

which displays Curie–Weiss behavior for the dielectric constant $\varepsilon$,

$$\varepsilon = 1 + 4\pi\chi \cong 4\pi\chi, \tag{2.15}$$

with either temperature at constant pressure $p_0$ or pressure at constant temperature $T_0$.

In the PE phase the constant pressure susceptibility $\chi_0$, given by Eq. (2.14) at $p = p_0$,

$$\chi_p^{-1} = \gamma_0(T - T_0), \tag{2.16}$$

yields

$$\varepsilon_p \cong 4\pi\chi_p = C/(T - T_0), \tag{2.17}$$

where the Curie constant $C = 4\pi/\gamma_0$. Similarly, the constant temperature susceptibility $\chi_T$ is

$$\chi_T^{-1} = Q(p - p_0), \tag{2.18}$$

with

$$\varepsilon_T \simeq C^*/(p - p_0), \tag{2.19}$$

where the modified Curie constant $C^* = 4\pi/Q$. The relationship between $C$ and $C^*$ is obtained by considering the change in $\alpha(p, T)$ with pressure. Since the transition is defined by $\alpha(p_0, T_0) = 0$, one has

$$\begin{aligned} d\alpha(p_0, T_0)/dp_0 &= \partial\alpha/\partial p_0 + (\partial\alpha/\partial T_0)(dT_0/dp_0) = 0 \\ &= -Q - \gamma_0(dT_0/dP_0), \end{aligned} \tag{2.20}$$

so that

$$dT_0/dp_0 = -Q/\gamma_0. \tag{2.21}$$

Comparing $C$ and $C^*$ one notes that the Curie constants are related by the pressure dependence of the transition temperature as

$$C/C^* = Q/\gamma_0 = -dT_0/dp_0. \tag{2.22}$$

A similar analysis yields Curie–Weiss behavior in the FE phase. Neglecting terms of order higher than $P^4$ in the free energy, Eq. (2.9) is solved for the nonzero value of $P_0$ as

$$P_0^2 = -\alpha/\xi, \tag{2.23}$$

which yields

$$\chi_{(T<T_0)}^{-1} = -2\alpha, \tag{2.24}$$

The temperature dependence of the dielectric constant for $T < T_0$ at constant $p = p_0$ is thus given by

$$\varepsilon \cong C'/(T_0 - T) \tag{2.25}$$

with $C' = C/2$. Examination of the pressure dependence of $\chi^{-1}$ at constant temperature yields analogous results.

Cochran[1,17] presented a lattice dynamical formulation of Landau's theory for continuous phase transitions. In this treatment he restricted the assumption that $\Delta\rho(\mathbf{r})$ belongs to a single irreducible representation to the change being a single mode of this irreducible representation. Thus, in terms of lattice dynamics, this implies that only *one* normal mode of a certain polarization and wave vector is involved in a second-order phase transition. The frequency of this mode vanishes precisely at the transition, and there is a one-to-one correspondence between the symmetry properties of the eigenvector of the soft mode and the static ionic displacements accompanying the transition.

For a first-order transition, on the other hand, there are no symmetry requirements; all that is necessary is that the free energies of the two phases be equal at the transition. If there is a "soft" mode associated with the transition, the frequency of this mode will remain finite at the transition. In some cases, while the transition is in fact first order, it appears to be nearly second order from a lattice dynamical viewpoint, and many of the considerations obtained from treatments of second-order phase transitions can be applied.

To understand soft mode transitions and the usefulness of high pressure in studying such transitions, it is necessary to examine the results of lattice dynamical theory and inquire into the conditions that can lead to soft mode behavior and the onset of a phase transition. Some insight can be gained by first considering a harmonic model.

## 3. Harmonic Model

Although many of the properties that we shall deal with later arise from anharmonic interactions, we first discuss the harmonic case and treat the anharmonic effects as perturbations on the harmonic results. The harmonic approximation provides a simple description of the atomic motions in the crystal and yields exact results not possible when more complicated potentials are considered.

In the harmonic approximation, periodic boundary conditions are

[17] W. Cochran, *in* "Structural Phase Transitions and Soft Modes" (E. Samuelsen, E. Andersen, and J. Feder, eds.), p. 1. Universitetsforlaget, Oslo, Norway, 1971.

assumed and the potential energy of the lattice $\Phi$ (in the adiabatic approximation) is written in terms of the atomic displacements $u$ as

$$\Phi = \frac{1}{2} \sum_{l_1 k_1 \alpha} \sum_{l_2 k_2 \beta} \Phi\alpha\beta(l_1k_1, l_2k_2) u_\alpha(l_1k_1) u_\beta(l_2k_2), \tag{3.1}$$

where the indices $k$ and $l$ designate the different atoms (nuclei) and unit cells, respectively, and

$$\Phi_{\alpha,\beta}(l_1 k_1, l_2 k_2) = \frac{\partial^2\Phi}{\partial x_\alpha(l_1 k_1)\, \partial x_\beta(l_2 k_2)} \tag{3.2}$$

is an effective force constant. Here $x_\alpha(l_1 k_1)$ . . . designate the rectangular coordinates ($\alpha$, $\beta$ = 1, 2, 3). The equations of motion have the form

$$m_k\left[\frac{\partial^2}{\partial t^2} u_\alpha(l_1k_1)\right] = -\sum_{l_2k_2\beta} \Phi_{\alpha\beta}(l_1k_1, l_2k_2) u_\beta(l_2k_2), \tag{3.3}$$

where $m_k$ is the mass of atom $k$. Their solutions yield the normal modes of vibration of the lattice. To relate the displacements $u(l_ik_i)$ to the motions associated with the normal modes, normal mode coordinates $Q(\mathbf{q}j)$ are defined by

$$u_\alpha(lk) = \sum_{(\mathbf{q}j)} \left[\frac{\hbar}{2\omega_0(\mathbf{q}j) m_k N}\right]^{1/2} \mathbf{e}_\alpha(k, \mathbf{q}\, j) Q(\mathbf{q}j) \exp[i\mathbf{q} \cdot \mathbf{r}(lk)]. \tag{3.4}$$

In Eq. (3.4) $\mathbf{q}$ is the wave vector ($q = 2\pi/\lambda$), $j$ designates the particular branch of the dispersion curves, $\omega_0(\mathbf{q}j)$ is the harmonic frequency of the mode ($\mathbf{q}\, j$), $N$ is the number of unit cells, $\mathbf{e}_\alpha(k, \mathbf{q}\, j)$ is the polarization vector of the normal mode, and $\mathbf{r}(l, k)$ is the position vector of atom $k$. Alternatively, each displacement $u_\alpha(lk)$ can be written as the sum of displacements as

$$u_\alpha(lk) = u_\alpha(k, \mathbf{q}j) \exp i[\mathbf{q} \cdot \mathbf{r}(lk) - \omega_0(\mathbf{q}j)t]. \tag{3.5}$$

For ionic crystals one chooses a dynamical model such as the modified rigid-ion model or the shell model. In these models the short-range interactions are usually considered to act between nearest neighbors only. The shell model[18] treats in a formal manner the effects of the electronic polarizabilities of the ions on the lattice vibrations by approximating each ion by a negative electron shell concentric with a positive ion core. The electronic polarizabilities, represented by the polarizabilities of the shells, modify both the short-range and the long-range interactions.

The equations of motion for such models are solved for the normal modes of vibration.[1,2] Among the particularly important modes for our

[18] B. G. Dick and A. W. Overhauser, *Phys. Rev.* [2] **112**, 90 (1958).

purposes are the long wavelength, or Brillouin zone-center (i.e., $\mathbf{q} = 0$), optic modes, especially the transverse optic (TO) modes; these modes have linear electric dipole moments in a polar crystal and make the primary contribution to the infrared polarization and static dielectric susceptibility. For the mode associated with an FE transition, positive and negative ions undergo opposite displacements along the polar axis. All ions of the same species in adjacent unit cells undergo identical displacements, which leads to a macroscopic net polarization. Thus the soft FE mode is by necessity a $\mathbf{q} = 0$ TO mode. We thus wish to examine the expression for the frequency of this mode.

The simplest case to consider which contains the important features of soft TO modes is that of a cubic diatomic ionic crystal with one (doubly degenerate) TO branch. For this case, solution of the equations of motion in the *harmonic approximation* for the $\mathbf{q} = 0$ TO and longitudinal optic (LO) mode frequencies $\omega_0$ yields[1,2]

$$\mu(\omega_0)^2_{\mathrm{TO}} = R_0' - \frac{4\pi(\varepsilon_\infty + 2)(Ze^*)^2}{9v}, \tag{3.6}$$

$$\mu(\omega_0)^2_{\mathrm{LO}} = R_0' - \frac{8\pi(\varepsilon_\infty + 2)(Ze^*)^2}{9v\varepsilon_\infty}, \tag{3.7}$$

where $\mu$ is the reduced mass for the ions, $\varepsilon_\infty$ the high frequency dielectric constant, $v$ the unit cell volume, $Ze^*$ the effective ionic charge, and $R_0'$ an effective short-range force constant which includes all of the appropriate short-range constants of the model. The second term on the right-hand sides of Eqs. (3.6) and (3.7) represents the contribution of the long-range Coulomb forces.

Note that $(\omega_0)_{\mathrm{TO}}$ is given by the difference between short-range and long-range forces. For ionic crystals such as NaCl the term $R_0'$ is about twice as large as the Coulomb term; $(\omega_0)_{\mathrm{TO}}$ is thus real and finite, and the crystal is stable with respect to this mode. Now consider a situation where the two forces are more nearly equal. As the Coulomb term increases and tends to cancel $R_0'$, $(\omega_0)_{\mathrm{TO}}$ decreases and tends to zero. The lattice thus becomes less stable with respect to the deformation associated with this mode. If the Coulomb term is larger than the short-range contribution, then $(\omega_0)^2_{\mathrm{TO}} < 0$ and the crystal is unstable in the harmonic approximation. This cancellation of forces leading to the decrease in $(\omega_0)_{\mathrm{TO}}$ does not imply any such cancellation in the case of $(\omega_0)_{\mathrm{LO}}$, which is given by the sum of the two terms.

In a strictly harmonic lattice, as assumed above, there is no interaction among phonons, and all phonon frequencies are temperature independent.

However, real crystals are never strictly harmonic; in fact they are highly anharmonic for cases that have a soft mode. We next examine some of the consequences of anharmonic interactions.

## 4. Anharmonic Interactions

Anharmonicities resulting from the interactions among the normal modes of vibration play an important role in the lattice dynamics of crystals. They are responsible for such crystal properties as thermal expansivity and temperature dependences of normal-mode frequencies. The temperature dependence of the frequency arises from two contributions. First, there is a pure-volume contribution associated with thermal expansion. Second, there is a volume-independent, or pure-temperature, contribution arising from higher-order anharmonic interactions. Isobaric measurements of the temperature dependence of phonon frequencies yield changes due to the combination of the two effects. However, measurements of both the pressure and temperature dependences of these frequencies along with a knowledge of the compressibility and thermal expansivity allow a separation of the pure-volume and pure-temperature contributions and, under suitable conditions, make it possible to determine the origin of the dominant anharmonic interaction.

The origin and sign of the pure-volume effect in ionic crystals can be intuitively understood. As the interionic distances decrease with decreasing temperature, the restoring forces between ions, and hence the frequencies, can be expected to increase. This is indeed found to be the case for normal ionic crystals. The origin of the pure-temperature effect, on the other hand, lies in the anharmonic vibrations of the ions about their equilibrium positions. To understand this effect, one must resort to the theory of anharmonic lattice dynamics. Unfortunately, detailed treatments of anharmonic lattice dynamics involve many-body effects and are very complex. Only approximate treatments which employ a perturbation expansion about a harmonic or quasi-harmonic basis have been given. These treatments have proved to be satisfactory for weakly anharmonic crystals. Extensions of these theories as well as other approaches (to be discussed later) have been used to treat soft modes.[19–21]

[19] R. A. Cowley, *Philos. Mag.* **11,** 673 (1965); *Adv. Phys.* **12,** 421 (1963).

[20] B. D. Silverman and R. I. Joseph, *Phys. Rev.* [2] **129,** 2062 (1963).

[21] A. A. Maradudin, *in* "Ferroelectricity," (E. F. Weller, ed.), p. 72. Elsevier, Amsterdam, 1967.

If the anharmonic crystal potential is expressed as a power series in the displacement of the ions from their equilibrium positions, it can be written in terms of the normal-mode coordinates introduced in Eq. (3.4) as

$$\begin{aligned}\Phi_{\text{anh}} = {} & [\omega_0(\lambda)^2 Q(\lambda) Q(-\lambda)] \\ & + \frac{1}{3!} \sum_{\lambda_1,\lambda_2,\lambda_3} V^{(3)}_{\lambda_1\lambda_2\lambda_3} Q(\lambda_1) Q(\lambda_2) Q(\lambda_3) \\ & + \frac{1}{4!} \sum_{\lambda_1,\lambda_2,\lambda_3,\lambda_4} V^{(4)}_{\lambda_1\lambda_2\lambda_3\lambda_4} Q(\lambda_1) Q(\lambda_2) Q(\lambda_3) Q(\lambda_4) \\ & + \cdots, \end{aligned} \tag{4.1}$$

where $\pm\lambda_i = (\pm q_i j_i)$. The anharmonic coefficients $V^{(n)}_{\lambda_1 \cdots \lambda_n}$ are functions of the anharmonic force constants $\Phi_{\alpha\beta\gamma}$ and the polarization and position vectors. Many-body techniques are then employed to solve for the normal modes.[19,22]

Following Cowley,[19] perturbation theory for a weakly anharmonic lattice allows us to express the renormalized frequency $\omega_T(\mathbf{q}j)$ as

$$\omega_T(\mathbf{q}j)^2 = \omega_0(\mathbf{q}j)^2 + 2\omega_0(\mathbf{q}j) D(\mathbf{q}jj', \Omega), \tag{4.2}$$

where $\omega_0(\mathbf{q}j)$ is the strictly harmonic frequency of the normal mode and $D(\mathbf{q}jj', \Omega)$ is the anharmonic contribution to the self-energy of the mode and is a function of the applied frequency $\Omega$. The self-energy is a complex quantity which can be written as

$$D(\mathbf{q}jj', \Omega) = \Delta(\mathbf{q}jj', \Omega) - i\Gamma(\mathbf{q}jj', \Omega), \tag{4.3}$$

where the real part $\Delta$ measures the anharmonic frequency shift and the imaginary part $\Gamma$ is the reciprocal of the phonon relaxation time. The real part can be written as

$$\Delta(\mathbf{q}jj', \Omega) = \Delta^{\mathrm{E}} + \Delta_3 + \Delta_4 + \cdots \equiv \Delta^{\mathrm{E}} + \Delta^{\mathrm{A}}. \tag{4.4}$$

The contribution $\Delta^{\mathrm{E}}$ involves the thermal strain $e^{\mathrm{T}}_{\alpha\beta}$ and represents the anharmonic frequency shift due to thermal expansion. It can be written as

$$\Delta^{\mathrm{E}} = \frac{2}{\hbar} \sum_{\alpha\beta} V_{\alpha\beta}(\lambda{-}\lambda') e^{\mathrm{T}}_{\alpha\beta}. \tag{4.5}$$

Of the higher-order anharmonicities, designated by $\Delta^{\mathrm{A}}$, we will explicitly consider only the anharmonic self-energy shift arising from the cubic $\Delta_3$

[22] A. A. Maradudin and A. E. Fein, *Phys. Rev.* [2] **128,** 2589 (1962).

and quartic $\Delta_4$ terms which contribute in the same order in the perturbation expansion. The leading term from cubic anharmonicity is

$$\Delta_3 = -\frac{18}{\hbar^2}\sum_{\lambda_1}\sum_{\lambda_2}|V(\lambda\lambda_1\lambda_2)|^2\left[\frac{n_1+n_2+1}{\Omega+\omega_1+\omega_2}-\frac{n_1+n_2+1}{\Omega-\omega_1-\omega_2}\right.$$
$$\left.+\frac{n_2-n_1}{\Omega+\omega_1-\omega_2}-\frac{n_2-n_1}{\Omega-\omega_1+\omega_2}\right], \qquad (4.6)$$

which results from the cubic interaction taken to second order in the perturbation expansion. The leading term in the quartic interaction is

$$\Delta_4 = \frac{12}{\hbar}\sum_{\lambda_1} V(\lambda^-\lambda'\lambda_1\lambda^-)[2n_1+1]. \qquad (4.7)$$

In Eqs. (4.6) and (4.7) $\omega_i = \omega(\mathbf{q}_i j_i)$ is the phonon frequency and $n_i = n(\mathbf{q}_i j_i) = [\exp(\hbar\omega_i/kT) - 1]^{-1}$ is the Bose–Einstein phonon occupation number.

The imaginary part $\Gamma$ of the anharmonic self-energy $D(\mathbf{q}jj', \Omega)$ is given by

$$\Gamma(\mathbf{q}jj', \Omega) = \frac{18\pi}{\hbar^2}\sum_{\lambda_1}\sum_{\lambda_2}|V(\lambda\lambda_1\lambda_2)|^2\{(n_1+n_2+1)$$
$$\times[\delta(\Omega-\omega_1-\omega_2)-\delta(\Omega+\omega_1+\omega_2)]$$
$$+(n_2-n_1)[\delta(\Omega-\omega_1+\omega_2)-\delta(\Omega+\omega_1-\omega_2)]\}. \qquad (4.8)$$

From Eq. (4.7) we observe that $\Delta_4$ is frequency ($\Omega$)-independent, and its contribution to the anharmonic self-energy can be either positive or negative depending on the sign of the quartic potential. The contribution $\Delta_3$, on the other hand, enters with a negative sign and involves the square of the cubic potential. This term is negative for low values of $\Omega$ but can become positive for large $\Omega$. For example, calculations by Cowley[19] have shown that $\Delta_3$ is negative for the $\mathbf{q} = 0$ TO soft mode of $SrTiO_3$ for $\Omega \leq \sim 14 \times 10^{12}$ sec$^{-1}$ and can become positive for higher frequencies. Under suitable conditions, the sign of the measured higher-order anharmonic frequency shift allows one to deduce the origin of the anharmonicity, assuming that the series for $\Delta^A$ converges rapidly. This assumption is believed to be valid, at least for crystals where the anharmonic contributions to the renormalized frequencies at all temperatures are much less than the harmonic contributions.

At sufficiently high temperatures we expect the thermal strain and the phonon occupation numbers to vary linearly with temperature so that

$$2\omega_0(\mathbf{q}j)D(\mathbf{q}jj', \Omega) = \alpha T, \qquad (4.9)$$

where $\alpha$ is a positive constant. We can then rewrite Eq. (4.2) as

$$\omega_T(\mathbf{q}j)^2 = \omega_0(\mathbf{q}j)^2 + \alpha T. \tag{4.10}$$

These results appear to be valid for weakly anharmonic crystals such as the alkali halides[19] as well as for crystals exhibiting weak soft-mode behavior as in the thallous halides[23] and $TiO_2$.[24] For these materials the harmonic frequency $\omega_0$ is always substantially larger than $\Delta$. However, the situation is more complicated for soft modes associated with phase transitions where $\Delta$ dominates. For crystals with finite transition temperatures, $\omega_0$ of the soft mode is expected to be imaginary, and the stabilization in the high-temperature phase is provided by $\Delta$. Thus a harmonic basis for this mode does not exist in principle. This fact was recognized in the original soft-mode treatments of Cochran,[1,2] who assumed that the soft-mode frequency $\omega_s$ was made real by the anharmonic interactions. Many authors[19-21,25] have since formally shown how this renormalization originates.

The general form of Eq. (4.10), which encompasses only the leading contributions from cubic and quartic anharmonicities, is confirmed by experiment for soft mode systems. At high temperature we rewrite Eq. (4.10) for the soft mode as

$$\omega_s^2 = \omega_0^2 + \alpha T, \tag{4.11}$$

where we have dropped the $(\mathbf{q}j)$ for convenience. For many crystals that exhibit displacive structural transitions, the measured temperature dependence of $\omega_s$ obeys the relation

$$\omega_s^2 = K(T - T_0), \tag{4.12}$$

where $K$ is a positive constant and $T_0$ is the (second-order) transition temperature. Comparing Eqs. (4.11) and (4.12), we can associate $K$ and $\alpha$ and

$$\omega_0^2 = -KT_0, \tag{4.13}$$

so that $\omega_0$ is imaginary for a finite $T_0$. From Eq. (4.11) we note that an estimate of $\omega_0^2$ can be obtained from extrapolating the linear $\omega_s^2(T)$ response back to 0 K.

It should be pointed out that the usual soft-mode description depicts $\omega_0^2 < 0$ for crystals with finite transition temperatures (where $\omega_s^2 \rightarrow 0$ at $T_0$), and $\omega_0^2$ as small and either negative or positive for a crystal with a mode

[23] R. P. Lowdnes, *Phys. Rev. Lett.* **27**, 1134 (1971); *J. Phys. C* [3] **4**, 3083 (1971).
[24] G. A. Samara and P. S. Peercy, *Phys. Rev. B: Solid State* [3] **7**, 1131 (1973).
[25] N. S. Gillis and T. R. Koehler, *Phys. Rev. B: Solid State* [3] **4**, 3971 (1971).

that softens with decreasing temperature although the structure remains stable with respect to the soft mode at 0 K. In the latter case if $\omega_0^2 < 0$, all of the stablization leading to a finite value of the measured response $\omega_s$ is provided by anharmonicities associated with zero-point motion. Designating these by $\Delta_0^A$, we then have at 0 K

$$\omega_s^2 = \omega_0^2 + \beta\Delta_0^A, \tag{4.14}$$

where $\beta$ is a constant.

This theory thus postulates that $\omega_0^2$ for the soft mode is negative as a result of the overcancellation of competing forces, so that the lattice is unstable in the harmonic approximation. Anharmonic interactions produce a frequency renormalization via $\Delta^A$ making the measured $\omega_s$ real and thereby stabilizing the lattice. For systems that exhibit structural phase transitions, the zero-point contribution to $\Delta^A$ is less than $|\omega_0^2|$; the high-temperature structure is therefore stabilized by the temperature-dependent contribution from anharmonic interactions. Starting in the high-temperature phase, this latter contribution decreases with decreasing temperature and ultimately $\omega_s \to 0$ as $T \to T_0$, the (second-order) transition temperature.

Thus far our discussion has dealt implicitly with the behavior of the soft mode in the high-temperature phase. In the low-temperature phase the soft-mode frequency is stabilized both by anharmonicities and, more importantly, by the nonzero thermal average $\langle Q(\mathbf{q}j)\rangle$ of the normal coordinates which describe the lattice distortion accompanying the transition from the high-temperature phase. In the case of a second-order transition, the thermal average of the soft-mode eigenvector is simply proportional to the order parameter.

## 5. Pressure Effects

We are now in a position to comment on the expected pressure effects on the soft-mode frequency and the transition temperature. To be specific, we consider first the case of an FE transition. For the model ferroelectric discussed earlier, the harmonic frequency $\omega_0$ of the soft FE mode is given by the difference between a short-range and a long-range force [Eq. (3.6)]. The negative value of $\omega_0^2$ results from the overcancellation of the short-range forces by the Coulomb term.

Since the Coulomb term varies with interionic distance $r$ as $1/r^3$, whereas the short-range forces vary as $1/r^n$, where $n$ has a large value ($\sim$10–11), for a given decrease in $r$ with pressure, $R_0'$ increases much more rapidly than does the Coulomb term leading to an increase in $\omega_0^2$. The effect is most dramatic when the two forces in Eq. (3.6) are nearly equal; then, a small change in $r$ causes a large fractional change in $\omega_0$.

The increase in $\omega_0^2$ (i.e., $\omega_0^2$ becoming less negative) with increasing pressure leads to an increase in $\omega_s$, which in turn leads to a decrease in $T_0$, as can be seen from Eq. (4.12). Ultimately, according to this model, at sufficiently high pressure the increase in the short-range forces $R_0'$ should make the harmonic contribution $\omega_0$ to the soft-mode frequency real and finite. In this case the crystal will become stable with respect to this mode at all temperatures, i.e., the FE transition should vanish and the PE phase remains stable to the lowest temperatures. Evidence for this behavior has been observed. This and a variety of other important pressure effects are dealt later in this review.

## 6. The Lyddane–Sachs–Teller (LST) Relation

An important relationship, first derived by Lyddane, Sachs, and Teller[26] for a cubic diatomic crystal, exists for ionic crystals between the high frequency ($\varepsilon_\infty$) and static ($\varepsilon$) dielectric constants and the frequencies of the longitudinal and transverse optic phonons in the limit of long wavelengths ($\mathbf{q} \approx 0$). For diatomic crystals, the result, often referred to as the LST relation, is

$$\frac{\varepsilon}{\varepsilon_\infty} = \left(\frac{\omega_{\mathrm{LO}}}{\omega_{\mathrm{TO}}}\right)^2. \tag{6.1}$$

Cochran[27] generalized the derivation of this relation to apply to diagonally cubic crystals containing any number of atoms per unit cell. The result is

$$\frac{\varepsilon}{\varepsilon_\infty} = \prod_i^{n-1} \frac{(\omega_{\mathrm{LO}})_i^2}{(\omega_{\mathrm{TO}})_i^2}, \tag{6.2}$$

where the product is over all the optic branches of the dispersion curves. Further extensions of the LST relation have been made to apply to uniaxial crystals[28,29] as well to any direction in a crystal of any symmetry.[30]

The LST relation can be derived from the macroscopic theory of lattice vibrations[13] and is thus model-independent. Although it is based on the

[26] R. H. Lyddane, R. G. Sachs, and E. Teller, *Phys. Rev.* [2] **59,** 673 (1941).
[27] W. Cochran, *Adv. Phys.* **18,** 157 (1969).
[28] L. Merten, *Z. Naturforsch., A* **15,** 47 (1960).
[29] A. S. Barker, *Phys. Rev.* [2] **136,** A1290 (1964).
[30] W. Cochran and R. A. Cowley, *J. Phys. Chem. Solids* **23,** 447 (1962).

harmonic and adiabatic approximations, it has proved quite accurate for simple ionic crystals such as the alkali halides. It also appears to work quite well for crystals that exhibit strongly anharmonic character (at least with respect to certain modes) such as TlCl,[23,31] $TiO_2$[24] and $KTaO_3$,[32] provided that *measured* values of the various quantities involved are used. This applicability of the LST relation undoubtedly results because the measured quantities represent the effective renormalized responses for the crystal.

The LST relation played a prime role in the early development of the soft mode theory of ferroelectricity. Fröhlich[33] was the first to suggest that as $\varepsilon \to \infty$, it follows from Eq. (6.1) that $\omega_{TO}$ should go to zero. Cochran[1] further suggested that the observed Curie–Weiss temperature dependence of $\varepsilon$ in the high-$T$ phase of an FE crystal [i.e., $\varepsilon = C/(T - T_0)$] should be associated with the $T$ dependence of the soft $\mathbf{q} \simeq 0$ TO mode. The soft FE mode should then vary with $T$ as indicated by Eq. (4.12). This has now been confirmed experimentally for many displacive ferroelectrics. Formally, the connection between $\epsilon$ and $\omega_s$ is made via the generalized LST relation, Eq. (6.2).

Experimentally, it is observed that essentially all of the temperature dependence of the phonons is associated with that of $\omega_s$, so that Eq. (6.2) can be rewritten as

$$\omega_s^2(T)\epsilon(T) = \epsilon_\infty\left[\prod_i \left(\omega_{LO}\right)_i^2 \Big/ \prod_i{}' \left(\omega_{TO}\right)_i^2\right] \approx B, \qquad (6.3)$$

where the product $\Pi_i'$ is taken over all the TO frequencies except $\omega_s$, and $B$ is a constant ($\equiv KC$) independent of temperature. The temperature dependence $\omega_s^2(T)\varepsilon(T) \approx B$ has been verified for many perovskite crystals; furthermore, the absolute value of $B$ is accurately given[34] by the middle term in Eq. (6.3).

## 7. Coupled Phonon Modes

In our discussion so far we have dealt with the simplest possible soft modes. In some crystals as the frequency of the soft mode changes, it may couple directly with other phonon modes of the crystal. Alternatively, the soft mode itself may be a coupled mode involving two or more fundamen-

[31] G. A. Samara, *Phys. Rev.* [2] **165,** 959 (1968).
[32] G. A. Samara and B. Morosin, *Phys. Rev. B: Solid State* [3] **8,** 1256 (1973).
[33] H. Frölich, *in* "Theory of Dielectrics," p. 159. Univ. Press Oxford, (Clarendon) London and New York, 1949.
[34] P. A. Fleury and J. M. Worlock, *Phys. Rev.* [2] **174,** 613 (1968).

tal excitations. It is necessary to consider these interactions explicitly in the analysis of inelastic light and neutron scattering from such coupled systems. For example, in many of the crystals of present interest ($BaTiO_3$, $KH_2PO_4$, etc.), the soft optic mode couples strongly to an acoustic mode through a piezoelectric interaction as $T$ approaches $T_c$. Similarly, interactions of the soft optic mode with other optic modes are important in SbSI, whereas in $KH_2PO_4$ and its isomorphs the soft mode is a coupled excitation between the proton motion and an optic mode of the lattice, and this excitation further couples to an acousitc mode near $T_c$. As will become clear later, pressure has been an important variable in understanding these coupled-mode interactions.

Several different types of coupled-mode systems have been investigated in different materials with different theoretical treatments given for each type of interaction. Scott[35] has shown that the various types of coupled-mode systems examined by inelastic light scattering measurements can be considered as special cases of unified treatment of coupled-mode systems. The spectrum of Stokes-shifted scattered light $S(\omega)$ is in general

$$S(\omega) \propto \mathrm{Im}[\chi(\omega)][\bar{n}(\omega) + 1], \tag{7.1}$$

where $\chi(\omega)$ is the complex susceptibility and $\bar{n}(\omega)$ is the Bose–Einstein population factor. For two modes $i$ and $j$ of strengths $P_i$ and $P_j$, $\mathrm{Im}[\chi(\omega)]$ is given by

$$\mathrm{Im}[\chi(\omega)] = \sum_{i,j} P_i P_j G_{ij}(\omega), \tag{7.2}$$

where the $G_{ij}(\omega)$ are the solutions of

$$\begin{bmatrix} \chi_1^{-1}(\omega) & \Delta^2 - i\omega\Gamma_{12} \\ \Delta^2 - i\omega\Gamma_{12} & \chi_2^{-1}(\omega) \end{bmatrix} \begin{bmatrix} G_{11} & G_{12} \\ G_{12} & G_{22} \end{bmatrix} = \begin{bmatrix} 1 & 0 \\ 0 & 1 \end{bmatrix}. \tag{7.3}$$

In Eq. (7.3), $\chi_j(\omega)$ is the decoupled susceptibility of mode $j$, and the modes 1 and 2 are coupled together through the completely general interaction $\Delta^2 - i\omega\Gamma_{12}$. These expressions can be readily extended to include more than two interacting modes by extending the sum in Eq. (7.2) over all the modes of interest. Such an extension was used by Lagakos and Cummins[36] to investigate the interaction of the lower-frequency branch of the coupled proton tunneling–optic mode system in $KH_2PO_4$ with an acoustic mode.

[35] J. F. Scott, *in* "Light Scattering in Solids" (M. Balkanski, ed.), p. 387. Flammarion Sciences, Paris, 1971.

[36] N. Lagakos and H. Z. Cummins, *Phys. Rev. B: Solid State* [3] **10**, 1063 (1974).

It should be noted that Eqs. (7.2) and (7.3) are overdetermined because of the general form of the interaction.[37] The choice of the phase of diagonalizing Eq. (7.3) is arbitrary—the data can be fit with a purely real interaction ($\Gamma_{12} = 0$), a purely imaginary interaction ($\Delta = 0$), or some combination of real and imaginary interaction (both $\Delta$ and $\Gamma_{12} \neq 0$). An additional constraint is required to obtain a unique form for the interaction. The form of the interaction is therefore based on the particular model with which one wishes to compare the measurements.

More physical insight can perhaps be obtained by considering a specific case. Here we examine the piezoelectric interaction between the soft optic mode and an acoustic mode which is observed in, for example, $KH_2PO_4$-type crystals[38] and $BaTiO_3$.[39,40] Various theoretical treatments have been given for such piezoelectric interactions. Fleury and Lazay[40] examined this interaction in $BaTiO_3$, and Reese *et al.*[41] gave a detailed phenomenological treatment of the coupled FE and acoustic modes in $KD_2PO_4$. We summarize these results in the following.

The equations of motion for the coupled system may be derived from a Lagrangian energy density. Assuming harmonic responses, one obtains the coupled equations

$$(1/\chi_o^0)P + aq(B/4\pi)X = F_P, \tag{7.4}$$

$$(aq/\rho)P + (1/\chi_a^0)X = F_x. \tag{7.5}$$

In Eqs. (7.4) and (7.5), $P$, $X$, $\rho$, and $q$ are the polarization, strain, density, and phonon wave vector, respectively; $B$ is a known constant that relates the optic mode frequency $\omega_o(T)$ to the clamped dielectric constant via the LST relation; $a$ is the relevant piezoelectric coefficient that couples the optic and acoustic modes; and $F_P$ ($F_X$) are the normalized driving forces for polarizations (strains). The decoupled susceptibilities are damped harmonic oscillator responses for the optic $\chi_o^0 \sim (\omega_o^2 - \omega^2 + i\omega\Gamma_o)^{-1}$ and acoustic $\chi_a^0 \sim (\omega_a^2 - \omega^2 + i\omega\Gamma_a)^{-1}$ modes. If the piezoelectric coefficient $a$ were zero, Eqs. (7.4) and (7.5) would reduce to two damped harmonic oscillator responses representing the optic and acoustic modes.

The different coupled susceptibilities (polarization response to strain or electric field, and strain response to stress or electric field) can be obtained from Eqs. (7.4) and (7.5), as shown by Reese *et al.*,[41] in terms of

[37] A. S. Barker and J. J. Hopfield, *Phys. Rev.* [2] **135,** A1732 (1964).

[38] E. M. Brody and H. Z. Cummins, *Phys. Rev. Lett.* **21,** 1263 (1968); **23,** 1039 (1969), and references therein.

[39] V. Dvorák, *Can. J. Phys.* **45,** 3909 (1967); *Phys. Rev.* [2] **167,** 525 (1968).

[40] P. A. Fleury and P. D. Lazay, *Phys. Rev. Lett.* **26,** 1331 (1971).

[41] R. L. Reese, I. J. Fritz, and H. Z. Cummins, *Phys. Rev. B: Solid State* [3] **7,** 4165 (1973).

these decoupled susceptibilities. In particular, the polarization response to electric field with zero stress is

$$\chi_{PP} = \left(\frac{P}{E}\right)_{\sigma=0} = \frac{\chi_o^0}{1 - A^2\chi_o^0\chi_a^0}, \tag{7.6}$$

where the coupling constant $A^2 = a^2q^2B/4\pi\rho$. Equation (7.6) is the response probed by inelastic light scattering measurements.

The treatment of this piezoelectric interaction can be shown to be a special case of the generalized coupled-mode picture discussed above.[41] If one recasts Eqs. (7.4) and (7.5) in the form of Eq. (7.3), then these equations correspond to two modes coupled through a real interaction $\Delta^2 = A$.

## III. Displacive Ferroelectric Transitions

### 8. Ferroelectricity and Incipient Ferroelectricity in Simple Diatomic Crystals

The theoretical arguments outlined in Section II for a soft-mode FE transition were first advanced and treated in some detail[1] for a model diatomic ionic crystal. As a result the question of the existence of FE behavior of this type in simple diatomic crystals has received considerable attention. The first indication that such behavior is indeed possible came from the inelastic neutron scattering studies of the phonon dispersion in SnTe.[42] This work showed the existence of a soft $\mathbf{q} \approx 0$ TO mode whose frequency decreased markely with decreasing $T$. More recent work on other IV–VI diatomic semiconductors (including mixed crystals) established the FE nature of these materials. However, direct measurements of the dielectric properties, which are the usually accepted signatures of ferroelectricity, are hampered by the relatively high electrical conductivity in these semiconductors. At present, there are no known diatomic dielectric crystals which undergo FE transitions; however, the cubic thallous halides exhibit soft-mode behavior similar to that observed in ferroelectrics for $T > T_c$. In this section we review some of the properties of the thallous halides and IV–VI crystals with emphasis on the pressure results.

#### a. *The Thallous Halides*

It has been known for quite some time that the cubic (*CsCl* structure) thallous halides exhibit high static dielectric constants $\varepsilon$ with anomalous

[42] G. S. Pawley, W. Cochran, R. A. Cowley, and G. Dolling, *Phys. Rev. Lett.* **17**, 753 (1966).

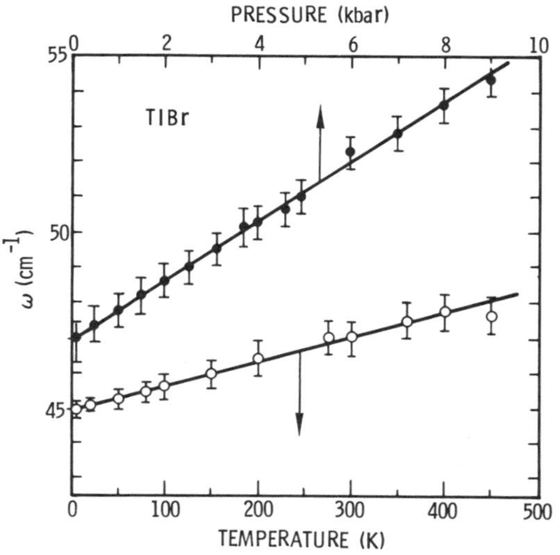


FIG. 1. Temperature and pressure dependences of the soft TO mode frequency in TlBr. [After R. P. Lowndes, *Phys. Rev. Lett.* **27**, 1134 (1971).]

negative temperature dependences. Furthermore, $\varepsilon$ obeys a Curie–Weiss law, $\varepsilon = C/(T - T_0)$, over a relatively wide temperature range.[31] However, no transitions are observed for these crystals down to <4 K.

Combined pressure and temperature measurements[31,43] show that the anomalous temperature dependence of $\varepsilon$ arises from pure-temperature (i.e., volume-independent) effects, which in turn arise from higher-order anharmonicities.

According to the theory of Section II, the above $\varepsilon(T)$ response implies that $\omega_{TO}(\mathbf{q} \approx 0)$ for these crystals should decrease with decreasing $T$ (unlike normal dielectrics such as the alkali halides). This behavior was confirmed experimentally.[43,44] The most detailed results are those of Lowndes[43] who measured the temperature and pressure dependences of $\omega_{TO}$ for both TlCl and TlBr. His results for TlBr are shown in Fig. 1, where it is seen that $\omega_{TO}$ increases with both temperature and pressure. Similar results were obtained for TlCl. Analysis of these results in terms of the theory of Section 4 shows that the anomalous $T$ dependence of this mode is dominated by the quartic contributions to the anharmonic self-energy. It was earlier argued that the anomalous temperature dependence of $\varepsilon$ for these crystals is also dominated by quartic anharmonicities.[31] Thus, the

[43] R. P. Lowndes, *Phys. Rev. Lett.* **27**, 1134 (1971); *Phys. Rev. B: Solid State* [3] **6**, 1490 (1972); **6**, 4667 (1972).

[44] E. R. Cowley and A. Okazaki, *Proc. R. Soc. London, Ser. A* **300**, 45 (1967).

experimental evidence indicates that the anomalous increase of $\varepsilon$ with decreasing $T$ is associated with the softening of $\omega_{\mathrm{TO}}(\mathbf{q} \approx 0)$. Although the softening is relatively weak compared with that observed for FE crystals, it is of the magnitude expected from the dielectric data via the LST relation, Eq. (6.1).

It is thus evident that the cubic thallous halides exhibit features characteristic of soft-mode ferroelectrics. However, these crystals fall far short of becoming ferroelectrics.

*b. SnTe and Isomorphous IV–VI Semiconductors*

It has been established[45,46] that GeTe transforms on cooling from the cubic *NaCl* structure to a face-centered rhombohedral structure ($R3m$-$C_{3v}^5$) at ~670 K. GeTe also forms a continuous series of solid solutions with SnTe and the cubic–rhombohedral transition temperature decreases nearly linearly with increasing SnTe concentration.[47] This transition is displacive; the rhombohedral phase is related to the NaCl phase by a small relative displacement of the Ge and Te sublattices in the [111] direction. The pattern of the displacements can be described by the distortion associated with a $q \sim 0$ TO mode with atomic motion along a [111] direction,[42] i.e., an FE mode. This suggests that the transitions in $Sn_{1-x}Ge_xTe$ may reflect an instability of the lattice with respect to this mode. Experimental evidence indicates that this is indeed the case.

Inelastic neutron scattering measurements on SnTe[42] and $Sn_{0.935}Ge_{0.065}Te$[48] show the existence of a soft $\mathbf{q} \approx 0$ TO mode in each case. The mixed crystal has a transition at ~80 K, however, the results show no evidence of a phase transition in SnTe, but suggest that the crystal is barely stable at 0 K.

Apparently, no inelastic neutron measurements exist on pure GeTe. However, Raman scattering results[49] confirm the FE nature of the low rhombohedral phase. In the cubic *NaCl* phase there are no Raman-active phonons, but in the rhombohedral phase the transverse polar E and $A_1$ phonons soften markedly on approaching $T_c$ from below. However, the softening is not complete since the transition is first order.

The isomorphous compound PbTe and the $Pb_{1-x}Ge_xTe$ alloys exhibit similar behavior. Although PbTe remains cubic and paraelectric down to

[45] J. Goldak, G. S. Barrett, D. Innes, and W. Yondelis, *J. Chem. Phys.* **44,** 3323 (1966).
[46] T. b. Zhudova and A. I. Zaslavskii, *Sov. Phys.—Crystallogr.* (*Engl. Transl.*) **12,** 28 (1967).
[47] J. N. Bierly, L. Muldawer, and O. Beckman, *Acta Metall.* **11,** 447 (1963).
[48] I. Lefkowitz, M. Shields, G. Dolling, W. J. L. Buyers, and R. A. Cowley, *Jpn. J. Phys.* **28,** Suppl. 249 (1970).
[49] G. Harbeke, E. F. Steigmeier, and R. K. Wehner, *Solid State Commun.* **8,** 1765 (1970).

at least 4 K,[50] the $Pb_{1-x}Ge_xTe$ alloys (for $x \gtrsim 0.01$) exhibit cubic-rhombohedral transitions at finite temperatures, and there is evidence that the rhombohedral phase is ferroelectric.[51]

There is thus evidence supporting the conclusion that GeTe and its alloys with SnTe and PbTe are diatomic ferroelectrics (FEs) (SnTe and PbTe being so-called incipient FEs, i.e., materials whose soft FE mode does not quite vanish at 0 K and $T_0 < 0$ K). Since these materials are highly conducting and covalent, the theory of Section II is not directly applicable. It is probable, however, that the stability of the FE mode is again determined by a delicate balance between long-range and short-range forces. If this is the case, this balance should be strongly dependent on the crystal volume with the frequency of the soft TO mode increasing and $T_c$ decreasing with increasing pressure. The transition temperature is indeed found to decrease with pressure. Measurements[52] on GeTe and two $Ge_{1-x}Sn_xTe$ alloys with $x = 0.06$ and 0.26 showed that their cubic-rhombohedral transition temperatures $T_c$ ($\simeq$670, 620, and 520 K, respectively, at 1 bar) are lowered to room temperature by the application of 35, 27, and 12 kbar, respectively.

High pressure studies on these semiconductors have also led to the discovery of a pressure-induced transition from the *NaCl* structure to a distorted *NaCl* (orthorhombic *Pnma*) structure. The transition was first observed in PbS, PbSe, and PbTe by volumetric[53] and then by electrical resistivity[54] techniques. The same transition was later observed in SnTe[55] by electrical resistivity and by x-ray diffraction which allowed identification of the high-pressure phase as orthorhombic. Similar work followed on SnTe and $Sn_{1-x}Ge_xTe$ alloys.[56,57]

Although the lattice dynamical origin of this pressure-induced transition is not known, the structure of the high pressure phase (orthorhombic *Pnma*) is identical to that of SnS and SnSe under normal conditions. Pawley[58] suggested that SnS and SnSe may be antiferroelectric (AFE). Significantly, the unit cell of SnTe and its isomorphs doubles at the

[50] R. T. Bate, D. L. Carter, and J. S. Wrobel, *Phys. Rev. Lett.* **25**, 159 (1970).
[51] G. A. Antcliffe, R. T. Bate, and D. D. Buss, *Solid State Commun.* **13**, 1003 (1973).
[52] S. S. Kabalkina, N. R. Serebryanaya, and L. F. Vereschagin, *Sov. Phys.—Solid State (Engl. Transl.)* **9**, 2527 (1968).
[53] P. W. Bridgman, *Proc. Am. Acad. Arts Sci.* **76**, 55 (1948).
[54] G. A. Samara and H. G. Drickamer, *J. Chem. Phys.* **37**, 1159 (1962).
[55] J. A. Kafalas and A. N. Mariano, *Science* **143**, 952 (1964).
[56] S. S. Kabalkina, N. R. Serebryanaya, and L. F. Vereschagin, *Sov. Phys.—Solid State (Engl. Transl.)* **10**, 574, (1969).
[57] G. A. Samara, unpublished.
[58] G. S. Pawley, *J. Phys., Colloq. (Orsay, Fr.) C* **4C**, Suppl. 29, C4-C145 (1968).

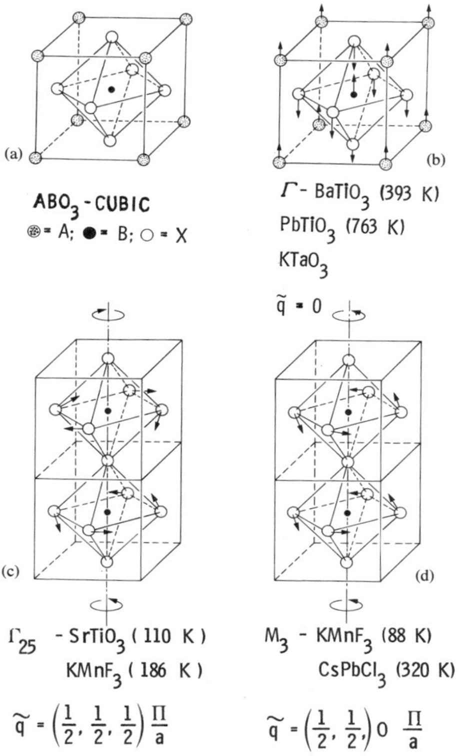


FIG. 2. Perovskite structure (a) and eigenvectors of three soft modes (b, c, d) observed for materials with this crystal structure. Examples of crystals exhibiting each of these soft modes and associated transition temperatures are given.

pressure-induced transition, as would be expected for AFE transition. Pawley,[58] using inelastic neutron scattering, observed that in SnTe at 1 bar, one of the optic phonon branches at the $(\pi/a, 0, 0)$ zone boundary decreases in frequency as the wave vector tends to the zone boundary. While it is possible that this observation is related to the pressure-induced transition, neutron scattering at high pressure is needed to prove definitely the connection.

## 9. Ferroelectricity in the Perovskite Structure

The cubic perovskite structure shown in Fig. 2a occupies a unique role in the study of soft modes. This structure provided the first detailed experimental evidence that beautifully confirmed the soft-mode concept and put in on a strong foundation. Many crystals having this structure exhibit temperature-induced structural phase transitions to FE, AFE, or normal dielectric phases, with some crystals exhibiting a succession of transitions. It is now established that these transitions are accompanied by the softening and ultimate condensation of certain phonon modes. In this and later sections we shall illustrate the nature of some of these modes and the effects of hydrostatic pressure on them. We begin by considering FE transitions.

Figure 2b shows the eigenvector of the zone-center ($\mathbf{q} = 0$) TO soft mode responsible for the FE transition in many $ABO_3$ oxides. This mode consists of the vibration of the positive A and B ions against the negative oxygen ions. In some crystals of this class, e.g., $KTaO_3$ and $SrTiO_3$, the mode consists predominantly of the vibration of the central B ion against a rigid oxygen octahedron, the A ions remaining effectively fixed. This is the motion that was first postulated by Slater and is referred to as the S1 mode.[59] In other $ABO_3$ crystals, e.g., $PbTiO_3$ and $BaTiO_3$, the dominant ionic motion consists of A ions vibrating against rigid $BO_6$ groups. This ionic motion can be decomposed into two distinct modes, S1 and S2.[59] The motion associated with either mode possesses a linear electric dipole moment which determines the dielectric properties. On approaching $T_c$, the frequency of the soft mode, and therefore the restoring force, decreases. Ultimately the restoring force vanishes and the structure elongates along the *c* axis, becoming tetragonal. In the lower-symmetry phase, the B ion is shifted slightly from the center of symmetry, and the crystal is noncentric and ferroelectric. This is perhaps the simplest and most beautiful kind of structural phase transition. The well-known cubic-to-tetragonal FE transitions in many perovskites are due to the softening of one of these modes. In some crystals, e.g., $KTaO_3$ and $SrTiO_3$, the mode softens with decreasing temperature but retains a small finite value ($\sim$10–20 $cm^{-1}$—a remarkably small frequency) at 4 K. As a result no transition occurs, but these crystals are just barely stable against this mode at these conditions and are thus often referred to as incipient FEs.

On the basis of the arguments advanced in Section II, the frequencies of these soft modes can be expected to increase with increasing pressure. As a consequence the transition temperatures should decrease. This is indeed found to be the case. A considerable amount of work has been reported on the effects of pressure on the properties of FE perovskites. Some of the early results were reviewed earlier.[9] In the present section we concentrate

[59] G. Shirane, J. D. Axe, J. Harada, and J. P. Remeika, *Phys. Rev. B: Solid State* [3] **2**, 155 (1970); *Acta Crystallogr., Sect. A* **26**, 608 (1970).

on some of the more recent results and emphasize the general characteristic features by reviewing the results on a few typical crystals.

*a. Incipient Ferroelectrics—$KTaO_3$*

Pure potassium tantalate, $KTaO_3$, has been thoroughly investigated and presents an interesting case to begin with. It has the ideal cubic perovskite structure and remains cubic and paraelectric down to the lowest temperatures; however, it is just barely stable against the FE mode at 4 K reflecting a delicate balance between the short-range and Coulomb forces. It is therefore possible to examine the behavior of the soft mode in this crystal over a wide temperature range while avoiding complications associated with phase transitions. $SrTiO_3$, which has also been thoroughly studied and exhibits qualitatively similar temperature and pressure behavior to $KTaO_3$ with respect to the zone-center FE soft mode, undergoes a structural phase transition at 105 K resulting from the condensation of a zone-boundary phonon (see Section IV). In the low-temperature tetragonal phase the static dielectric constant $\varepsilon$ becomes anisotropic. Although this anisotropy, which increases as the order parameter increases below the transition, has generally been ignored, Sakudo and Unoki[60] have shown that for single-domain tetragonal $SrTiO_3$ the anisotropy in $\varepsilon$ is $\simeq 2$ at 4 K.

The FE soft mode in $KTaO_3$ is underdamped and well defined, and its temperature dependence at 1 bar has been studied by far-infrared,[61,62] Raman,[63] and inelastic neutron scattering[64] techniques. (Although $KTaO_3$ has no first-order Raman modes, the application of an electric field causes the FE mode to become Raman active.) The temperature and pressure dependences of $\varepsilon$ have also been studied by several authors[32,65–70] with general agreement among the various results.

[60] T. Sakudo and H. Unoki, *Phys. Rev. Lett.* **26,** 851 (1971).

[61] R. C. Miller and W. G. Spitzer, *Phys. Rev.* [2] **129,** 94 (1963).

[62] C. H. Perry and T. F. McNelly, *Phys. Rev.* [2] **154,** 456 (1967).

[63] P. A. Fleury and J. M. Worlock, *Phys. Rev.* [2] **174,** 613 (1968).

[64] G. Shirane, R. Nathans, and V. J. Minkiewicz, *Phys. Rev.* [2] **157,** 396 (1967).

[65] J. K. Hulm, B. T. Matthias, and E. A. Long, *Phys. Rev.* [2] **79,** 885 (1949).

[66] S. H. Wemple, *Phys. Rev.* [2] **137,** A1575 (1965).

[67] D. G. Demurov and Yu. N. Venevtsev, *Sov. Phys.—Solid State (Engl. Transl.)* **13,** 553 (1971).

[68] S. H. Wemple, A. Jayaraman, and M. DiDomenico, *Phys. Rev. Lett.* **17,** 142 (1966).

[69] W. R. Abel, *Phys. Rev. B: Solid State* [3] **4,** 2696 (1971).

[70] R. P. Lowndes and A. Rastogi, *J. Phys. C* [3] **6,** 932 (1973).

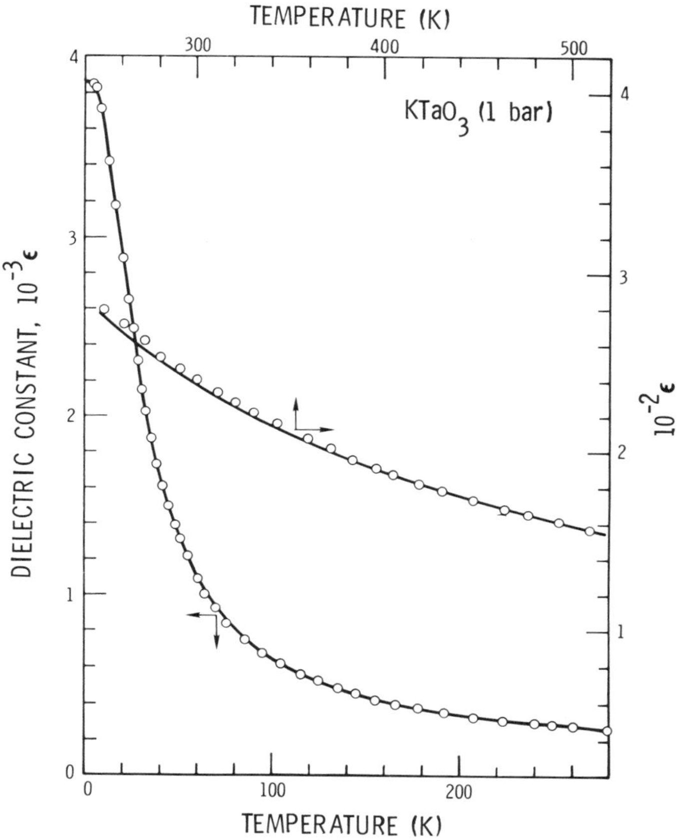


FIG. 3. Temperature dependence of the dielectric constant of $KTaO_3$. The solid curve is a fit to Eq. (9.1). After G. A. Samara and B. Morosin, *Phys. Rev. B: Solid State* [3] **8**, 1256 (1973).

(*i*) *The Static Dielectric Constant* $\varepsilon$. The $T$ dependence of $\varepsilon$ of $KTaO_3$ at 1 bar is shown in Fig. 3. Above ~50 K the data can be well fit by the modified Curie–Weiss law

$$\varepsilon = A + C/(T - T_0), \tag{9.1}$$

where $A$ is a temperature-independent constant that represents the contribution to $\varepsilon$ from the electronic polarization and lattice modes other than the FE mode. Anharmonic lattice dynamical treatments of displacive FEs lead to an expression of this form for $T > T_0$.[20] At low temperatures

($<50$ K) $\varepsilon$ increases less rapidly than is predicted by Eq. (9.1), and this deviation has been attributed by Barrett[71] to quantum effects on the ionic polarizabilities of the crystal, and by Silverman and Joseph[20] to the stabilization of the bare phonon by the phonon cloud due to zero-point fluctuations.

The solid line in Fig. 3 is a least squares fit of the data to the expression

$$\varepsilon = A + \frac{C}{\frac{1}{2}T_1 \coth (T_1/2T) - T_o}, \tag{9.2}$$

where $A = 47.5$, $C = 5.45 \times 10^4$ K, $T_1 = 56.9$ K, and $T_0 = 13.1$ K. This expression was first derived (with $A = 0$) by Barrett[71] and was used by Abel[69] to describe the deviations from the simple Curie–Weiss law at both low and high $T$. Note that Eq. (9.2) reduces to $\varepsilon = A + C(\frac{1}{2}T_1 - T_0) \equiv$ constant as $T \rightarrow 0$ K and to $\varepsilon = A + C/(T - T_0)$ [i.e., Eq. (9.1)] at high $T$. In Eq. (9.2) $T_1$ is the temperature below which deviations from the simple Curie–Weiss law occur. More precisely, in Barrett's model the lowest quantum level for the ion has an energy of $kT_1$ for $T < T_1$, all of the ions are in their lowest energy states and further reduction in $T$ causes no change in $\varepsilon$. For this model one would expect $T_1$ to increase with pressure since decreasing the volume should raise the energy of the lowest quantum level, and hence $T_1$. Indeed, Abel[69] finds this to be the case for $KTaO_3$, where $dT_1/dp \simeq 1.5$ K/kbar. It should be emphasized, however, that whereas Eq. (9.2) describes the behavior of $\varepsilon(T)$ over a wide temperature range, it predicts relations between the pressure derivatives of the various parameters that are not completely consistent with the experimental results.[69] Nevertheless, Eq. (9.2) has proved useful when $A$, $C$, $T_1$, and $T_0$ are treated merely as empirically determined constants. Finally, we note that the $\varepsilon(T)$ data can also be fit over the whole temperature range by the expression given by Silverman and Joseph[20] determined from a perturbation lattice dynamic treatment. However, their result involves many parameters and renders physical interpretation difficult.

The pressure dependence of $\varepsilon$ of $KTaO_3$ has been studied extensively at temperatures ranging from 4 to 500 K.[32,69,70] The temperature dependence of the initial logarithmic derivatives $(\partial \ln \varepsilon/\partial p)_T$ and $(\partial \ln \varepsilon/\partial T)_{1\text{ bar}}$ are shown in Fig. 4. Note that $(\partial \ln \varepsilon/\partial p)_T$ is large at low $T$ ($-39$%/kbar at 4 K) and decreases (in magnitude) rapidly and monotonically with increasing $T$ reaching a value of $-1.36$%/kbar at 450 K. The temperature derivative $(\partial \ln \varepsilon/\partial T)$, on the other hand, equals 0 at 0 K, reaches a sharp

[71] J. H. Barrett, *Phys. Rev.* [2] **86**, 118 (1952).

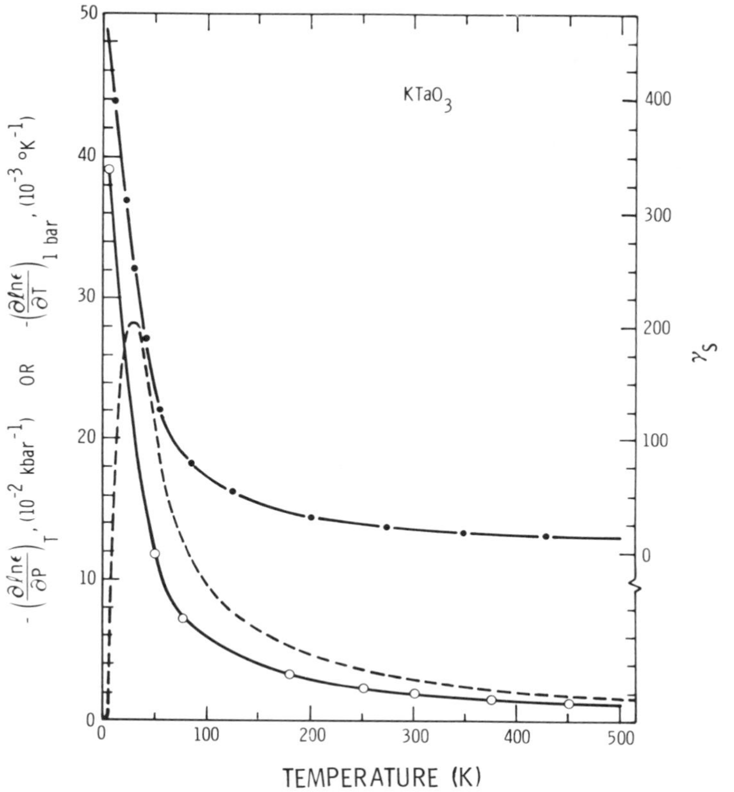


FIG. 4. Temperature dependences of the initial pressure and temperature derivatives of the dielectric constant of $KTaO_3$ and of the Grüneisen parameter of the soft mode. [After G. A. Samara and B. Morosin, *Phys. Rev. B: Solid State* [3] **8**, 1256 (1973).] (—○—) $(\partial \ln \varepsilon/\partial P)_T$; (----) $(\partial \ln \varepsilon/\partial T)_{1\ \mathrm{bar}}$; (—●—) $\gamma_s$.

maximum (in magnitude) of $-2.8\%/\mathrm{K}$ at ~30 K, and then decreases monotonically with increasing $T$.

Over a wide pressure range $\varepsilon(p)$ is found to obey a modified Curie–Weiss law with pressure,[32,68] i.e., Eq. (2.19), where $C^*$ and $p_0$ are constants at any given $T$. This expression is valid for many crystals in the PE phase.[9]

(*ii*) *The Soft Mode Frequency* $\omega_s$. The temperature dependence of $\omega_s$ for $KTaO_3$, plotted as $\omega_s^2$ versus $T$, is shown in Fig. 5. Typical reported error bars are indicated. Yamada and Shirane's[72] results on $SrTiO_3$ are

[72] Y. Yamada and G. Shirane, *Jpn. J. Phys.* **26**, 396 (1969).

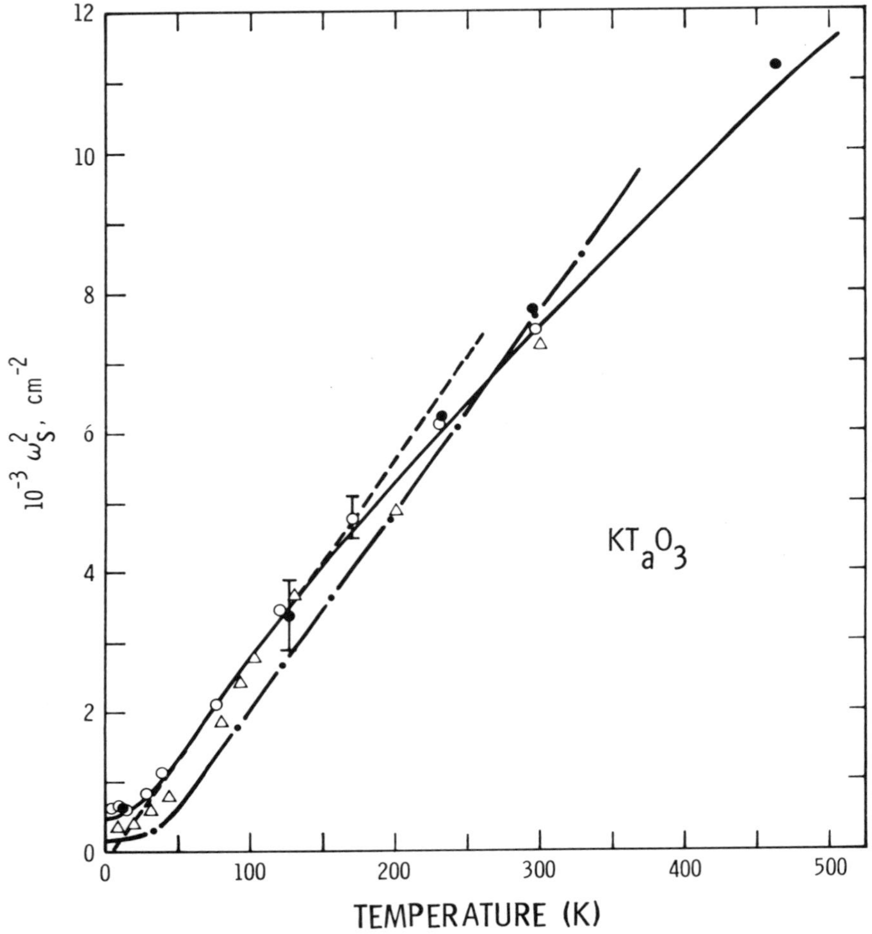


FIG. 5. Temperature dependence of the soft TO mode frequencies in $KTaO_3$ and $SrTiO_3$ measured by different techniques. [After G. A. Samara and B. Morosin, *Phys. Rev. B: Solid State* [3] **8**, 1256 (1973).] (○) Neutron; (●) infrared; (△) Raman; (——) $1.81 \times 10^6/\varepsilon$ $SrTiO_3$; (—●—) Y. Yamada and G. Shirane, *Jpn. J. Phys.* **26**, 396 (1969).

also shown for comparison. Note the very low values of $\omega_s$ at 4 K for both crystals. For $KTaO_3$ at 4 K, $\omega_s = 21 \pm 4\,\text{cm}^{-1}$ [$= (6.3 \pm 1.2) \times 10^{11}\,\text{sec}^{-1}$]. If the strictly harmonic frequency $\omega_0$ of the soft mode were truly imaginary, one would have to conclude that at 4 K $\omega_s$ is stabilized by anharmonic interactions associated with zero-point motion.

As discussed in Section 4 an estimate of $\omega_0^2$ can be obtained from extrapolating the linear portion of the $\omega_s^2(T)$ response back to 0 K. Unlike $SrTiO_3$, where $\omega_s^2(T)$ is linear over a substantial temperature range above ~50 K, in $KTaO_3$ an approximately linear region, denoted by the dotted line in Fig. 5, extends only over the relatively small region from ~30 to ~125 K. Extrapolation of this linear response back to 0 K yields an intercept $\equiv \omega_0^2 = -200 \pm 200\ \mathrm{cm}^{-2}$. Although this approach of evaluating $\omega_0^2$ involves considerable uncertainty, we can conclude that for $KTaO_3$ $\omega_0^2 \lessapprox 0$. The stabilization of $\omega_s$ at 0 K is thus provided by zero-point fluctuations which contribute a frequency shift of something slightly over $21 \pm 4\ \mathrm{cm}^{-1}$. The fact that $\omega_0^2 \approx 0$ emphasizes the almost perfect cancellation between the short-range and Coulomb interactions in $KTaO_3$ at low temperatures.

The large increase in $\omega_s$ with increasing $T$ in Fig. 5 is responsible for the $\varepsilon(T)$ behavior in Fig. 3 since, as noted earlier, the two quantities are related by an LST relationship [Eq. (6.3)]. That Eq. (6.3) holds very well $KTaO_3$ is shown by the solid line in Fig. 5 which is deduced from the $\varepsilon(T)$,[32] with the constant ($B = 1.81 \times 10^6\ \mathrm{cm}^{-2}$) evaluated solely from the known room temperature values of $\varepsilon = 239$ and $\omega_s = 87\ \mathrm{cm}^{-1}$. $B$ can also be evaluated from the middle term in Eq. (6.3) as discussed in Section 6. Within the uncertainties in the directly measured values for $\omega_s$, the agreement between the solid line and the data is excellent. It is thus seen that measurements of $\varepsilon$ give direct information on $\omega_s$. This fact has been used[32] to deduce the effect of pressure on $\omega_s$ from the $\varepsilon(p)$ data, since $\omega_s(p)$ has not been measured directly for $KTaO_3$. From Eq. (6.3) it follows that

$$(\partial \ln \omega_s/\partial p)_T = -\tfrac{1}{2}(\partial \ln \varepsilon/\partial p)_T \tag{9.3}$$

and

$$(\partial \ln \omega_s/\partial T)_p = -\tfrac{1}{2}(\partial \ln \varepsilon/\partial T)_p. \tag{9.4}$$

The pressure and temperature dependences of $\omega_s$ are then readily deduced from the results in Fig. 4.

It is found that $\omega_s$ increases with pressure, as expected for the FE mode, at a rate of ~20%/kbar at 4 K. This rate of increase decreases monotonically with increasing temperature reaching a value of ~1%/kbar at 400 K. This is the expected behavior for the soft mode.

(*iii*) *Evaluation of the Self-Energy Shifts.* The measured temperature dependence of $\omega_s$ (or $\varepsilon$) at constant pressure for a cubic crystal derives from two contributions: (*a*) the contribution associated with the change in crystal volume $V$, i.e., the explicit volume effect; and (*b*) the explicit temperature dependence which would occur even if the volume of the sample were to remain fixed. Writing $\omega_s = \omega_s(V, T)$ we then have

$$\left(\frac{\partial \ln \omega_s}{\partial T}\right)_p = \left(\frac{\partial \ln V}{\partial T}\right)_p \left(\frac{\partial \ln \omega_s}{\partial \ln V}\right)_T + \left(\frac{\partial \ln \omega_s}{\partial T}\right)_V$$

$$= -\frac{\beta}{\varkappa}\left(\frac{\partial \ln \omega_s}{\partial p}\right)_T + \left(\frac{\partial \ln \omega_s}{\partial T}\right)_V. \quad (9.5)$$

Thus, the pure-temperature contribution $(\partial \ln \omega_s/\partial T)_V$ can be evaluated from the isobaric and isothermal derivatives and a knowledge of the volume thermal expansivity, $\beta \equiv (\partial \ln V/\partial T)_p$, and compressibility, $\varkappa \equiv -(\partial \ln V/\partial p)_T$. This has been done for $KTaO_3$, and the results show that the pure-volume and pure-temperature contributions are opposite in sign, but the pure-temperature effect, $(\partial \ln \omega_s/\partial T)_V$, dominates the isobaric temperature derivative of $\omega_s$ at all temperatures examined, as expected (Section 4).[32,70]

This conclusion is made more explicit by evaluating the pure-volume and pure-temperature contributions to $\omega_s$ at every temperature. Rewriting Eq. (9.5) (deleting the subscript s in $\omega_s$ for convenience),

$$(\Delta\omega_T)_p = -(\Delta\omega_p)_T + (\Delta\omega_T)_V, \quad (9.6)$$

$(\Delta\omega_T)_p$ is the measured change in $\omega_s$ on raising the temperature from 0 to $T$ K at constant pressure (1 bar); $(\Delta\omega_T)_V$ is the change in $\omega_s$ caused by raising the temperature from 0 to $T$ K at constant volume, that being the volume of the crystal at 0 K and 1 bar; $-(\Delta\omega_p)_T$ is the change in $\omega_s$ caused by raising the pressure at a constant temperature $T$ from 1 bar to a value $p$ sufficient to produce a volume change equal in magnitude to that caused by raising the temperature from 0 to $T$ K at 1 bar. The term $(\Delta\omega_p)_T$ enters with a negative sign, since the signs of $\beta$ and $\varkappa$ of $KTaO_3$ are normal, i.e., $\Delta V$ is positive on heating and negative on compression, and it is readily evaluated from $\omega_s(p)$ data and the known $\beta$ and $\varkappa$. The temperature dependences of $(\Delta\omega_T)_V$ and $(\Delta\omega_p)_T$ are given in Fig. 6. These results illustrate that the pure-temperature effect dominates the measured $\omega_s(T)$ at all temperatures. The pure-volume effect produces a negative frequency shift which partially cancels the pure-temperature effect.

Let us now examine the implications of these results in terms of the theory developed in Section 4. The origin and sign of the thermal expansion contribution $(\Delta\omega_p)_T$ are easily understood qualitatively. As the interatomic distances increase with increasing $T$, the restoring force for the FE mode, which is a vibration of the Ta ion against the oxygen octahedron, and hence the frequency $\omega_s$, decreases. This same pure-volume effect causes $\omega_s$ to increase with pressure. The effect can be treated more formally in terms of quasi-harmonic lattice dynamics, and the ability to evaluate $(\Delta\omega_p)_T$ as a function of $T$ over a wide temperature

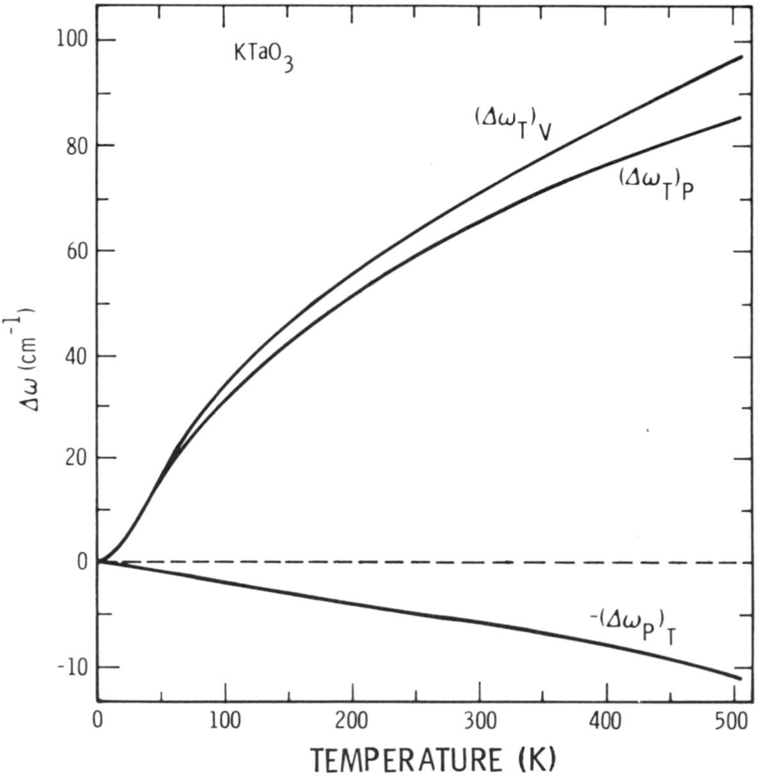


FIG. 6. Decomposition of the measured isobaric temperature dependence of the soft TO mode frequency in $KTaO_3$ into the pure volume and pure temperature contributions. [After G. A. Samara and B. Morosin, *Phys. Rev. B: Solid State* [3] **8,** 1256 (1973).] Note the change of scale for the negative ordinate.

range can be used for obtaining better values of specific dynamical model parameters.[73,74]

In $KTaO_3$ as well as in other displacive or incipient FEs the predominant soft FE mode frequency shift at 1 bar arises from the pure-temperature contribution $(\Delta\omega_T)_V$. The origin of this term lies in the anharmonic vibrations of the ions about their equilibrium positions, and it is responsible for the high temperature stabilization of $\omega_s$. Interpreted in terms of both the perturbation and self-consistent phonon treatments, the positive sign of $(\Delta\omega_T)_V$ implies that this stabilization is dominated by quartic anharmonicities.

[73] W. G. Stirling, *J. Phys. C* [3] **5,** 2711 (1972).

[74] A. D. Bruce and R. A. Cowley, *J. Phys. C* [3] **6,** 2422 (1973).

One of the important results of the anharmonic theory of Section 4 is that, within the approximations made, $\omega_s^2$ should vary linearly with temperature over some suitably high temperature range, i.e., above the Debye temperature, $\theta_D$. However, it has long been recognized that for many members of the perovskite family, $\omega_s^2(T)$ varies linearly over a wide temperature range well below $\theta_D$. This behavior is illustrated by the data for $SrTiO_3$ ($\omega_D \approx 1000$ K) in Fig. 5 which show the linear region extending down to ~50 K. This feature is not explained by perturbation theory; however, Gillis and Koehler[25] found that by treating all modes self-consistently, the density of soft modes with $\hbar\omega \ll kT$ is weighted more heavily at low temperatures than in the perturbation treatments. These authors have also observed that the temperature range over which the linear $\omega_s^2(T)$ response obtains decreases with decreasing quartic anharmonicity. We should also note that the linear $\omega_s^2(T)$ response is obtained in lowest-order treatments. When higher-order anharmonic interactions are included, $\omega_s^2(T)$ is no longer linear with temperature. Interpreted in terms of Gillis and Koehler's results,[25] the fact that the linear $\omega_s^2(T)$ region in $KTaO_3$ is much smaller than that in $SrTiO_3$ probably implies that the quartic anharmonicities play a less dominant role in $KTaO_3$; furthermore, the nonlinear $\omega_s^2(T)$ response at high temperatures suggests that lowest-order anharmonic treatments are not adequate for $KTaO_3$.

*(iv). Stress-Induced Ferroelectricity.* As mentioned earlier, $KTaO_3$ and $SrTiO_3$ remain paraelectric down to the lowest temperatures (<1 K) and the application of hydrostatic pressure increases $\omega_s$ which further stabilizes the PE phases. However, application of sufficient uniaxial stress along certain crystal directions can induce ferroelectricity. This was first investigated by Burke and Pressley[75] in $SrTiO_3$ and then by Uwe and Sakudo on $KTaO_3$[10,76] and $SrTiO_3$.[11,77] $KTaO_3$ is ideal for these studies because in the absence of stress it is cubic at all temperatures, whereas $SrTiO_3$ is tetragonal below 105 K. Space limitations do not allow us to discuss the uniaxial results here. Uwe and Sakudo[10,11] provide an excellent account of the soft-mode aspects of this interesting work.

### b. *Ferroelectrics with Finite $T_c$ and the Vanishing of the Ferroelectricity at High Pressure*

*(i) $K_{1-x}Na_xTaO_3$.* Although pure $KTaO_3$ remains cubic at all temperatures, substitution of Na for K yields mixed crystals which exhibit FE

[75] W. J. Burke and R. J. Pressley, *Solid State Commun.* **9,** 191 (1971).

[76] H. Hwe, H. Unoki, Y. Fujii, and T. Sakudo, *Solid State Commun.* **13,** 737 (1973); H. Uwe and T. Sakudo, *Jpn. J. Phys.* **38,** 183 (1975).

[77] T. Sakudo and H. Uwe, *Ferroelectrics* **8,** 857 (1974).

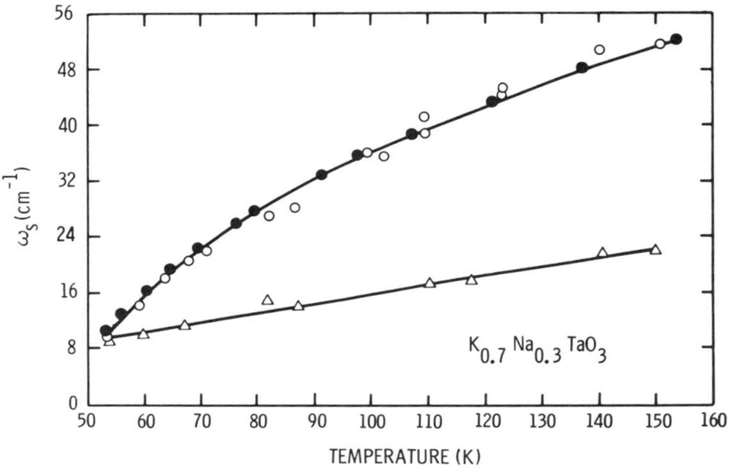


FIG. 7. Temperature dependence of the soft TO mode frequency and linewidth in $K_{0.7}Na_{0.3}TaO_3$ measured using electric field-induced Raman scattering. The frequency evaluated from dielectric measurements is also shown for comparison. [After T. G. Davis, *Phys. Rev. B: Solid State* [3] **5**, 2530 (1972).] (○) Soft mode frequency; (△) mode linewidth; (●) $1.27 \times 10^3\ \varepsilon^{-1/2}$.

transitions. For compositions with ≤30 at.% Na there is a single second-order transition from a high temperature cubic PE phase to a tetragonal FE phase.[78] Davis[78] has investigated the temperature dependence of the soft mode in the cubic PE phase for several compositions using electric field-induced Raman scattering. In all cases investigated the mode was underdamped and was followed through the transition. Typical results on the soft mode frequency and linewidth are shown in Fig. 7 where also values of $\omega_s$ deduced from the static dielectric constant via Eq. (6.3) are shown. Equation (6.3) is well obeyed for each of the compositions investigated.[78]

Figure 8 shows the square of the soft-mode frequency versus temperature at different pressures in the cubic PE phase for a crystal with $x = 0.28$. These data were deduced from combined field-induced Raman scattering[78] (available only at 1 bar) and static dielectric constant[79] measurements. Note that the $\omega_s^2(T)$ response is linear over a substantial temperature range. According to Eq. (4.11), the extrapolation of this linear response to 0 K gives an estimate of the harmonic frequency $\omega_0$ of the soft mode. At atmospheric pressure (1 bar) $\omega_0^2$ is negative and $\omega_0$ is

[78] T. G. Davis, *Phys. Rev. B: Solid State* [3] **5**, 2530 (1972).
[79] G. A. Samara, *Ferroelectrics* **7**, 221 (1974).

therefore imaginary. Several important pressure effects are evident. At constant temperature the measured soft-mode frequency $\omega_s$ increases markedly with pressure, the Curie temperature decreases, and $\omega_0^2$ changes from negative to positive at sufficiently high pressure, so that $\omega_0$ becomes real and finite. When $\omega_0 > 0$ the crystal becomes stable with respect to the soft mode at all temperatures and the transition vanishes. At the lowest temperatures, zero-point anharmonicities help to further stabilize the crystal and contribute to the deviation from linearity of the $\omega_s^2(T)$ response in Fig. 8.

The vanishing of the FE phase is shown in Fig. 9, where $\varepsilon(T)$ is plotted at various pressures. $T_c$ decreases with an initial slope of $-4.5$ K/kbar. There is no transition above 10 kbar. These effects can be expected to be qualitatively typical of the behavior of other perovskites as well as other displacive soft-mode FEs.

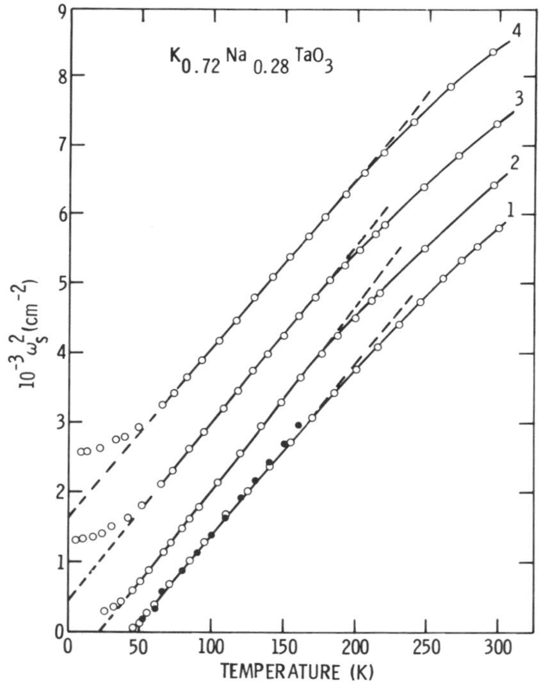


FIG. 8. Temperature dependence of the soft TO mode frequency determined from dielectric measurements in $K_{0.72}Na_{0.28}TaO_3$ for various pressures. [After G. A. Samara, *Ferroelectrics* **7**, 221 (1971).] (1) 1 bar; (2) 6 kbar; (3) 15 kbar; (4) 25 kbar.

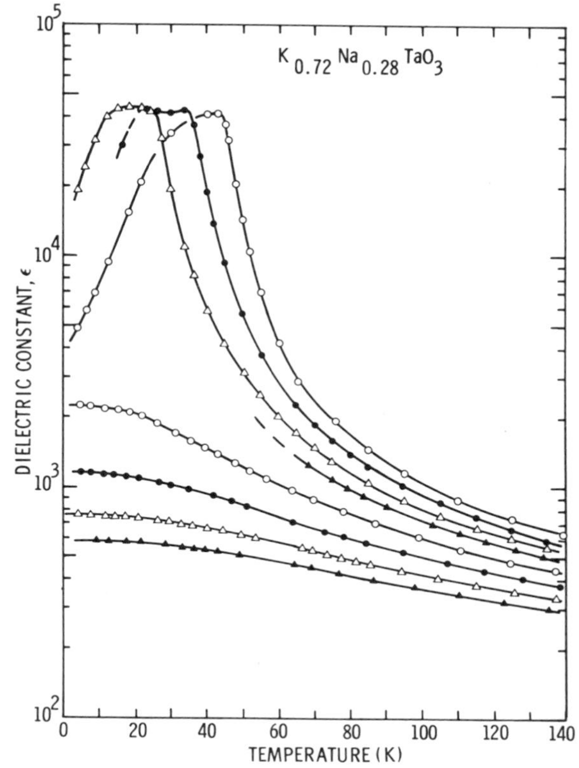


FIG. 9. Temperature dependence of the static dielectric constant of $K_{0.72}N_{0.28}TaO_3$ at various pressures illustrating the suppression of the FE phase. For upper curves: (○) 1 bar; (●) 2 kbar; (△) 4; (▲) 6; for lower curves: (○) 10 kbar; (●) 15; (△) 20; (▲) 25. [After G. A. Samara, *Ferroelectrics* **7**, 221 (1974).]

(*ii*) *$PbTiO_3$*. $PbTiO_3$ is a classic displacive FE whose dielectric and soft-mode properties have been widely investigated and found to follow the soft-mode picture. On cooling, the crystal transforms from the cubic PE phase to a tetragonal FE phase at ~765 K. The tetragonal distortion is large with $c/a \simeq 1.06$.[16] (For example, $BaTiO_3$ exhibits a similar transition at ~393 K with $c/a \simeq 1.01$.) The transition in $PbTiO_3$ is strongly first order with discontinuous changes in physical properties and thermal hysteresis. Unlike $BaTiO_3$, $PbTiO_3$ apparently does not undergo other transitions upon further cooling.

The effects of pressure on the dielectric properties[80] and the Raman-

[80] G. A. Samara, *Ferroelectrics* **2**, 277 (1971), and references therein.

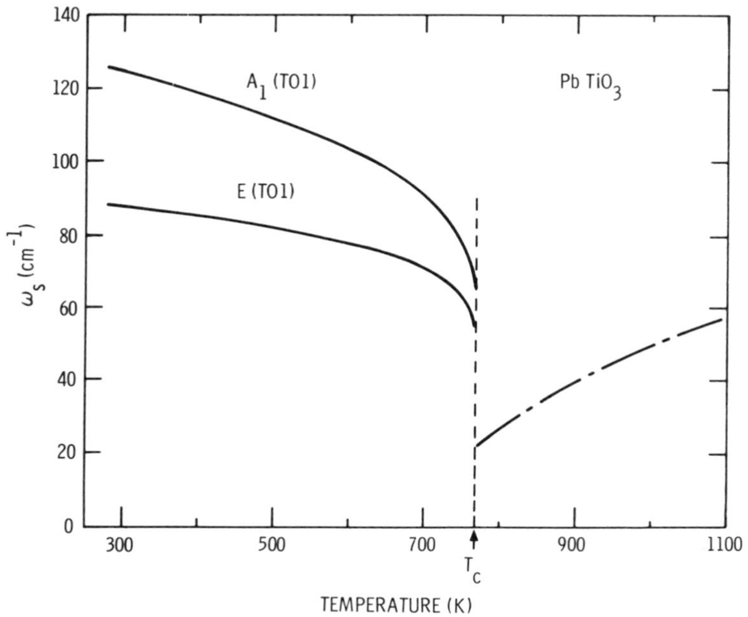


FIG. 10. Temperature dependence of the soft TO mode frequency in $PbTiO_3$ measured by neutron[59] and Raman[82] scattering techniques. —, Raman; ---, neutron.

active phonons[81] in $PbTiO_3$ have been reported. The dielectric results are qualitatively similar to those observed for $BaTiO_3$ and have been reviewed earlier,[9] the main pressure effects being (*a*) a large decrease in $T_c$ ($dT_c/dp = -8.4$ K/kbar), (*b*) an increase in the peak value of $\varepsilon$, and (*c*) a decrease in the temperature hysteresis at $T_c$. Results *b* and *c* suggest that the transition may become second order at sufficiently high pressure.[80]

The temperature and pressure dependences of $\varepsilon$ of $PbTiO_3$ in the cubic phase can be understood in terms of the behavior of the soft FE mode as discussed earlier for $K_{1-x}Ta_xO_3$. The temperature dependence of the soft-mode frequency at 1 bar has been studied by inelastic neutron scattering[59] and Raman[82] spectroscopy. Unlike $BaTiO_3$, where the soft mode in the PE phase is overdamped, in $PbTiO_3$ the mode is underdamped and well defined. Figure 10 shows the $\omega_s(T)$ dependence at 1 bar. The fact that $\omega_s$ remains finite at $T_c$ is consistent with the transition being first order.

[81] F. Cerdeira, W. B. Holzapfel, and D. Bauerle, *Phys. Rev. B: Solid State* [3] **11,** 1188 (1975).

[82] G. Burns and B. A. Scott, *Phys. Rev. B: Solid State* [3] **7,** 3088 (1973).

The cubic perovskite structure (space group $O_h^1$-$Pm3m$) has twelve $\mathbf{q} \cong 0$ optic phonon modes which transforms as the $3T_{1u} + T_{2u}$ irreducible representation of the point group $O_h$. The triply degenerate $T_{1u}$ ($\Gamma_{15}$) modes are IR active but not Raman active. At $\mathbf{q} \approx 0$ the electrostatic interactions lower the degeneracy splitting each of the $T_{1u}$ modes into a doubly degenerate TO mode and a singly degenerate LO mode. The soft FE mode is one of these TO modes. The $T_{2u}$ ($\Gamma_{25}$) mode is also triply degenerate but is optically silent. In the tetragonal ($C_{4v}'$-$P4mm$) phase the irreducible representations become $T_{1u} \rightarrow A_1 + E$ and $T_{2u} \rightarrow B_1 + E$ of the point group $c_{4v}$. All of these modes are Raman active and all but $B_1$ are IR active.

The effect of pressure on the frequencies of the TO modes in the $C_{4v}$ FE phase at room temperature have been reported by Cerdeira *et al.*[81] and are shown in Fig. 11, where the labeling follows that given by Burns and Scott.[82] These results show a variety of features, but perhaps most important is that the separation of the pairs of modes $A_1$(TO3) : $E$(TO4) and $A_1$(TO2) : $E$(TO2) decreases appreciably with increasing pressure. Since increasing pressure in the FE phase causes the crystal to approach the transition, these results identify the above pairs as belonging to cubic $T_{1u}$ representations which are split by the tetragonal anisotropy in the FE phase. The mode labeled $E$(TO3) + $B_1$ is identified as originating from the silent $T_{2u}$ mode of the cubic phase, which is apparently not split by the tetragonal distortion. The mode labeled $E$(TO1) is the soft mode. Its frequency decreases rapidly with pressure. Cerdeira *et al.*[81] find that the square of this frequency decreases linearly with pressure and extrapolates to zero at ~90 kbar at room temperature. This pressure is comparable to the pressure where the frequencies of the aforementioned $A_1$ : $E$ pairs are expected to cross (see Fig. 11).

Although the frequency of the soft mode as well as the frequencies of several other modes in the FE phase decrease with pressure, the normal behavior is for phonon frequencies to increase with pressure. In the FE phase, however, the soft mode is stabilized largely by the nonzero value of the order parameter. The order parameter (which is proportional to the spontaneous ionic displacements), and hence the stabilization of the frequencies, increases as the system moves away from the transition into the FE phase. Pressure reduces the order parameter by pushing the crystal closer to the transition, and thus decreases the frequencies (Fig. 11). These results are consistent with the measurements of the pressure dependence of the lattice parameters by Ikeda *et al.*,[83] which show that

[83] T. Ikeda, K. Inoue, and A. Nakaue, *Proc. Int. Conf. High Pressure, 4th, 1974* p. 236 (1975).

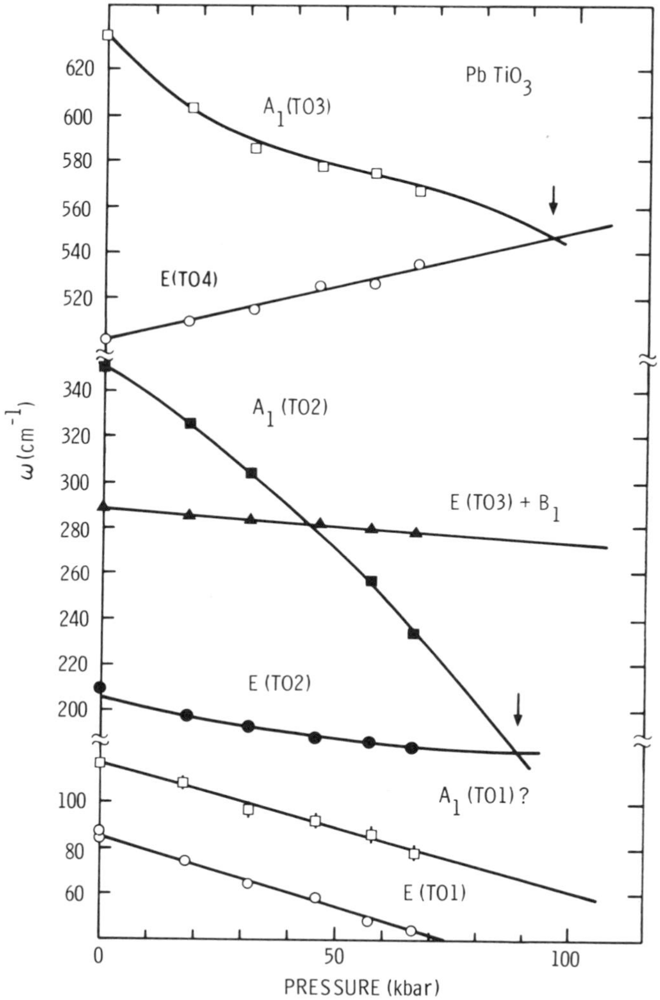


FIG. 11. Pressure dependences of the frequencies of the TO modes in the FE phase of $PbTiO_3$ at 295 K. [After F. Cerdeira, W. G. Holzapfel, and D. Bauerle, *Phys. Rev. B: Solid State* [3] **11,** 1188 (1975).]

*c*/*a* and the spontaneous strain decrease markedly with pressure. The spontaneous strain is simply related to the square of the spontaneous polarization.

Finally, we note that the Curie constant $C$ and the spontaneous polarization $P$ can be related to the dynamic properties of the soft FE mode. Cochran[2] has shown that for a displacive FE $C$ is given by

$$C = 4\pi v P^2(T - T_0)/\omega_s^2 \sum_i m_i \langle u_i^2 \rangle, \tag{9.7}$$

where $v$ is the unit cell volume, $\omega_s$ the soft-mode frequency, $m_i$ the mass of ion $i$, and $\langle u_i^2 \rangle$ its rms displacement. Combining Eqs. (2.17) and (9.7) yields

$$P^2 = \frac{\varepsilon \omega_s^2}{4\pi v} \sum_i m_i \langle u_i^2 \rangle. \tag{9.8}$$

The validity of Eq. (9.8) has been tested for $BaTiO_3$ and $PbTiO_3$.[80,84] For $BaTiO_3$ the experimental value of $C$ agrees well with the value calculated from Eq. (9.8)—specifically $C_{exp} = (1.5 \pm 0.3) \times 10^5$ K compared with $C_{calc} = (1.2 \pm 0.3) \times 10^5$ K. The agreement is much poorer for $PbTiO_3$. If we take $\omega_s^2 \varepsilon$ to be a constant, then the pressure and temperature dependences of the quantity $\Sigma\ m_i \langle u_i^2 \rangle$, which is dominated by the displacements associated with the soft mode, are simply related to the measured dependences of the polarization which are known.[9,80]

(*iii*) *Other Ferroelectric Perovskites.* Two other important members of the FE perovskite family are $BaTiO_3$ and $KNbO_3$. These materials have remarkably similar properties. On cooling they exhibit the following sequence of phase transitions[16]: cubic PE to tetragonal FE (polar axis along the $c$ axis) to an orthorhombic FE phase (polar axis parallel to a face diagonal) and finally to a rhombohedral FE phase (polar axis along a body diagonal). The soft modes in these crystals have been studied by inelastic light and neutron scattering.[85-87] In the cubic phase the soft-mode frequency exhibits the expected temperature dependence, i.e., $\omega_s^2 = A(T - T_0)$, but in both crystals the mode is overdamped even at $T \gg T_0$ and the damping is highly anisotropic.[86,87]

The effects of pressure on the dielectric properties of $BaTiO_3$ have been thoroughly investigated and were reviewed earlier.[9,80] An important feature of the response of $BaTiO_3$ is the coupling that occurs between the soft FE mode and an acoustic mode of the same symmetry as the frequency of the FE mode decreases into the domain of the acoustic mode on approaching the transition. The temperature and pressure dependences of this coupling have been investigated in the FE tetragonal phase on

[84] W. Cochran, *Adv. Phys.* **18,** 157 (1969).

[85] M. DiDomenico, S. P. S. Porto, and S. Wemple, *Phys. Rev. Lett.* **19,** 855 (1967).

[86] G. Shirane, J. D. Axe, J. Harada, and A. Linz, *Phys. Rev. B: Solid State* [3] **2,** 3651 (1970).

[87] A. C. Nunes, J. D. Axe, and G. Shirane, *Ferroelectrics* **2,** 291 (1971).

approaching the tetragonal-orthorhombic transition at 280 K. These results along with coupled-mode interactions in other crystals will be discussed in Section VII.

Apparently there are no pressure measurements on $KNbO_3$, presumably because suitable crystals have not been readily available and the transition temperatures are considerably higher than in $BaTiO_3$ (thus making the pressure experiments more difficult). However, the pressure effects are expected to be very similar to those in $BaTiO_3$, as has been recently demonstrated by results on mixed $K_{1-x}Na_xNbO_3$ samples.[88]

### 10. Other Displacive Ferroelectrics

#### *a. SbSI and Its Isomorphs*

Antimony sulfo-iodide (SbSI) and some of its isomorphs (e.g., SbSBr and BiSI) exhibit an unusual combination of strongly coupled photoconductive, semiconductive, and ferroelectric properties which have been subjects of much recent interest.[89] Of these crystals SbSI has been the most widely studied. In its FE phase, SbSI is the most strongly piezoelectric crystal known.

The spatial arrangement of the atoms in SbSI is shown in Fig. 12a. The structure is orthorhombic with four molecules per unit cell and with space group $D_{2h}^{16}$-*Pnam* above and $C_{2v}^{9}$-$Pna2_1$ below the FE transition temperature $T_c \approx 290$ K. The structure is highly anisotropic consisting of doubly linked chains extending along the *c* axis with weak coupling between adjacent double chains. The transition is first order involving relative displacements of the Sb and S ions along the *c* axis. The pattern of these displacements relative to the iodine ions is shown in Fig. 12b. Although the lattice dynamics of the crystal are not fully known at present, Raman scattering[90–92] and IR reflectivity[92,93] measurements at atmospheric pressure have established that the transition is driven by the softening of the frequency $\omega_s$ of a long wavelength TO mode with the displacements of

[88] G. A. Samara, to be published.

[89] See, e.g., M. Balkanski, M. K. Teng, S. M. Shapiro, and M. K. Ziolkiewicz, *Phys. Status Solidi B* **44,** 355 (1971), for pertinent references.

[90] E. F. Steigmeier, G. Harbeke, and R. K. Wehner, *in* "Light Scattering in Solids" (M. Balkanski, ed.), p. 397. Flammarion Sciences, Paris, 1971.

[91] M. Balkanski, M. K. Teng, M. Massot, and S. M. Shapiro, *in* "Light Scattering in Solids" (M. Balkanski, ed.), p. 392. Flammarion Sciences, Paris, 1971.

[92] D. K. Agrawal and C. H. Perry, *Phys. Rev. B: Solid State* [3] **4,** 1893 (1971).

[93] F. Sugawara and T. Nakamura, *J. Phys. Chem. Solids* **33,** 1665 (1972), and references therein.

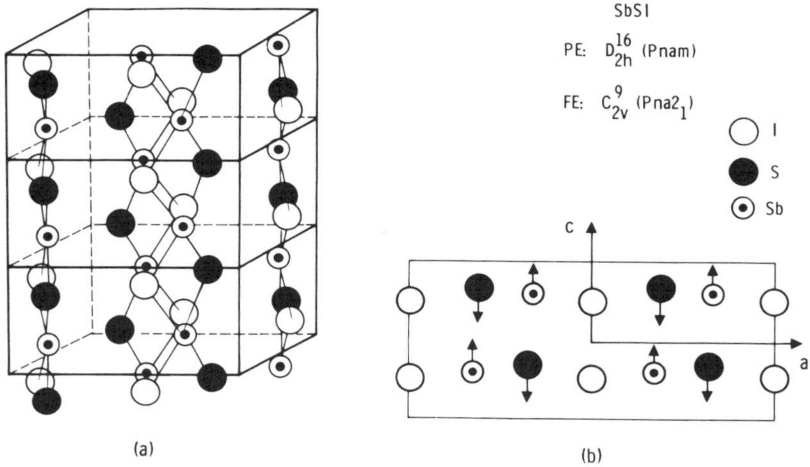


FIG. 12. (a) Structure of SbSI and (b) atomic motion associated with the soft TO mode. [After M. Balkanski, M. K. Teng, S. M. Shapiro, M. K. Ziolkiewicz, *Phys. Status Solidi B* **44**, 355 (1971).]

$Sb^{3+}$ and $S^{2-}$ ions indicated in Fig. 12b. These measurements have also established that the temperature dependence of $\omega_s$ is responsible for the $\varepsilon_c(T)$ response in both the PE and FE phases.

Figure 13 shows the temperature dependence of $\omega_s$ at 1 bar. In the FE phase the soft mode is both Raman and IR active, whereas in the PE phase the soft mode is only IR active (symmetry $A_{2u}$). Because IR reflectivity measurements are difficult at very low frequencies, $\omega_s$ is not known as accurately in the PE phase as in the FE phase.

Early reports[94] suggested that on further cooling a second-order transition occurs at ~233 K with the space group becoming $C_2^2$-$P2_1$. Evidence for this transition, however, was not conclusive. More recent dielectric, Raman, and x-ray measurements fail to reveal any evidence for this transition.[92,95]

The strong electromechanical coupling in SbSI suggests that its electrical properties should be very pressure sensitive, and this has led to a number of studies dealing with the effects of pressure on its FE proper-

[94] See, e.g., V. N. Nosov, *Sov. Phys.—Crystallogr. (Engl. Transl.)* **13,** 273 (1968); E. I. Gerzanich and V. M. Fridkin, *Sov. Phys.—Solid State (Engl. Transl.)* **10,** 2452 (1969); D. M. Bercha, I. V. Bercha, V. Yu. Slivka, I. D. Turyanitsa, and D. V. Chepur, *ibid.* **11,** 1356 (1969).

[95] G. A. Samara, *Ferroelectrics* **9,** 209 (1975), and references therein.

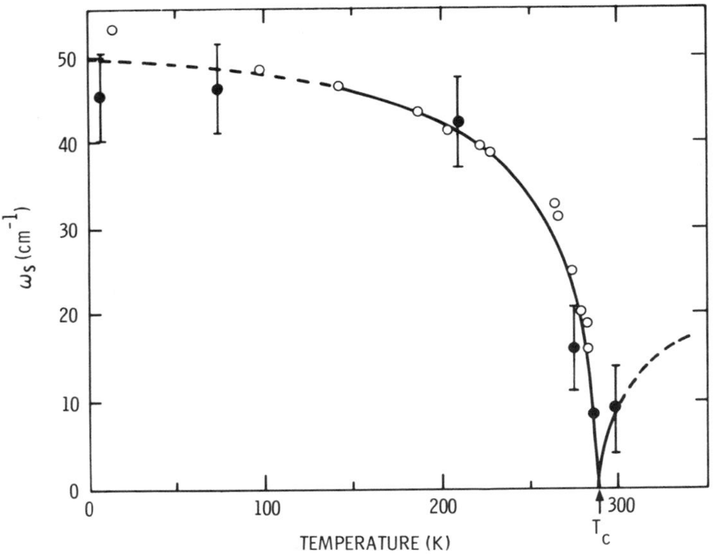


FIG. 13. Temperature dependence of the soft TO mode frequency in SbSI. [After D. K. Agrawal and C. H. Perry, *Phys. Rev. B: Solid State* [3] **4,** 1893 (1971).] (●) Infrared data, (○) Raman data.

ties,[95] energy gap,[96] and Raman spectra.[97,98] Extremely large pressure effects were observed. Some of the early results were discussed previously,[9] and here we shall restrict our comments to the more recent dielectric and Raman results.

The dielectric constant results[95] are qualitatively similar to those of $K_{1-x}Na_xTaO_3$ discussed in Section 9,b and show that the FE transition vanishes above 8 kbar. Figure 14 shows the pressure dependence of $T_c$. The initial shift is linear with $dT_c/dp = -39.6 \pm 0.6$ K/kbar. At high pressure $T_c(p)$ becomes nonlinear, but much or all of this nonlinearity is probably associated with the nonlinear pressure dependence of the volume of SbSI. The slope $dT_c/dp$ appears finite as $T_c \rightarrow 0$ K, similar to the behavior of the perovskites, and this, as we have seen, can be understood in terms of the soft-mode picture. It is in marked contrast with the behavior of $KH_2PO_4$-type crystals to be discussed in Section V.

[96] V. M. Fridkin, K. Gulyamov, V. A. Lyakhovitskaya, V. N. Nosov, and N. A. Tikhomirova, *Sov. Phys.—Solid State* (*Engl. Transl.*) **8,** 1510 (1966).

[97] M. K. Teng, M. Balkanski, and M. Massot, *Phys. Rev. B: Solid State* [3] **3,** 1031 (1972).

[98] P. S. Peercy, *Phys. Rev. Lett.* **35,** 1581 (1975); *Proc. Int. Conf. Low Lying Vibrational Modes Their Relationship Supercond. Ferroelectr., Ferroelectrics* **16,** 193 (1977).

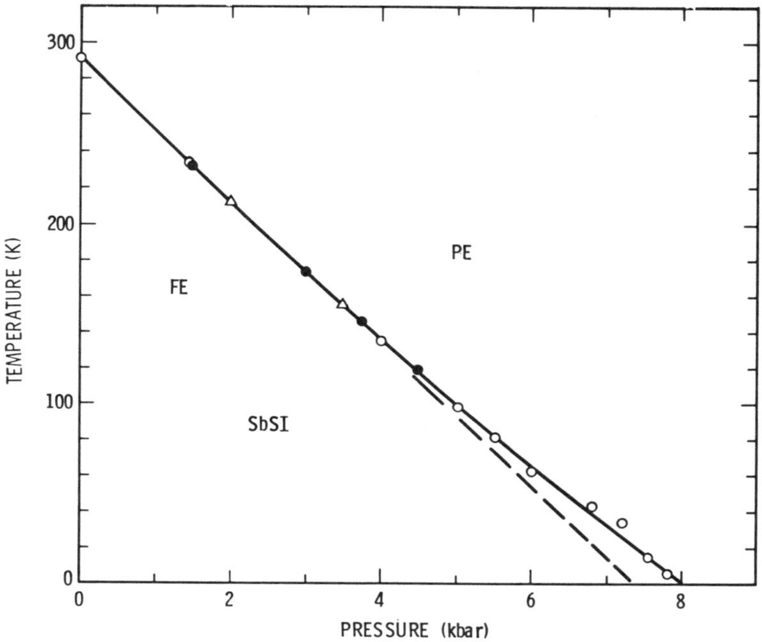


FIG. 14. Pressure dependence of the transition temperature of SbSI. [After G. A. Samara, *Ferroelectrics* **9**, 209 (1975).] Different symbols represent different samples.

Figure 15 shows $\varepsilon_c^{-1}(T)$ results in the PE phase at different pressures. The response obeys the simple Curie–Weiss law, Eq. (2.17), over a substantial temperature range. Deviations from this intermediate temperature approximation at high temperature reflect (*a*) the increased importance of relatively temperature-independent contributions arising from the large electronic polarization ($\varepsilon_\infty = n^2 \simeq 25$) as well as from lattice polarization associated with modes other than the FE mode, and (*b*) high-order anharmonic contributions to $\omega_s$. Low-temperature deviations from the Curie–Weiss law are observed below ~60 K (not shown) and are associated with anharmonic quantum effects. Both the Curie constant $C$ and the Curie–Weiss temperature decrease with pressure in a manner similar to that observed for the perovskites.[95]

The $\varepsilon_c^{-1}(T)$ response in Fig. 15 also represents $B'\omega_s^2(T)$, where to a very good approximation $B'$ should be a constant independent of pressure and temperature. Although $B'$ can be determined from the known values of $\varepsilon_c$ and $\omega_s$ at any given temperature, the absolute values of $\omega(T)$ are uncertain. For example, Agrawal and Perry[92] give $\omega_s \simeq 9\ \text{cm}^{-1}$ at 1 bar and 298 K,

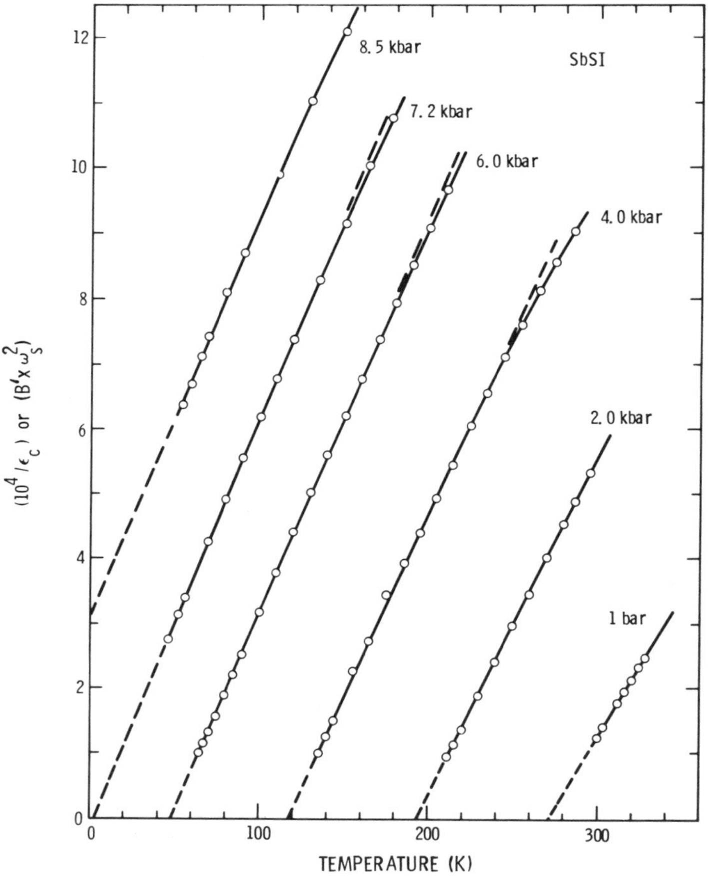


FIG. 15. Plots of the inverse of the $c$-axis dielectric constant of SbSI illustrating the Curie–Weiss behavior exhibited by SbSI over a wide temperature range. [After G. A. Samara, *Ferroelectrics* **9**, 209 (1975).]

whereas Sugawara and Nakamura[93] give $\omega_s \simeq 4\ \text{cm}^{-1}$ at the same conditions. Despite this difficulty, $\varepsilon_c^{-1}(T, p)$ should accurately reflect the $\omega_s^2(T, p)$ response. Representing the linear response in Fig. 15 by Eq. (4.11), it is seen that for pressures $\leq 7.2$ kbar $\omega_0$ is imaginary. At higher pressures $\omega_0$ becomes real and finite, which implies that the lattice becomes stable with respect to this mode at all temperatures; i.e., the FE transition vanishes.

An important feature of the results in Fig. 14 is that the value of $dT_c/dp$ is about the same when $T_c \simeq 290$ K (the 1 bar value) as it is when $T_c \lesssim 10$

K. Since the magnitude of the anharmonic contributions must have decreased drastically between 290 and 10 K, the experimental evidence indicates that $dT_c/dp$ is dominated by changes in the harmonic interactions, i.e., by changes in $\omega_0$, as noted above. This result, which obtains also for the perovskites and is expected to hold for other soft-mode FEs, serves to emphasize an important point. Whereas at finite temperatures all of the stabilization of the soft-mode frequency of a FE is provided by anharmonic interactions [the $\alpha T$ term in Eq. (4.11)], the most interesting and unusual features of the soft mode are not so much a consequence of the large anharmonicities as they are a result of the extremely small (or imaginary) harmonic frequency. This point was emphasized early in the development of the soft-mode concept.[1,4]

As mentioned earlier, the soft FE mode in SbSI is Raman active only in the FE phase. Its temperature dependence has been studied by several groups,[90–92] and its pressure dependence was studied by Balkanski *et al.*[97] and more recently by Peercy,[98] whose work emphasized the behavior of the mode very near the transition. Some of Peercy's results are shown in Fig. 16 for $T = 119$ K. At this temperature, the transition occurs at a pressure $p_c$ of 4.54 kbar. Starting at a slightly higher pressure (4.59 kbar in Fig. 16) where the crystal is in the PE state and decreasing the pressure below $p_c$, one sees in Fig. 16 the beautiful evolution of the soft mode. Below $p_c$ (i.e., in the FE phase), the soft mode first appears as a shoulder on the laser line; as the pressure is further decreased, the mode moves smoothly to higher frequency as an underdamped excitation in the FE phase. At pressures such that $\omega_s \simeq 30$ cm$^{-1}$, the soft mode interacts with another optic mode of the crystal. As shown in Fig. 16, this latter mode is initially quite weak, but as the frequency of the soft mode approaches it, there is an intensity transfer between the modes. It is also evident that at still lower pressures than in Fig. 16, the soft mode interacts with another optic mode at $\sim$40 cm$^{-1}$. This latter mode, which was part of a degenerate doublet in the PE phase, splits smoothly into two lines as the pressure is decreased. The pressure dependences of the various mode frequencies are shown in Fig. 17 illustrating this splitting and the mode couplings. It can be seen from these data that the interactions between the modes are quite weak with level repulsions of $\lesssim 2$ cm$^{-1}$ in the case of both couplings. These mode couplings will be discussed in more detail in Section VII.

In Fig. 16 it is seen that the soft mode is underdamped and well resolved near the transition at 119 K. At this temperature, the mode is much less heavily damped than at the atmospheric pressure transition temperature of 292 K. Thus, by pressure the transition has been moved into a temperature regime which has made it possible to interpret unambiguously the soft mode behavior and couplings shown in Fig. 17. This feature

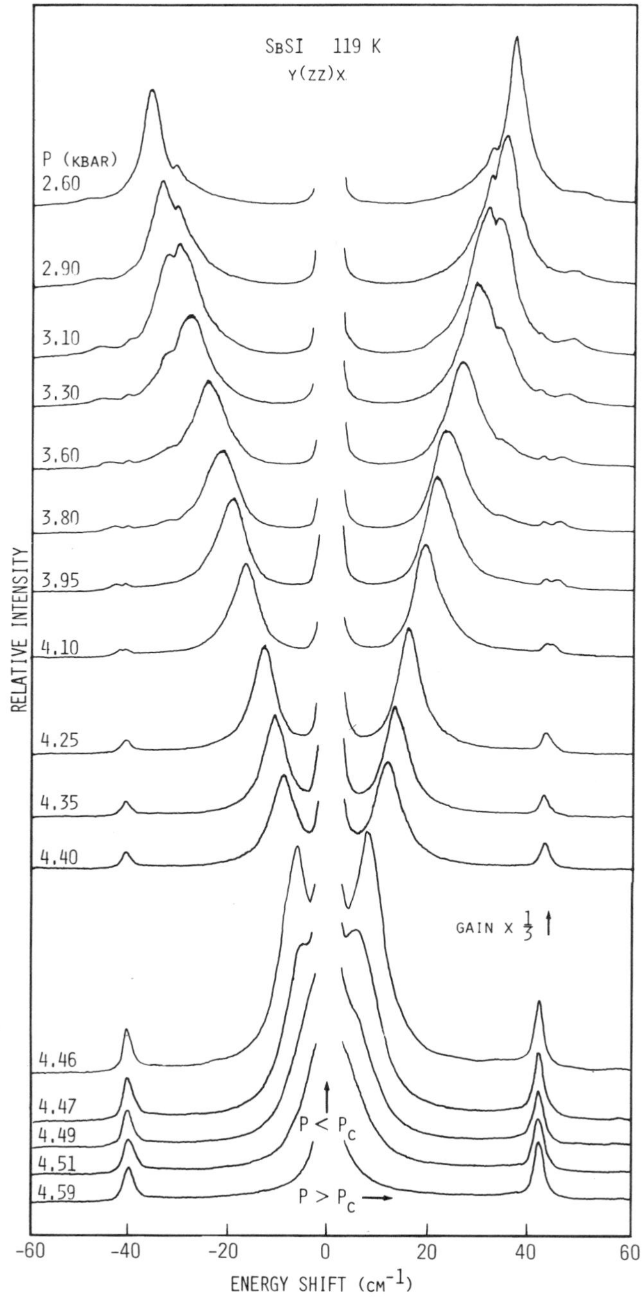


FIG. 16. Raman spectra showing the pressure dependence of the soft TO mode in the FE phase of SbSI near the transition pressure of 4.54 kbar at 119 K. [After P. S. Peercy, *Phys. Rev. Lett.* **35**, 1581 (1975).]

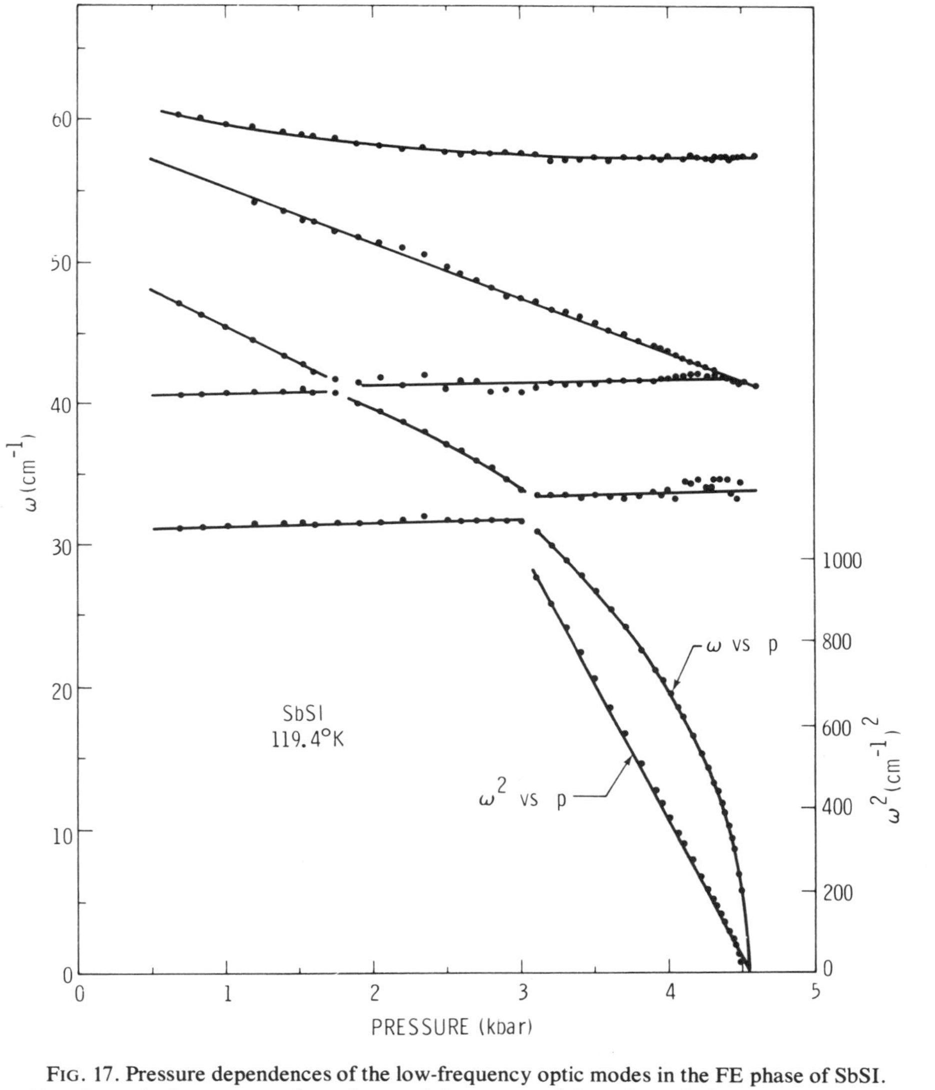


FIG. 17. Pressure dependences of the low-frequency optic modes in the FE phase of SbSI. Also shown is the pressure dependence of the square of the soft-mode frequency near the transition pressure. [After P. S. Peercy, *Phys. Rev. Lett.* **35**, 1581 (1975).]

illustrates an additional benefit of the use of pressure as a variable in the study of soft modes.

The soft mode in SbSI exhibits mean-field behavior with pressure. This can be inferred from the $\varepsilon(p, T)$ results[95] and is further illustrated in Fig. 17 where $\omega_s^2$ is plotted versus pressure. These results demonstrate that $\omega_s^2$

$\propto (p_c - p)$ for pressures sufficiently close to $p_c$ so that the soft mode does not couple with other optic modes. This pressure dependence of $\omega_s$, which holds for other soft-mode FEs, is, therefore, in agreement with the Landau treatment presented in Section 2.

### b. $Pb_5Ge_3O_{11}$

Lead germanate, $Pb_5Ge_3O_{11}$ (or $5PbO \cdot 3GeO_2$), is a relatively newly discovered[99,100] FE having a transition temperature of $T_c = 450$ K at 1 bar. The space group is trigonal $C_3^{\prime}$-$P3$ in the FE phase and becomes hexagonal $C_{3h}^{\prime}$-$P\bar{6}$ in the nonpolar PE phase. There is a soft FE mode associated with the transition. The mode has been studied in the FE phase by Raman scattering[101] measurements which show evidence that the mode interacts with other optic modes. There is also evidence for coupling between this mode and acoustic modes leading to anomalies in the elastic constants at $T_c$.[102] There are apparently no direct detailed measurements of the soft mode in the PE phase. In this phase, the mode is not Raman active, but it is IR active as can be inferred from the observed Curie–Weiss behavior of the dielectric constant $\varepsilon_{33}$.[100] The Curie constant is $C = 1.2 \times 10^4$ K. Recent neutron scattering results show evidence for a broad overdamped soft mode above $T_c$.[103]

The effects of pressure up to 10 kbar on the dielectric constant $\varepsilon_{33}$ and $T_c$ have been reported.[104,105] The results are qualitatively similar to those observed for perovskites (see Section 9). Gesi and Ozawa[104] observed a superlinear decrease of $T_c$ with pressure, the initial slope being $-6.7$ K/kbar. Kirk *et al.*,[105] on the other hand, report a linear decrease of $T_c$ up to 10 kbar with slope of $-15.5$ K/kbar. It is not clear why there is this large quantitative difference between the two results; a reexamination of this pressure effect is desirable. In any case, however, the decrease of $T_c$ with pressure is undoubtedly associated with an increase in the soft-mode frequency in the PE phase with pressure.

### c. $Gd_2(MoO_4)_3$

Gadalinium molybdate and its isomorphous rare earth molybdates, $R_2(MoO_4)_3$ (where R is Gd, Sm, Eu, Tb, or Dy), constitute an important

[99] H. Iwasaki, K. Sugii, T. Yamada, and N. Niizeki, *Appl. Phys. Lett.* **18,** 444 (1971).
[100] S. Nanamatsu, H. Sugiyama, K. Doi, and Y. Kondo, *Jpn. J. Phys.* **31,** 616 (1971).
[101] J. F. Ryan and K. Hisano, *J. Phys. C* [3] **6,** 566 (1973).
[102] G. R. Barsch, L. J. Bonczar, and R. E. Newnham, *Phys. Status Solidi B* **29,** 241 (1975).
[103] R. A. Cowley, J. D. Axe, and M. Iizumi, *Phys. Rev. Lett.* **36,** 806 (1976).
[104] K. Gesi and K. Ozawa, *Jpn. J. Appl. Phys.* **13,** 897 (1974).
[105] J. L. Kirk, L. E. Cross, and J. P. Dougherty, *Ferroelectrics* **11,** 439 (1976).

class of ferroelectrics which have been subjects of much interest. They undergo FE transitions in the 420–460 K range.[106] In the PE phase, the crystals are tetragonal, space group $D_{2d}^{3}$–$P_4\bar{2}_1m$, whereas in the FE phase, the symmetry is orthorhombic, space group $C_{2v}^{8}$–$Pba2$. One of the most striking properties of these crystals, and one that contrasts with the behavior of most FEs, is that only a very small dielectric constant anomaly is observed along the polar orthorhombic $c$ axis at the transition, and this anomaly disappears when the crystal is clamped, i.e., for measurements above the mechanical resonance frequencies.[107] The transition is also accompanied by a doubling of the unit cell.[108–110] These facts emphasize that the order parameter cannot be a homogeneous polarization (or any other homogeneous parameter) and, therefore, these crystals are unlike other FEs.

Indeed this has been shown to be the case. Theoretical analyses[108,109] and neutron scattering results[110] have shown that, unlike other displacive FE transitions which involve soft zone-center polar phonons, the transition in the molybdates is driven by a soft zone-boundary optic phonon at the $M(\frac{1}{2}, \frac{1}{2}, 0)$ point of the Brillouin zone. The ionic displacements associated with this mode (primarily counter-rotations of $MoO_4$ tetrahedra) are antiferrodistortive and cannot directly produce the observed spontaneous polarization $P_s$ in the FE phase. Rather, $P_s$ and the accompanying spontaneous strain $u_s$ arise indirectly as a result of anharmonic coupling between the soft-mode displacements and zone-center optic and acoustic phonon displacements. The ferroelectricity below $T_c$ is thus accidental, being a consequence of this coupling. Because $P_s$ is not the primary order parameter for the transition, the molybdates belong to a class of so-called *improper* ferroelectrics.[109]

Because the soft mode in these crystals is a nonpolar zone-boundary phonon, we shall discuss and compare the pressure effects on these crystals with those of other crystals with transitions associated with short wavelength optic phonons. This is done in Section IV.

### 11. The Grüneisen Parameter of the Soft Ferroelectric Mode

In addition to the static susceptibility, many other physical properties of crystals possessing a soft FE mode are influenced by this mode. For

[106] H. J. Borchardt and P. E. Bierstedt, *J. Appl. Phys.* **38,** 2057 (1967).

[107] L. E. Cross, A. Fouskova, and S. E. Cummins, *Phys. Rev. Lett.* **22,** 812 (1968).

[108] E. Pytte, *Solid State Commun.* **8,** 2101 (1970).

[109] J. Petzelt and V. Dvorak, *Phys. Status Solidi B* **46,** 413 (1971).

[110] J. D. Axe, B. Dorner and G. Shirane, *Phys. Rev. Lett.* **26,** 519 (1971); *Phys. Rev. B: Solid State* [3] **6,** 1950 (1972), and references therein.

example, the ultrasonic properties,[111] thermal conductivity,[112] critical scattering,[84] and nature of the phase transition[113] are dominated by interactions with the FE mode. In treating many of these properties it is often necessary to know the volume (or strain) dependence of the soft-mode frequency, i.e., the mode Grüneisen parameter $\gamma_s$. This is especially true for theoretical treatments of ultrasonic attenuation and velocity where $\gamma_s$ enters as one of the principal parameters. $\gamma_s$ can be determined from measurements of the shift of $\omega_s$ with pressure by either optical (IR or Raman) or inelastic neutron spectroscopy. The pressure dependence of $\omega_s$ can also be determined from the pressure dependence of $\varepsilon$.

The soft-mode Grüneisen parameter is defined by[114]

$$\gamma_s \equiv -\left(\frac{\partial \ln \omega_s}{\partial \ln V}\right)_T = \frac{1}{\varkappa}\left(\frac{\partial \ln \omega_s}{\partial p}\right)_T \tag{11.1}$$

$$\cong -\frac{1}{2\varkappa}\left(\frac{\partial \ln \varepsilon}{\partial p}\right)_T, \tag{11.2}$$

where $\varkappa$ is the isothermal volume compressibility, and Eq. (11.2) follows from Eq. (6.3). For most FEs the isothermal pressure dependence of $\varepsilon$ in the PE phase obeys the relationship $\varepsilon = C^*/(p - p_o)$ over a wide pressure range [Eq. (2.19)]. Since $C^*$ and $p_0$ are constants at a given temperature,

$$(\partial \ln \varepsilon/\partial p)_T = -\varepsilon/C^*, \tag{11.3}$$

which yields

$$\gamma_s = \varepsilon/2\varkappa C^*. \tag{11.4}$$

Thus, $\gamma_s$ can be evaluated from easily measurable quantities.

Both $\varkappa$ and $C^*$ are very weak functions of $T$; e.g., for $SrTiO_3$ $(\partial \ln C^*/\partial T) \simeq -1.3 \times 10^{-4}\ K^{-1}$, whereas $(\alpha \ln \varkappa/\partial T) \simeq +1.8 \times 10^{-4}\ K^{-1}$.[114] Because the temperature dependences of $\varkappa$ and $C^*$ very nearly cancel each other, the temperature dependence of $\gamma_s$ primarily reflects that of $\varepsilon$. Combining Eqs. (2.17) and (11.4) yields

$$\gamma_s = \left(\frac{C}{2\varkappa C^*}\right)\left(\frac{1}{T - T_0}\right) \simeq \frac{K}{T - T_0}, \tag{11.5}$$

where $K$ is a constant; i.e., $\gamma_s$ obeys the Curie–Weiss law and diverges as $T \to T_0$.

[111] H. H. Barrett, *Phys. Rev.* [2] **178,** 743 (1969).
[112] R. E. Nettleton, *Phys. Rev.* [2] **140,** A1453 (1965).
[113] P. C. Kwok and P. B. Miller, *Phys. Rev.* [2] **151,** 387 (1966).
[114] G. A. Samara, *Ferroelectrics* **2,** 277 (1971).

The data necessary to evaluate $\gamma_s$ and its temperature dependence for a number of crystals are given in Table I. Of the crystals listed, $SrTiO_3$ and $KTaO_3$ are incipient FEs for which $\varepsilon(T)$ deviates from a simple Curie–Weiss behavior at low and high temperatures (Section 9a). In regions where such deviations occur, $\gamma_s$ should be evaluated directly from Eq. (11.2) and not from the data in Table I. $BaTiO_3$ and $PbTiO_3$ exhibit classic FE transitions, and Eq. (11.4) applies. $PbZrO_3$ and $PbHfO_3$ on the other hand, undergo PE → AFE transitions, but they also possess soft FE modes that are responsible for the large polarizabilities and Curie–Weiss behavior of their $\varepsilon$'s in the PE phase (Section IV). Equation (11.4) applies well to them. Figure 4 illustrates the $\gamma_s(T)$ response for $KTaO_3$. This response is to be contrasted with the nearly temperature-independent optic- and acoustic-mode Grüneisen parameters (~1–3) normally observed for most ionic crystals. The large values and strong temperature dependence of $\gamma_s$ are a manifestation of the unique properties of the soft mode whose frequency $\omega_s$ becomes increasingly smaller as $T \rightarrow T_0$.

It is also possible to determine $\gamma_s$ from the phenomenological theory of ferroelectricity which relates the electrostrictive constants $g_{ij}$ to the static susceptibility $\chi_0$. For the perovskites, the appropriate result is[114]

$$\gamma_s = \tfrac{1}{2}\chi_0[\tfrac{1}{3}g_{11} + \tfrac{2}{3}g_{12}]. \quad (11.6)$$

The electrostrictive constants are generally temperature independent; thus, for large $\chi_0$, Eq. (11.6) has the same form as Eq. (11.4).

Accurate values of the electrostrictive constants are known for only a few crystals of interest. For $SrTiO_3$[114] $g_{11} \simeq 5$ and $g_{12} \ll g_{11}$ and is thus

TABLE I. VALUES OF THE ISOTHERMAL COMPRESSIBILITY $\varkappa$, THE TRANSITION TEMPERATURE $T_c$, THE CURIE CONSTANT $C$, THE CURIE TEMPERATURE $T_0$, AND THE CONSTANTS $C^*$ AND $K$ APPEARING IN EQS. (11.4) AND (11.5) FOR A NUMBER OF CUBIC PEROVSKITE CRYSTALS[a]

| Crystal | $\varkappa$ ($10^{-4}$ kbar$^{-1}$) | $T_c$ (K) | $C$ ($10^5$ K) | $T_0$ (K) | $C^*$ ($10^4$ kbar) | $K$ ($10^3$ K) |
|---|---|---|---|---|---|---|
| $SrTiO_3$ | 5.81 | — | 0.83 | 36 | 1.26 | 5.67 |
| $KTaO_3$ | 5.05 | — | 0.57[b] | 4 | 1.23 | 4.59[b] |
| $BaTiO_3$ | 7.8 | 395 | 1.60 | 385 | 3.58 | 2.86 |
| $PbTiO_3$ | 11.5 | 765 | 4.1 | 722 | 4.55 | 3.92 |
| $PbZrO_3$ | 13.4 | 507 | 1.60 | 475 | 1.00 | 5.97 |
| $PbHfO_3$ | 14.2 | 476 | 1.65 | 378 | ~1.1 | ~5.3 |
| SbSI[c] | ~34 | 292 | 2.26 | 272 | 0.6 | 5.5 |

[a] After G. A. Samara, *Ferroelectrics* **2**, 277 (1971).
[b] See the discussion in the text.
[c] Data from G. A. Samara, *Ferroelectrics* **9**, 209 (1975), and references therein.

negligible. Equation (11.6) then yields $\gamma_s \simeq \frac{1}{2}\chi_0(\frac{5}{3})c = 0.066\varepsilon$ compared to $\gamma_s = 0.068\varepsilon$ obtained from Eq. (11.4) and the data in Table I. The agreement is excellent considering the uncertainty in $g_{11}$.

## IV. Displacive Antiferroelectric and Antiferrodistortive Transitions: Soft Short-Wavelength Optic Phonons

Because of the antiparallel displacement of similar ions in adjacent unit cells, Cochran[2] was the first to suggest that a transition to an AFE state can be associated with an instability of a mode whose wavelength is on the order of a lattice parameter, e.g., a zone-boundary mode. The existence of soft zone-boundary modes responsible for phase transitions has now been established experimentally for several crystals. It is not, of course, necessary that all transitions driven by soft zone-boundary modes involve AFE phases; in fact, the best understood cases do not. Here again, the perovskite structure has provided the best examples. Figures 2c and 2d show the eigenvectors of two other soft modes in this structure. These modes consist of rotations of the oxygen octahedra about axes through the central B ions as shown. The mode of symmetry $R_{25}$ corresponds to a phonon at the $R$ point, or corner, of the Brillouin zone (shown in Fig. 18) and involves the counterrotations of adjacent octahedra. Mode $M_3$ corresponds to a phonon at the edge of the Brillouin zone, and here adjacent octahedra rotate similarly about the indicated axis, but this rotation is opposite in sense from those about adjacent parallel axes. The order parameters for the associated transitions are the angles of rotation

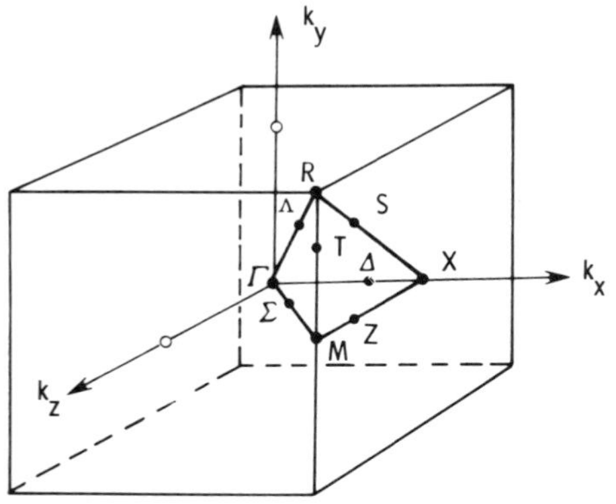


FIG. 18. Schematic illustration of the unit cell for the perovskite structure labeling the symmetry of points of interest in soft-mode systems.

of the octahedra. Because these zone-boundary modes have no net dipole moments, no dielectric constant anomalies are generally associated with their softening when the phase transitions are second order. When the transitions are first order, on the other hand, the accompanying lattice strains will often lead to small discontinuities in the dielectric constant.

One or both of these two modes are known to soften with decreasing temperature and lead to second-order (or nearly so) phase transitions in a number of perovskites. We shall discuss some examples later in this section.

## 12. The Antiferroelectric Perovskites

Experimentally an interesting situation is observed in $PbZrO_3$, $PbHfO_3$, $NaNbO_3$, and a number of other perovskites. The situation is best illustrated by the dielectric response at 1 bar of $PbZrO_3$ shown in Fig. 19. In the high-temperature phase $PbZrO_3$ is cubic and its dielectric constant obeys a Curie–Weiss law, as is true of the isomorphous crystal $PbTiO_3$.[80] However, unlike $PbTiO_3$, which transforms on cooling to a FE phase, $PbZrO_3$ transforms at 505 K to an orthorhombic AFE phase.

Since according to the LST relationship, a large $\varepsilon$ is associated with the existence of a low-frequency $\mathbf{q} = 0$ TO mode, the above behavior suggests that in $PbZrO_3$ there are two low-frequency and nearly degenerate temperature-dependent modes: a FE mode that determines the large $\varepsilon$ and an AFE mode associated with the transition. On lowering the temperature, the crystal becomes unstable with respect to the AFE mode at the AFE transition temperature $T_a$ before the instability due to the FE mode is reached at $T_0$. While the validity of this picture can in principle be checked by measuring the temperature dependence of the phonon dispersion curves, data are apparently not available because of the lack of suitable single crystals. In the absence of such data, high-pressure measurements provided the first confirmation that the large $\epsilon$ anomaly at the AFE transition and the transition itself are indeed determined by two independent modes.[115] Whereas for $PbTiO_3$ both the transition temperature $T_c$ and $T_0$ decrease with pressure and the dielectric anomaly is enhanced,[80] the behavior of $PbZrO_3$, shown in Fig. 19, is quite different. For $PbZrO_3$, $T_0$ decreases corresponding to an increase in the FE mode frequency, whereas $T_a$ increases corresponding to a softening of the AFE mode frequency. As a result, the dielectric anomaly decreases dramatically.

[115] G. A. Samara, *Phys. Rev. B: Solid State* [3] **1**, 3777 (1970).

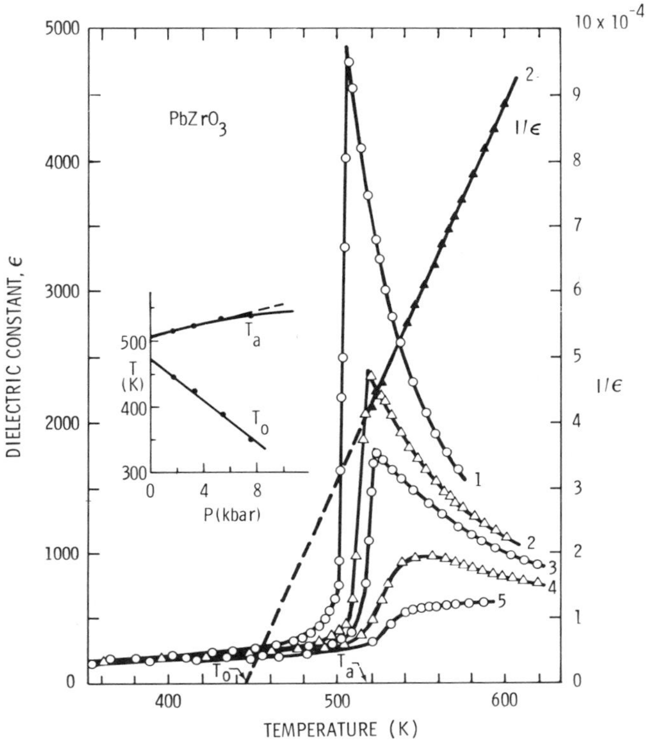


FIG. 19. Temperature dependence of the static dielectric constant of $PbZrO_3$, which exhibits an AFE phase transition, at various pressures. The inset shows the pressure dependences of the AFE transition temperature $T_a$ and the Curie–Weiss temperature $T_0$. (1) $P = 1$ bar; (2) 1.90 kbar; (3) 3.40 kbar; (4) 7.60 kbar; (5) 11.40 kbar. [After G. A. Samara, *Phys. Rev. B, Solid State* [3] **1,** 3777 (1970).]

In the PE phase of $PbZrO_3$ the pressure dependence of $\varepsilon$, and thereby the FE soft-mode frequency $\omega_s$, is very similar to that for the FE perovskites.[115] Specifically, at constant temperature $\varepsilon$ obeys the expression $\varepsilon = C^*/(p - p_0)$, and the data allow the evaluation of the soft FE mode Grüneisen parameter $\gamma_s$ (Table I).

In the absence of direct phonon measurements to identify the AFE mode, some insight into the nature of this mode can be gained by examining the displacements of the various ions that occur at the PE → AFE transition. Since

$$\omega(\mathbf{q}, j)^2 \propto 1/\langle Q(\mathbf{q}, j)\rangle, \tag{12.1}$$

where $\langle Q(\mathbf{q}, j)\rangle$ is the thermal expectation value of the normal-mode coordinate, the pattern of the displacements associated with the soft mode will be "frozen in" the new phase.[113] In $PbZrO_3$, x-ray and neutron diffraction studies[116] have shown that the atomic motion associated with the transition to the orthorhombic AFE phase involves primarily antiparallel displacements of the Pb ions along one of the cubic $\langle 110 \rangle$ directions and antiparallel displacements of the oxygen ions in the (001) plane.

Cochran and Zia[117] examined the measured displacements and concluded that several modes may be involved in the transition (which is first order). Using the notation $(\mathbf{q}, R)$ to identify a wave vector $\mathbf{q}$ for which the pattern of displacements is described by the irreducible representation $R$, they find that at least four values of $(\mathbf{q}, R)$ are represented in the low-temperature orthorhombic phase: $(000, \Gamma_{15})$, $(\frac{1}{4}\frac{1}{4}0, \Sigma_3)$, $(\frac{1}{2}\frac{1}{2}0, M_5')$, and $(\frac{1}{2}\frac{1}{2}\frac{1}{2}, \Gamma_{25})$. The first of these is associated with the $\mathbf{q} = 0$ FE mode. The fact that the symmetry change involves more than one irreducible representation is consistent with the transition being first order. Without further information it is difficult to assess the relative contribution of each of these modes to the transition. However, since the unit cell is enlarged by a factor of 8 at the cubic or orthorhombic transition, the softening of $\omega(\Sigma_3)$ may be an important factor in the $PbZrO_3$ transition.

Another important AFE perovskite is $PbHfO_3$. Although the pressure dependence of the dielectric properties of this crystal are qualitatively similar to those of $PbZrO_3$, the crystal exhibits two AFE transitions at 1 bar. A third AFE phase becomes stable at high pressure.[115] All of the AFE transition temperatures increase with pressure corresponding to the softening of their respective AFE modes with pressure. Here again there is no direct information concerning the nature of these modes. In the PE phase the effects of pressure on the dielectric properties are similar to those for $PbZrO_3$ (see Table I).

$NaNbO_3$ also exhibits AFE behavior. No pressure results are available on this crystal possibly because the PE → AFE transition temperature is high and the crystal exhibits a complicated sequence of at least five phase transitions between 635 and 915 K. There have been some recent advances into the nature of the lattice dynamical origins of some of these transitions. Specifically, the cubic-tetragonal transition at 915 K and the tetragonal-orthorhombic transition at 845 K are now believed to result, respectively, from the condensation of the $M_3$ phonon and a phonon in the tetragonal phase that derives from the $\Gamma_{25}$ mode of the cubic phase.[118,119] We expect both transition temperatures to increase with pressure.

[116] F. Jona, G. Shirane, F. Mazzi, and R. Pepinsky, *Phys. Rev.* [2] **105,** 849 (1957).

[117] W. Cochran and A. Zia, *Phys. Status Solidi* **25,** 273 (1968).

[118] A. M. Glazer and H. D. Megaw, *Philos. Mag.* **25,** 1119 (1972); M. Ahtee *et al., ibid.* **26,** 995 (1972).

[119] K. Ishida and G. Honjo, *J. Phys. Soc. Jpn.* **34,** 1279 (1973).

## 13. Antiferrodistortive Transitions in $SrTiO_3$ and Its Isomorphs

It has long been recognized that $SrTiO_3$ undergoes a structural phase transition at ~105 K. This transition, from the high-temperature cubic phase to a tetragonal ($D_{4h}^{18}$-$I4/mcm$) phase, involves a doubling of the unit cell and is accompanied by large anomalies in the elastic properties but not in the dielectric constant. Fleury *et al.*[120] attributed the dynamical origin of the transition to the softening of the $R_{25}$ zone-boundary phonon (Fig. 2c) of the cubic phase. Because of the large wave vector associated with this mode in the cubic phase, it is not amenable to study by light scattering techniques. Light scattering can be used to investigate this mode in the tetragonal phase, however, because the zone boundary of the cubic phase becomes the zone center of the tetragonal phase when the unit cell doubles.

The $SrTiO_3$ transition is an ideal second-order transition which has been studied in considerable detail. The understanding of this transition has also led to the understanding of several similar transitions in other perovskites, e.g., $LaAlO_3$, $KMnF_3$, and $CsPbCl_3$. In $LaAlO_3$ the transition (527 K) is also caused by condensation of the $R_{25}$ phonon at the $R$ point, but the low-temperature phase is rhombohedral ($D_{3d}^{6}$-$R\bar{3}c$) and not tetragonal as in $SrTiO_3$.[121] This difference is a consequence of the fact that the $R_{25}$ mode is threefold degenerate in the cubic phase; therefore three independent linear combinations of ionic displacements are possible, namely rotations of oxygen octahedra about [100], [110], and [111] axes, resulting in tetragonal, orthorhombic, and rhombohedral structures, respectively. The distortion that occurs in a given crystal is determined by the anharmonic forces and energetics of the particular crystal.

In $KMnF_3$ there are two structural phase transitions.[122] The first, at 184 K, results from the condensation of the $R_{25}$ mode, whereas the second, at ~90 K, results from the condensation of the $M_3$ zone-boundary phonon (Fig. 2d). This sequence is reversed in $CsPbCl_3$ which exhibits three successive phase transitions at 320, 315, and 311 K.[123] The first transition results from the condensation of the $M_3$ mode, whereas the second and third transitions result from the condensation in the tetragonal phase of modes which derive from the triply degenerate $R_{25}$ mode of the cubic phase.

[120] P. A. Fleury, J. F. Scott, and J. M. Worlock, *Phys. Rev. Lett.* **21,** 16 (1968).

[121] J. D. Axe, G. Shirane, and K. A. Muller, *Phys. Rev.* [2] **183,** 820 (1969).

[122] V. J. Minkiewicz and G. Shirane, *J. Phys. Soc. Jpn.* **26,** 674 (1969); *Solid State Commun.* **8,** 1941 (1970).

[123] Y. Fujii, S. Hoshino, Y. Yamada, and G. Shirane, *Phys. Rev. B: Solid State* [3] **9,** 4549 (1974).

Apparently no direct pressure measurements on the behavior of the soft-mode frequency have been made for any of the above crystals. However, the effects of pressure on the transition temperatures have been reported for several cases. The results are summarized in Table II. In all

TABLE II. PRESSURE DEPENDENCE OF THE TRANSITION TEMPERATURE $T_c$ FOR A VARIETY OF CRYSTALS THAT EXHIBIT DISPLACIVE STRUCTURAL PHASE TRANSITIONS[a]

| Crystal | Symmetry of high-temperature phase | Soft mode (symmetry) | $T_c$[b] (K) | $dT_c/dP$ (K/kbar) | Reference |
|---|---|---|---|---|---|
| $BaMnF_4$ | Orthogonal-$C_{2v}^{12}$ | z.b.[c] | 247.3 | 3.3 | Samara *et al.*[e] |
| $Gd_2(MoO_4)_3$ | Tetragonal-$D_{2d}^{3}$ | z.b. | 435 | 29.5 | Shirokov *et al.*[f] |
| $(NH_4)_2Cd_2(SO_4)_3$ | Cubic-$T^4$ | z.b. | 92 | 3.3 | Glogarova *et al.*[g] |
| $SrTiO_3$ | Cubic-$O_h^1$ | z.b. ($R_{25}$) | 110 | 1.7 | Sorge *et al.*[h] |
| | | z.c. ($\Gamma_{15}$) | 36 | −14.0 | Samara[i] |
| $KMnF_3$ | Cubic-$O_h^1$ | z.b. ($R_{25}$) | 186 | 3.0 | Okai and Yoshimoto[j] |
| $CsPbCl_3$ | Cubic-$O_h$ | z.b. ($M_3$) | 320 | 7.6 | Gesi *et al.*[k] |
| | | z.b. ($R_{25}$)[d] | 315 | 5.2 | Gesi *et al.*[k] |
| | | z.b. ($R_{25}$)[d] | 311 | 5.4 | Gesi *et al.*[k] |
| $PbZrO_3$ | Cubic-$O_h^1$ | z.b. | 507 | 4.5 | Samara[l] |
| | | z.c. ($\Gamma_{15}$) | 475 | −16.0 | Samara[l] |
| $PbHfO_3$ | Cubic-$O_h^1$ | z.b. | 434 | 5.9 | Samara[l] |
| | | z.c. ($\Gamma_{15}$) | 378 | −10.0 | Samara[l] |
| $BaTiO_3$ | Cubic-$O_h^1$ | z.c. ($\Gamma_{15}$) | 393 | −5.2 | Samara[i] |
| $PbTiO_3$ | Cubic-$O_h^1$ | z.c. ($\Gamma_{15}$) | 765 | −8.4 | Samara[i] |
| SbSI | Orthogonal-$D_{2h}^{16}$ | z.c. | 292 | −37.0 | Samara[m] |
| $Pb_5Ge_3O_{11}$ | Hexagonal-$C_{3h}^1$ | z.c. | 450 | −6.7 | Gesi and Ozawa[n] |

[a] After G. A. Samara, *Comments Solid State Phys.* **8,** 13 (1977).
[b] $T_c$ is either the actual transition temperature of the Curie–Weiss temperature or deduced from the static susceptibility.
[c] There is some evidence that this soft mode is a short-wavelength phonon incommensurate with the lattice (see text). z.b., Zone-boundary; z.c., zone-center.
[d] These modes derive from the $R_{25}$ mode of the cubic phase.
[e] G. A. Samara *et al.*, *Phys. Rev. Lett.* **35,** 1767 (1975).
[f] A. M. Shirokov *et al.*, *Sov. Phys. Solid State* (*Engl. Transl.*) **13,** 2610 (1972).
[g] M. Glogarova *et al.*, *Phys. Status Solidi B* **53,** 369 (1972).
[h] G. Sorge *et al.*, *Phys. Status Solidi* **37,** K17 (1970).
[i] G. A. Samara, *Ferroelectrics* **2,** 277 (1971).
[j] B. Okai and J. Yoshimoto, *J. Phys. Soc. Jpn.* **34,** 873 (1973).
[k] K. Gesi *et al.*, *J. Phys. Soc. Jpn.* **38,** 463 (1975).
[l] G. A. Samara, *Phys. Rev. B: Solid State* [3] **1,** 3777 (1970).
[m] G. A. Samara, *Ferroelectrics* **9,** 209 (1975), and references therein.
[n] K. Gesi and K. Ozawa, *Jpn. J. Appl. Phys.* **13,** 897 (1974).

cases the transition temperature $T_c$ *increases* with pressure, in contrast to the behavior of transitions associated with the $\mathbf{q} = 0$ FE mode for which $T_c$ and $T_0$ *decrease* with pressure. This contrast is highlighted in Table II for $SrTiO_3$ which has both kinds of soft modes. The interpretation of this behavior is discussed later (Section 15).

## 14. Antiferrodistortive Transitions in $Gd_2(MoO_4)_3$ and $BaMnF_4$

Table II also summarizes results on $PbZrO_3$, $PbHfO_3$, and two other crystals which do not belong to the perovskite family, namely $Gd_2(MoO_4)_3$ and $BaMnF_4$, but which undergo transitions resulting from the softening of short-wavelength optic phonons.

As already noted (Section 10), in $Gd_2(MoO_4)_3$ and its isomorphs the cell-doubling phase transition results from the condensation of a doubly degenerate phonon at the $(\frac{1}{2}, \frac{1}{2}, 0)$ $M$ point of the high-temperature phase. Neutron scattering results[110] have shown that the square of the soft-mode frequency, $\omega_s^2$, decreases linearly with decreasing temperature and $\omega_s$ becomes very small, but not zero, at $T_c$ since the transition is first order. In Raman scattering in the low temperature orthorhombic phase there is a soft $A_1$(TO) mode with a peak at 47 $cm^{-1}$ at room temperature in $Gd_2(MoO_4)_3$.[124] This peak shifts to lower frequencies and broadens on heating and ultimately the response becomes overdamped near $T_c$. Ganguly *et al.*[125] studied the effects of uniaxial stress along [100] on this Raman-active mode at several temperatures within 20 K below $T_c$. An increase in peak frequency and narrowing of the Raman line is observed with increasing stress. A Lorentzian line shape analysis of the data showed that the force constant parameter increased slightly with stress but the damping constant decreased markedly. These results indicate that $T_c$ increases with this stress. Apparently, the only hydrostatic pressure measurements available are those on the shift of $T_c$.[126,127] The initial slope $dT_c/dp$ is given in Table II. The increase of $T_c$ with pressure can be interpreted as resulting from the softening of $\omega_s$ in the PE phase.

$BaMnF_4$ is orthorhombic (space group $C_{2v}^{12}$-$A2_1am$) at room temperature, the structure consisting of layered sheets of linked $MnF_6$ octahedra with the Ba ions located between the layers.[128] On cooling, the crystal

[124] P. A. Fleury, *Solid State Commun.* **8,** 601 (1970).
[125] B. N. Ganguly, F. G. Ullman, R. D. Kirby, and J. R. Hardy, *Phys. Rev. B: Solid State* [3] **12,** 3783 (1975).
[126] A. M. Shirokov, V. P. Mylov, A. I. Baranov, and T. M. Prokhortseva, *Sov. Phys.—Solid State (Engl. Transl.)* **13,** 2610 (1972).
[127] G. A. Samara, *in* "High Pressure and Low Temperature Physics" (C. W. Chu and J. A. Woollam, eds.), p. 255, Plenum, New York, 1978.
[128] E. T. Keve, S. C. Abrahams, and J. L. Bernstein, *J. Chem. Phys.* **51,** 4828 (1969).

undergoes an apparently second-order structural transition at 247 K.[129–132] Raman scattering results[130] indicated that the transition is associated with a zone-boundary soft optical phonon. This conclusion was based on measurements below $T_c$ which showed a soft optic mode whose frequency vanished at $T_c$ with no evidence for the mode above $T_c$. Since all optic modes of the high-temperature phase's $C_{2v}$ point group are Raman active, it was concluded that the soft mode below $T_c$ corresponds to a zone-boundary phonon of the high-temperature phase, and thus the transition involves doubling of the primitive unit cell below $T_c$. On this basis, it has been suggested[133,134] that the low-temperature structure should be monoclinic (space group $P2_1$), but this remains unconfirmed. Single-crystal x-ray diffraction studies[135] show no definite evidence for the transition, thus emphasizing its subtle nature. A mechanism for the transition, based on the known Raman, ultrasonic, and other data, has been proposed.[133] It involves rotations of the $MnF_6$ octahedra about the orthorhombic $b$ axis, with a pair of phonons at the Brillouin zone $S$ points $\pi(0, 1/b, \pm 1/c)$ being the soft modes. Though consistent with the then available data, this mechanism is not strictly correct. Recent inelastic neutron scattering work[136] indicates that the soft mode is more complicated than was suggested by the Raman work. Specifically, the neutron results suggest that the wavelength of the soft mode is incommensurate with the lattice. At any rate, it is certain that the transition involves a short-wavelength soft optic phonon that most likely involves rotations of adjacent $MnF_6$ octahedra. In this latter regard the $BaMnF_4$ transition has some similarity to the transitions in $SrTiO_3$ and $Gd_2(MoO_4)_3$, and also similarly, its transition temperature increases with pressure (Table II).

## 15. Theoretical Treatment

Table II reveals some important features which we now wish to dwell on. For some of the crystals in the table including $Gd_2(MoO_4)_3$ and $BaMnF_4$ the transitions are associated with soft short-wavelength optic

[129] E. G. Spencer, H. J. Guggenheim, and G. J. Kominiak, *Appl. Phys. Lett.* **17**, 300 (1970).
[130] J. F. Ryan and J. F. Scott, *Solid State Commun.* **14**, 5 (1974); *in* "Light Scattering in Solids" (M. Balkanski, R. C. C. Leite, and S. P. S. Porto, ed.). Wiley, New York, 1976.
[131] I. J. Fritz, *Phys. Lett. A* **51**, 219 (1975).
[132] G. A. Samara and P. M. Richards, *Phys. Rev. B: Solid State* [3] **14**, 5073 (1976).
[133] I. J. Fritz, *Phys. Rev. Lett.* **35**, 1511 (1975).
[134] V. Dvorak, *Phys. Status Solidi B* **71**, 269 (1975).
[135] B. Morosin, private communication.
[136] S. M. Shapiro, R. A. Cowley, D. E. Cox, M. Eibschütz, and H. J. Guggenheim, *in* "Neutron Scattering" (R. M. Moon, ed.). Gatlinburg, Tennessee, 1976.

phonons, whereas for others, the transitions (FE) involve soft zone-center ($\mathbf{q} = 0$) TO phonons. In some cases (e.g., $SrTiO_3$, $PbZrO_3$, and $PbHfO_3$) the crystals possess both types of soft modes. Note that in all cases where the transition involves a zone-center FE TO mode $T_c$ *decreases* with increasing pressure, whereas for those transitions involving short-wavelength optic modes $T_c$ *increases* with increasing pressure. This appears to be a general result for relatively simple displacive phase transitions involving soft optic phonons.[137–139] This result has important implications relating to the nature of the balance of forces that leads to the softening of the modes in the two cases.

From Eq. (4.12) the increase of $T_c$ with pressure for those transitions associated with soft short-wavelength optic phonons can be interpreted as resulting from a decrease in the soft-mode frequency with pressure. Thus, for soft short-wavelength optic phonons, the mode frequency in the high-temperature phase softens with both decreasing temperature and increasing pressure—both influences leading to a decrease in unit cell volume. This is perhaps not too surprising since for these soft modes the interaction between lattice strain and the order parameter of the transition appears to be a key feature.[140] Phenomenological theory employing optical soft mode and elastic strain coordinates has been developed[5,139] that can account for the observed properties (at least for the perovskites), including the increase of $T_c$ with pressure.[141]

Space limitations do not allow us to go into the details of the theory here. It suffices to indicate that in this theory the (configurational) free energy is written as a Landau series expansion in terms of the order parameters plus terms corresponding to (*a*) the interactions between the

[137] G. A. Samara, T. Sakudo, and K. Yoshimitsu, *Phys. Rev. Lett.* **35**, 1767 (1975).

[138] G. A. Samara, *Comments Solid State Phys.* **8**, 13 (1977).

[139] The case of Co-I-boracite ($Co_3B_7O_{13}I$) should be mentioned here. This material is believed to be an improper ferroelectric and its transition temperature decreases with pressure [see J. Fousek, F. Smutny, C. Frenzel, and E. Hegenbarth, *Ferroelectrics* **4**, 23 (1972)]. On the surface, this result appears to be an exception to the generalizations discussed above. However, it should be emphasized that our knowledge of the nature of the soft mode and crystal structure of the boracites is not complete. In addition, several of these material, including Co-I-boracite, exhibit simultaneous ferromagnetic and ferroelectric order parameters with coupling between the two phenomena, but the exact nature of the coupling is not clear. Furthermore, the transition in Co-I-boracite is strongly first order involving large lattice strains and thermal hysteresis and these features complicate the description of the transition in terms of a soft-mode picture. Thus, there are many unanswered questions which need to be resolved before the pressure results on Co-I-boracite can be put in the proper perspective.

[140] See, e.g., J. C. Slonczewski and H. Thomas, *Phys. Rev. B: Solid State* [3] **1**, 3599 (1970).

[141] K. Fossheim and B. Berre, *Phys. Rev. B: Solid State* [3] **5**, 3292 (1972).

order parameters and elastic strains, (*b*) elastic energy, and (*c*) interactions between applied stress and elastic strain (see Section 2). For the soft zone-boundary phonon transitions in the perovskites, the most general case requires three order parameters $Q_i$ corresponding to the three possible rotations of the octahedra about the cubic axes.[140,141]

From the lattice dynamical point of view, there appears to be a physical explanation for the above behavior of $T_c$ and $\omega_s$ for antiferrodistortive transitions that contrasts with that of the FE soft-mode transitions described earlier (Section III). The clue is suggested by the results of lattice dynamical calculations. Such calculations are apparently available in sufficient detail only for $SrTiO_3$, but it is believed that the results are qualitatively true for other soft zone-boundary optic phonons.[138]

Stirling[73] has studied three *harmonic* models: (*a*) rigid ion, (*b*) rigid shell, and (*c*) deformable shell. He employed measured phonon dispersion curves plus other auxiliary data to describe the frequencies and eigenvectors of the normal modes of $SrTiO_3$ throughout the whole zone within the framework of these models. For all models, the short-range forces are dominated by Ti–O and Sr–O interactions. Bruce and Cowley[74] extended these results by assuming that the anharmonic forces are also dominated by these interactions. They evaluated the various parameters for the anharmonic shell model using the available experimental data. In all of the above models, the parameters implicit in the coefficients of the Hamiltonian are related to the real-space expansion of the potential energy, and hence to the derivatives of the potential describing the interactions among the atoms.

For our present purposes, it suffices to examine the results of the harmonic models. The characteristic frequencies of the $T_2$ phonon branch along the $(\frac{1}{2}\frac{1}{2}\zeta)$ direction, which is compatible with the modes $M_3$ and $R_{25}$ at $\zeta = 0$ and $\frac{1}{2}$, respectively, are given by

$$\mu_x \omega^2(T_2) = [a_1 + b_1 + b_2 + 4b_3] + \frac{(Z_x e^*)^2}{v} [C(\zeta)], \qquad (15.1)$$

where $\mu$ is the reduced mass, $(Z_x e^*)$ is the effective ionic charge, and $v$ is the volume. The term $C(\zeta)$ represents the electrostatic force constant. The terms in the first square bracket on the right-hand side represent the various short-range force constants for the model and are given by[73,74]

$$\left(\frac{e^2}{2v}\right) a_i = \left(\frac{\partial^2 V_i}{\partial r^2}\right)_{\parallel} \qquad (15.2)$$

$$\left(\frac{e^2}{2v}\right) b_i = \left(\frac{\partial^2 V_i}{\partial r^2}\right)_{\perp} \qquad (15.3)$$

where the subscript $i = 1, 2, 3$ refers to the Sr–O, Ti–O, and O–O interactions, respectively, $e$ is the electronic charge, $v$ is the volume, and $V_i$ are the appropriate ionic potentials for the various interactions. The short-range forces are assumed to be axially symmetric, and the forces are taken to be central and act between nearest neighbors.

Stirling[73] calculated the various short-range force constants for the different harmonic models. He finds that in all cases $b_2$ is large and *negative* and is by far the dominant term in the sum in the square brackets in Eq. (15.1); i.e., the sum $[a_1 + b_1 + b_2 + 4b_3]$ of the short-range force constants is negative. Qualitatively similar results were obtained from rigid-ion model calculations by Cowley.[4] These results and their implications do not appear to have been emphasized until recently.[137,138]

It thus appears that for the $\Gamma_{25}$ and $M_3$ soft modes the balance of forces leading to an imaginary harmonic frequency (i.e., $\omega_0^2 < 0$) is due to an overcancellation of forces by the short-range interactions that are negative (i.e., attractive) as opposed to the case of the zone-center FE mode where the overcancellation is by the Coulomb forces. In other words, for the zone-boundary soft modes, the roles of the short-range and electrostatic forces are reversed from what they are for the $\mathbf{q} = 0$ FE mode. These results provide a ready explanation for the increase of $T_c$ with pressure for the zone-boundary phonon transition.[138] Since increasing pressure increases the magnitude of the short-range forces more rapidly than the magnitude of electrostatic forces, $\omega_0^2$ becomes more negative. Therefore, higher temperatures are required to provide the necessary stabilization of the high-temperature phase by anharmonic interactions, i.e., a higher $T_c$.

The reversal in the roles of the short-range and long-range forces is also seen in the recent results of Yoshimitsu.[142] Yoshimitsu analyzed the phonon dispersion relations for $SrTiO_3$ in the [111] direction using a harmonic rigid-ion model. He expresses the dynamical matrix $D$ of the lattice vibrations as

$$D = D_{\mathrm{d}} + D_{\mathcal{S}}, \tag{15.4}$$

where $D_{\mathrm{d}}$ and $D_{\mathfrak{S}}$ denote the contributions to $D$ from the dipole–dipole (electrostatic) and the short-range interactions, respectively. The normal modes are obtained by the usual diagonalization of $D$ with an appropriate matrix $U$, i.e.,

$$\mathrm{UDU}^{-1} = \omega_i^2 \delta_{ij}, \tag{15.5}$$

[142] K. Yoshimitsu, *Prog. Theor. Phys.* **54,** 583 (1975).

where $\omega_i$ are the eigenfrequencies. $D_d$ and $D_{\mathscr{S}}$ are given by

$$UD_d U^{-1} = [\omega_d^2(ij)] \tag{15.6}$$

$$UD_{\mathscr{S}} U^{-1} = [\omega_{\mathscr{S}}^2(ij)], \tag{15.7}$$

where $\omega_d^2(ij)$ and $\omega_{\mathscr{S}}^2(ij)$ are the $(ij)$ components of the matrices and satisfy the relations

$$\omega_d^2(ij) + \omega_{\mathscr{S}}^2(ij) = \begin{cases} 0, & i \neq j \quad (15.8a) \\ \omega_i^2, & i = j \quad (15.8b) \end{cases}$$

Equations (15.8) indicate that the observed mode frequency squared $\omega_i^2$ is expressed as the sum of two contributions: one from the dipolar interactions and the other from the short-range interactions.

Using the results of one of Cowley's[4] rigid-ion calculations (Model I), Yoshimitsu determined $U$ that diagonalizes $D$ and with this $U$ transformed $D_d$ and $D_{\mathscr{S}}$ to obtain $\omega_d^2(ii)$ and $\omega_{\mathscr{S}}^2(ii)$. His results are shown in Fig. 20. Figure 20a shows the dispersion curves in the [111] direction calculated by Cowley. Figure 20b and c show the decomposition of these curves according to Eqs. (15.8) into the two contributions $\omega_{\mathscr{S}}^2(ii)$—solid line, and $\omega_d^2(ii)$—dashed line. Note specifically the branches labeled a and e corresponding, respectively, to the $\Gamma_{15}$ soft FE mode at $\mathbf{q} = 0$ and to the $R_{25}$ soft mode at the zone boundary. It is clear that the roles of $\omega_{\mathscr{S}}^2(ii)$ and $\omega_d^2(ii)$ are reversed in the two cases. For $\Gamma_{15}$ the soft mode is stabilized by the short-range interactions [i.e., by $\omega_{\mathscr{S}}^2(ii)$] and destabilized by the dipolar interactions. The reverse is true for $R_{25}$.

It has been suggested[138] that the aforementioned reversal of roles of short-range and electrostatic forces from what they are for zone-center FE phonons obtains not only in $SrTiO_3$, but rather, it may be the general rule for displacive transitions involving soft short-wavelength optic phonons. The pressure results in Table II provide support for this hypothesis. In fact, pressure is a unique and all important variable in this regard. By applying pressure, we simply reduce the interionic separation and thereby change the balance between the competing forces.

An interesting observation based on the above discussion is that with increasingly higher pressure, and as the ions get closer and closer together, the short-range interactions could not be expected to continue to be attractive. Ultimately they must become repulsive. If this were so, a reversal in the sign of $dT_c/dP$ might be expected at sufficiently high pressure. Such an effect has been observed in $BaMnF_4$, where $T_c$ first increases with pressure, reaches a maximum at ~10 kbar, and then decreases.[132] It should be noted, however, that in $BaMnF_4$ the wavelength of the soft mode is not commensurate with the lattice,[136] so that the sign

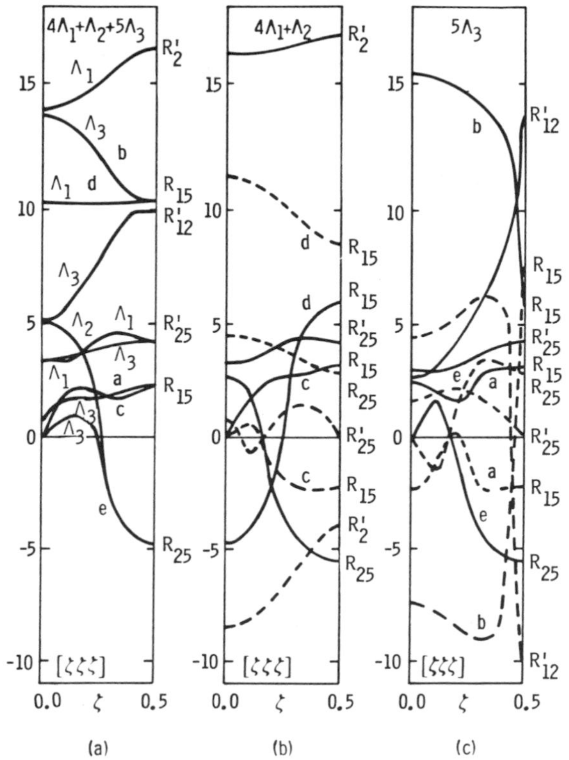

FIG. 20. The phonon dispersion curves of $SrTiO_3$ in the [111] direction. Part (a) gives the results of rigid-ion model calculations, whereas (b) and (c) show decomposition of each curve into contributions from short-range (solid line) and dipole–dipole (dashed line) interactions. [After K. Yoshimitsu, *Prog. Theor. Phys.* **54**, 583 (1975).] The normal modes are designated by the irreducible representations of the space group of the crystal. Along [111] the little group at a general point is $3m$, and its irreducible representations are $4A_1 + A_2 + 5A_3$. At the zone boundary the little group symmetry is increased to $m3m$, and the irreducible representations are $R_2' + R_{12}' + R_{25} + R_{25}' + 2R_{15}$.

reversal of $dT_c/dP$ might in some way be related to this incommensurability.

The apparent generalization of behavior and conclusions described above are based on the contrast between results of soft FE modes and on soft short-wavelength optic modes. For the FE case, the explanations are undoubtedly correct; whereas, for the short-wavelength optic modes, the explanations are somewhat speculative, although they find support in calculations on $SrTiO_3$. Clearly more work is needed here to evaluate fully the general validity and full implications of these results.

It should be emphasized that the list of zone-boundary transitions in Table II includes AFE, i.e., antipolar, transitions (e.g., $PbZrO_3$ and $PbHfO_3$) as well as transitions that do not involve soft polar modes (e.g., $M_3$ and $R_{25}$). Although there are differences between the two cases, both involve zone-boundary optic phonons and both behave qualitatively the same under pressure, thus suggesting that the dominant interactions are probably similar in the two cases. It should also be noted that neither the $C(\zeta)$ function in Eq. (15.1) nor its $r$ dependence are the same as the corresponding term for the FE case [see Eq. (3.6)] or its $r$ dependence. In the case of FE transitions, there are strong electric dipole forces not present in the zone-boundary transitions. Furthermore, it is not necessary that $C(\zeta)$ vary exactly as $r^{-3}$, but its dependence on $r$ will certainly be different and weaker than that of the short-range interactions.

## V. Potassium Dihydrogen Phosphate and Its Isomorphs

### 16. General Properties

There is probably no crystal class for which pressure studies have yielded more important information than potassium dihydrogen phosphate ($KH_2PO_4$ or KDP) and its isomorphs. These materials comprise a large class of FE and AFE crystals. Ferroelectrics isomorphous to KDP can be obtained by substituting deuterium for hydrogen, the alkali metal Rb for K, as well as $AsO_4$ for $PO_4$, whereas AFE behavior is obtained when $NH_4^+$ ions are substituted for the alkali metal ions. The properties of these materials can be varied widely by these various substitutions.

The crystals structure of KDP, the prototype crystal of the FE members of this class, is shown in Fig. 21.[143] In the high-temperature PE phase the crystal is tetragonal (space group $I\bar{4}2d$). The unit cell contains four K–$PO_4$ groups, and each tetrahedral $PO_4$ group is connected to four neighboring $PO_4$ groups by O–H$\cdots$O bonds lying very nearly in the basal plane. The protons are disordered on the O–H$\cdots$O bonds,[144,145] occupying positions on either side of the bond centers. On cooling the crystal transforms at 123 K to an orthorhombic FE phase (space group $Fdd2$) in which the protons order, and spontaneous polarization sets in along the $c$ axis. An important feature to note is that this polarization is perpendicular to the proton motion which is essentially in the basal plane. Thus the FE properties cannot be accounted for by proton motion alone.

[143] J. West, *Z. Kristallogr., Kristallgeom., Kristallphys., Kristallchem.* **74,** 306 (1930).

[144] G. E. Bacon and R. S. Pease, *Proc. R. Soc. London, Ser. A* **220,** 397 (1953); **230,** 359 (1955).

[145] B. C. Frazer and R. Pepinsky, *Acta Crystallogr.* **6,** 273 (1953).

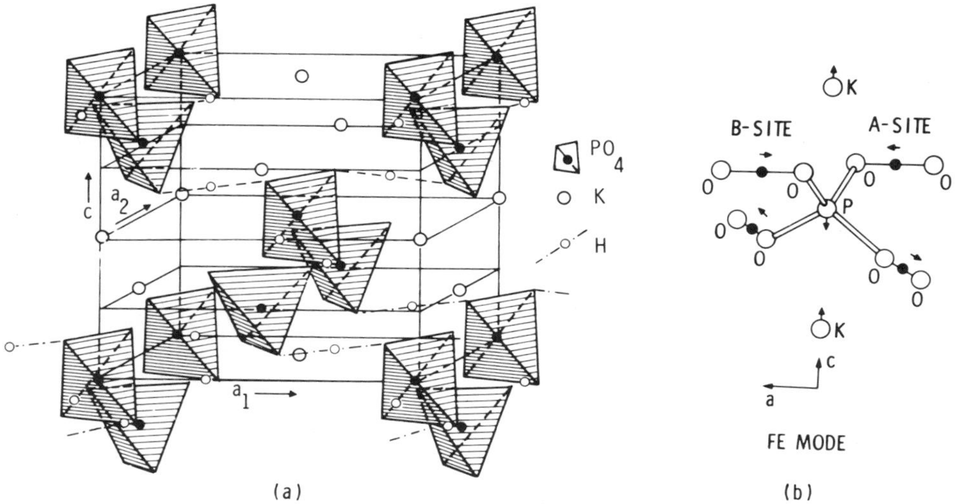


FIG. 21. (a) Structure of $KH_2PO_4$. [After J. West, *Z. Kristallogr., Kristallgeom., Kristallphys., Kristallchem.* **74,** 306 (1930).] (b) The atomic motion associated with the soft mode. [After W. Cochran, *Adv. Phys.* **20,** 401 (1961).]

The AFE members of this class, exemplified by ammonium dihydrogen phosphate ($NH_4H_2PO_4$ or ADP) are isomorphous with KDP in the high-temperature PE phase but on cooling transform to orthorhombic (space group $P2_12_12_1$) AFE phases. For ADP, $T_c = 148$ K.

One of the most important aspects of the properties of these crystals is the role played by the proton. The importance of this role is evidenced by the large changes in both static and dynamic properties observed on replacing the protons by deuterons. Thus, for example, $T_c$, the Curie constant $C$, and the saturation polarization $P_s$ of KDP increase very nearly linearly by approximately 80, 40, and 20%, respectively, on complete deuteration.[146]

The phase transition in KDP-type materials has been studied in great detail both experimentally and theoretically. Slater[147] originally proposed that the protons move between two minima of a double-well potential on the O–H···O bonds, and the phase transition results from the ordering of the protons in one of these minima. The proton motion is coupled to an optic mode of the lattice; when the protons order, the K–$PO_4$ units distort to produce the spontaneous polarization along the $c$ axis. Theoretical

[146] G. A. Samara, *Ferroelectrics* **5,** 25 (1973).

[147] J. C. Slater, *J. Chem. Phys.* **9,** 16 (1941).

treatments have been given emphasizing either the proton motion[148–151] or the lattice motion,[2,84] and Kobayashi[152] extended these microscopic treatments to consider explicitly coupling between the proton motion and an optic mode of the lattice. Similar treatments of both the proton motion models and coupled proton motion–optic mode models have subsequently been given by Elliot and co-workers.[153] Blinc and Zeks[154] have given a comprehensive discussion of the various models for the phase transition in KDP-type crystals. In these models proton motion between the two potential minima along the O–H···O bonds is taken to be via tunneling. Tunneling is attractive because it provides a ready qualitative explanation for the isotope effect on $T_c$—the much lower transition temperature for the undeuterated crystals being associated with the much higher tunneling probability for the proton as compared to the deuteron.

The effects of hydrostatic pressure on the properties of KDP-type crystals have been reported by several groups.[155–166] Several of these investigations have illustrated the special importance of pressure as a variable for studying these crystals. For example, Samara[159] demonstrated that the FE phase could be suppressed in KDP, and Peercy[160] and later Lowndes *et al.*[162,163] demonstrated that the soft mode could be made

148 R. Blinc, *J. Chem. Solids* **13,** 204 (1959).
149 P. G. deGennes, *Solid State Commun.* **1,** 132 (1963).
150 R. Brout, K. A. Müller, and H. Thomas, *Solid State Commun.* **4,** 507 (1966).
151 M. Tokunaga and T. Matsubara, *Prog. Theor. Phys.* **35,** 581 (1966); M. Tokunaga, *Ferroelectrics* **1,** 195 (1970).
152 K. K. Kobayashi, *J. Phys. Soc. Jpn.* **24,** 497 (1968).
153 R. J. Elliott and A. P. Young, *Ferroelectrics* **7,** 23 (1974); *J. Phys. C* [3] **7,** 2721 (1974).
154 R. Blinc and B. Zeks, *Adv. Phys.* **29,** 693 (1972).
155 H. Umebayashi, B. C. Frazer, G. Shirane, and W. B. Daniels, *Solid State Commun.* **5,** 591 (1967); J. Skalyo, Jr., B. C. Frazer, and G. Shirane, *J. Phys. Chem. Solids* **30,** 2045 (1969).
156 G. A. Samara, *Phys. Lett. A* **25,** 664 (1967).
157 C. Frenzel, B. Pietrass, and E. Hegenbarth, *Phys. Status Solidi A* **2,** 273 (1970); E. Hegenbarth and S. Ullwer, *Cryogenics* **7,** 306 (1971).
158 B. Morosin and G. A. Samara, *Ferroelectrics* **3,** 49 (1971).
159 G. A. Samara, *Phys. Rev. Lett.* **27,** 103 (1971).
160 P. S. Peercy, *Phys. Rev. Lett.* **31,** 380 (1973).
161 P. S. Peercy and G. A. Samara, *Phys. Rev. B: Solid State* [3] **8,** 2033 (1973).
162 R. P. Lowndes, R. C. Leung, W. B. Spillman, and N. E. Tornberg, *Proc. Int. Conf. High Press. 4th, 1974* p. 260 (1975); *in* "Light Scattering in Solids" (M. Balkanski, R. C. C. Leite, and S. P. S. Porto, eds.), p. 796. Wiley, New York, 1976.
163 R. P. Lowndes, N. E. Tornberg, and R. C. Leung, *Phys. Rev. B: Solid State* [3] **10,** 911 (1974).
164 P. S. Peercy, *Solid State Commun.* **16,** 439 (1975).
165 P. S. Peercy, *Phys. Rev. B: Solid State* [3] **12,** 2725 (1975).
166 G. A. Samara, *Ferroelectrics* **22,** 925 (1979).

underdamped in the PE phase by the application of pressure. These results reemphasized the important role of the hydrogen bond in these materials and demonstrated that the proton motion must be considered as a collective, propagating excitation rather than a single particle-like diffusive excitation. Pressure was also used to demonstrate that the proton motion remains coupled to an optic mode of the lattice in the FE phase.[164] In what follows we shall briefly outline the theoretical situation (developed largely for KDP) and then review and discuss the important features of the pressure results. Much of this work has been on KDP and its FE isomorphs with relatively much less effort devoted to ADP and its AFE isomorphs. In the latter crystals the transition is strongly first order, whereas in the former crystals the transition is very nearly second order. Space limitations restrict our discussion almost exclusively to KDP with a few remarks on ADP.

## 17. Theoretical Model for the Transition

The picture that has emerged for the behavior of KDP is that in the PE phase the protons tunnel between two minima of a double-well potential along the O–H···O bonds with the proton motion coupled to the K–$PO_4$ $\mathbf{q} = 0$ TO mode of the lattice along the $c$ axis. The transition is then accompanied by the preferential ordering of the protons in the minima of the double-well potentials and by a distortion of the K–$PO_4$ sublattices as the optic mode "freezes in." The atomic motion associated with this soft mode is shown in Fig. 21b. In the Kobayashi treatment[152] of the coupled modes, the transition reflects an instability of the lower frequency branch $\omega_-$ of the coupled proton tunneling–optic mode system. In reality the transition does not occur when $\omega_-$ vanishes, but when a transverse acoustic (TA) mode coupled piezoelectrically to $\omega_-$ vanishes. As $\omega_-$ decreases on approaching the transition, a level-anticrossing effect forces the TA mode frequency to decrease and approach zero, thereby precipitating the transition. This coupling produces a difference between the free (uncoupled) and clamped transition temperatures.[41] We shall discuss this coupling and the effect of pressure on it in Section VII, but for the main part of the discussion that follows it is not necessary to take this coupling into account, since the FE properties of the crystals are dominated by the proton tunneling–TO mode interaction.

Blinc[148] demonstrated that the proton motion could be considered as a tunneling motion, and de Gennes[49] originally showed that the individual tunneling of the protons between two minima of a double-well potential on the O–H···O bonds would lead to collective excitations of the proton motion when proton–proton interactions are considered. In the coupled mode treatment, the proton Hamiltonian $H_p$, which leads to soft-mode behavior when considered independently,[150] is coupled to a lattice contribution $H_L$ through a proton–lattice interaction $H_{PL}$. The resulting Hamiltonian $H$ is[152]

$$H = H_P + H_L + H_{PL}. \tag{17.1}$$

The proton contribution $H_P$ is cast in the form of an Ising model in a transverse field and solved independently of the lattice. This solution corresponds to the tunneling model and yields a collective proton tunneling frequency $\Omega_0$ given by

$$\Omega_0^2 = 2\Omega(2\Omega - J\langle S^0\rangle), \tag{17.2}$$

with $2\Omega = \Delta E/\hbar$, where $\Delta E$ is the tunneling splitting of the lowest energy level of the potential wells on the O–H···O bonds, $J$ is the proton–proton interaction, and

$$\langle S^0\rangle = \tfrac{1}{2}\tanh(\Omega/kT). \tag{17.3}$$

Assuming that $\Omega_0 \to 0$ at the transition temperature $T_0$ of the proton system (in the absence of explicit proton–lattice interactions) yields the expression

$$4\Omega/J = \tanh(\Omega/kT_0) \tag{17.4}$$

for $T_0$. The temperature dependence of the collective proton mode frequency $\Omega_0$ is obtained by combining Eqs. (17.2)–(17.4) to yield

$$\Omega_0^2 = 4\Omega^2\left(1 - \frac{\tanh(\Omega/kT)}{\tanh(\Omega/kT_0)}\right). \tag{17.5}$$

The tunneling Hamiltonian also yields an expression for the dielectric constant $\varepsilon$ in the PE phase of the form[152]

$$\varepsilon(T) = \varepsilon_L + \frac{4N\mu^2\tanh(\Omega/kT)}{4\Omega - J\tanh(\Omega/kT)}, \tag{17.6}$$

where $\varepsilon_L$ is the "background" lattice dielectric constant. In Eq. (17.6), $N$ is the number of dipoles per unit volume, and $\mu$ is the dipole moment. Expanding Eq. (17.6) about the transition temperature $T_c$ yields a Curie–Weiss law of the form

$$\varepsilon(T) \cong \varepsilon_L + \frac{C}{T - T_c}, \tag{17.7}$$

with the Curie constant $C$ given by

$$C = \frac{16N\mu^2 k T_c^2}{J^2 - 16\Omega^2}. \tag{17.8}$$

The spontaneous polarization $P$ is given in the tunneling model by

$$P = 2N\mu\langle S^z\rangle, \tag{17.9}$$

where $\langle S^z\rangle$ is the thermal expectation value of the $z$ component of the pseudospins and describes the ordering protons. In the PE phase, $\langle S^z\rangle = 0$ so that $P = 0$; at $T \ll T_c$, the spins do not completely align ($\langle S^z\rangle < \frac{1}{2}$) because a disordering field due to a finite tunneling probability remains even at $T = 0$. At $T \ll T_c$, the saturated spontaneous polarization $P_s$ is given by

$$P_s = N\mu(1 - 16\Omega^2/J^2)^{1/2}. \tag{17.10}$$

The remaining contributions to $H$ given in Eq. (17.1) are solved as follows: The lattice contribution $H_L$ is solved in the harmonic approximation (see Section 3), and the proton–lattice interaction $H_{PL}$ is taken to be bilinear in the proton and phonon coordinates. In the further approximation that the protons interact with only the lowest-frequency TO mode of the lattice of frequency $\omega_j$, Kobayashi[152] obtained the coupled proton tunneling–optic mode frequencies of the form

$$\omega_\pm^2 = \tfrac{1}{2}(\omega_j^2 + \Omega_0^2) \pm \{[\tfrac{1}{2}(\omega_j^2 - \Omega_0^2)]^2 + 8\langle S^0\rangle\Omega|G_i|^2\}^{1/2}, \tag{17.11}$$

where $|G_i|^2$ is the proton–lattice interaction. The soft mode of the system is $\omega_-$, and the transition occurs when $\omega_- \to 0$ in this model, where the interaction between the soft mode and the acoustic modes is neglected. Assuming $\omega_- = 0$ at $T = T_-$,* Eqs. (17.3), (17.4), and (17.6) yield an expression for $T_-$ given by

$$4\Omega/\tilde{J} = \tanh(\Omega/kT_-), \tag{17.12}$$

where

$$\tilde{J} = J + 4|G_i|^2/\omega_j^2 \equiv J + 4G.^{166a} \tag{17.13}$$

The proton–lattice interaction thus has the effect of enhancing the dipolar proton–proton interaction $J$, and the coupled-mode model becomes similar to the tunneling model if $J$ is merely replaced by $\tilde{J}$. In fact, it has been shown by Blinc and co-workers[154] that not only is Eq. (17.12) formally equivalent to the pure tunneling result [Eq. (17.4)], but also that all mean field results for the tunneling model are recovered in the coupled-

[166a] In this section four "transition" temperatures are used. These are (*a*) the temperature where $\omega_-$ vanishes, $T_-$; (*b*) the observed transition temperature $T_c$; (*c*) the observed Curie–Weiss temperature $T_c'$; and (*d*) the ordering temperature of the proton system in the absence of coupling to the lattice, $T_0$.

mode model if $J$ is replaced by $\tilde{J}$. The coupled-mode model may therefore be considered as a modified tunneling model.

## 18. Soft Mode in the Paraelectric Phase

The soft mode in KDP was first observed in Raman scattering by Kaminow and Damen,[167] and Katiyar, Ryan, and Scott[168] noted that the soft mode was coupled to an optic mode of the lattice in their Raman measurements on the isomorphous arsenates.

In all crystals of the KDP class the soft mode is heavily overdamped for all temperatures throughout the PE phase at atmospheric pressure, and the signal appears as a wing on the laser line. A question thus arises as to whether the response should be considered as a collective, propagating excitation or as a single particle-like relaxational, or diffusive, excitation. High pressure experiments have clarified this long-standing question. It has been found that the response in KDP can be made underdamped at room temperature by the application of hydrostatic pressure, and this response must therefore be considered as a collective, propagating excitation.[160] This behavior is illustrated in Fig. 22, which shows data taken at various pressures at 140 K. Whereas the low-pressure spectra peak at $\omega = 0$, for $P \gtrsim 4$ kbar the spectral response peaks at $\omega \neq 0$, indicating that the mode is less than critically damped at high pressure. Similar behavior is observed for the isomorphous $RbH_2PO_4$[169] and $KH_2AsO_4$[163] indicating that the conclusion that the soft-mode response must be considered as a propagating excitation appears to be generally valid for KDP-type crystals. Analogous conclusions cannot be drawn for deuterated members of this crystal class, however, since the response does not become underdamped within the pressure range of present measurements (see Section 22).

Measurements of the temperature dependence of the coupled-mode response in KDP at high pressure demonstrate that not only is the response underdamped at room temperature, but it is also underdamped throughout most of the PE phase at modest pressure.[165]

In addition to the question concerning the nature of the response, several authors have stressed the difficulties encountered in obtaining reliable, independent values for the spectral parameters from data in which the response is heavily overdamped. Pressure results have made it possible to overcome this difficulty. Data such as those shown in Fig. 22 may be used to obtain more reliable values for the spectral parameters

[167] I. P. Kaminow and T. C. Damen, *Phys. Rev. Lett.* **20,** 1105 (1968).
[168] R. S. Katiyar, J. F. Ryan, and J. F. Scott, *Phys. Rev. B: Solid State* [3] **4,** 2685 (1971).
[169] P. S. Peercy, unpublished.

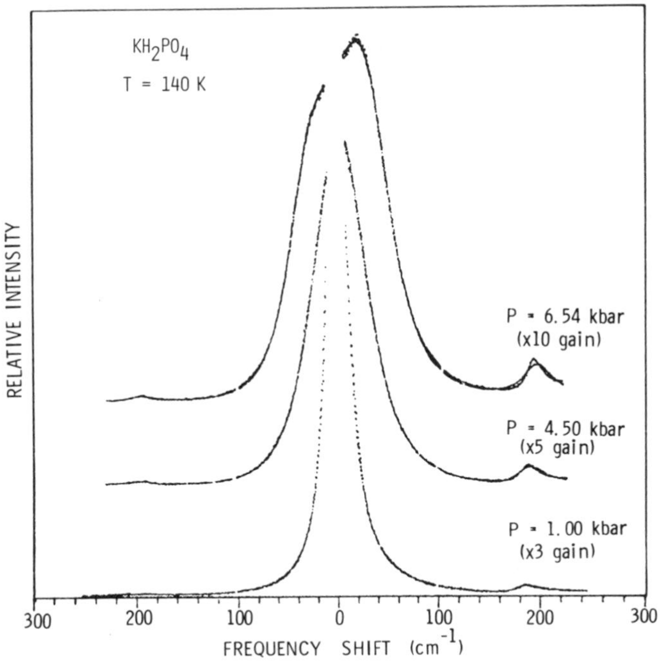


FIG. 22. Raman spectra of the soft mode in $KH_2PO_4$ at various pressures at 140 K illustrating that the soft mode becomes underdamped at high pressure. [After P. S. Peercy, *Phys. Rev. B: Solid State* [3] **12,** 2725 (1975).]

such as mode frequencies and dampings than can be obtained from data at atmospheric pressure where the response is heavily overdamped at all temperatures. These parameters may then be used to examine microscopic models for the phase transition in these materials. We now turn attention to this examination.

In the discussion of coupled-mode systems (Section 7) we noted that the general equations given for two coupled mods were overdetermined [Eqs. (7.2) and (7.3)] and the choice of diagonalization was determined by the model with which the data are to be compared. It should be emphasized that the data of Fig. 22 can be fit by Eqs. (7.2) and (7.3) using either a real interaction ($\Gamma_{ab} = 0$) or a dissipative interaction ($\Delta^2 = 0$). Whereas the fits one obtains are identical, the physical significance of the parameters are different. If the data are fit assuming a real interaction, one is assuming there are two modes of frequencies $\omega_a$ and $\omega_b$, which do not necessarily correspond in position to the measured response frequencies, that are coupled together through the real interaction $\Delta^2$ to give the measured response frequencies $\omega_{\pm}$. If, on the other hand, the data are fit assuming $\Delta^2 = 0$, then one obtains the measured response frequencies $\omega_{\pm}$. Furthermore, if one fits the data assuming $\Gamma_{ab} = 0$ to obtain the mode frequencies $\omega_{a,b}$, dampings $\Gamma_{a,b}$, and interaction $\Delta^2$ and substitutes these quantities back into the complex susceptibility $\chi(\omega)$, the normal responses $\omega_{\pm}$ of the system can be obtained from $\chi(\omega)$. In the limit of small damping and a real coupling between $\omega_a$ and $\omega_b$ the normal modes $\omega_{\pm}$ of the system are

$$\omega_{\pm}^2 = \tfrac{1}{2}(\omega_b^2 + \omega_a^2) \pm \{[\tfrac{1}{2}(\omega_b^2 - \omega_a^2)]^2 + \Delta^4\}. \quad (18.1)$$

It can be immediately seen that Eq. (18.1) is formally identical to Eq. (17.11) of the Kobayashi model. To examine the spectra within the framework of the Kobayashi coupled proton tunneling–optic mode model of the phase transition, the spectra must therefore be fit assuming a real interaction ($\Gamma_{ab} = 0$), and one must make the identifications $\omega_b = \omega_j$, $\omega_a = \Omega_0$, and $\Delta^4 = 4|G_i|^2\Omega \tanh \beta\Omega$. To examine the observed responses of the system, however, one should consider the frequencies $\omega_{\pm}$.

The pressure dependences of $\omega_a$ and $\omega_b$ are shown in Fig. 23 for KDP at room temperature, and the parameters and their pressure derivatives are summarized in Table III. One of the surprising results of these measurements is the small pressure dependence of the "proton" mode $\omega_a$; in fact, $\omega_a$ and $\omega_b$ have comparable pressure dependences. The primary effect of pressure is to reduce the damping $\Gamma_a$ of this mode. The pressure dependences of $\Gamma_a$ is shown in Fig. 24. It should be noted that the damping of the proton mode is expected to depend on the proton–proton interaction $J$. Thus a decrease in $\Gamma_a$ is indicative of a decrease in $J$. These comments will be made more quantitative below.

The lower curve in Fig. 24 gives the pressure dependence of the relaxation rate $\tau_a^{-1} \equiv \omega_a^2/\Gamma_a$; $\tau_a^{-1}$ displays the largest pressure dependence of the parameters measured, increasing by a factor of ~2 by 10 kbar.

The coupling $\Delta$ between $\omega_a$ and $\omega_b$ does not display an appreciable pressure dependence throughout the pressure range investigated. Although $\Delta$ appears to increase with pressure, the increase is only ~2 or 3 standard deviations and is not considered significant.

The temperature and pressure dependences of the coupled responses $\omega_{\pm}$ of KDP are shown in Figs. 25 and 23. The higher-frequency branch $\omega_+$ displays a normal temperature dependence, increasing in frequency as the temperature is decreased, whereas $\omega_-$ exhibits a normal soft-mode temperature dependence, $\omega_-^2 = 46.8(T - T_-)\ \mathrm{cm}^{-2}$, over a temperature range of more than 100 K above the transition temperature $T_c$. The difference between $T_c$ and the temperature at which $\omega_-$ vanishes is $(T_c - T_-) \sim 5$ K. This difference reflects primarily the difference between the

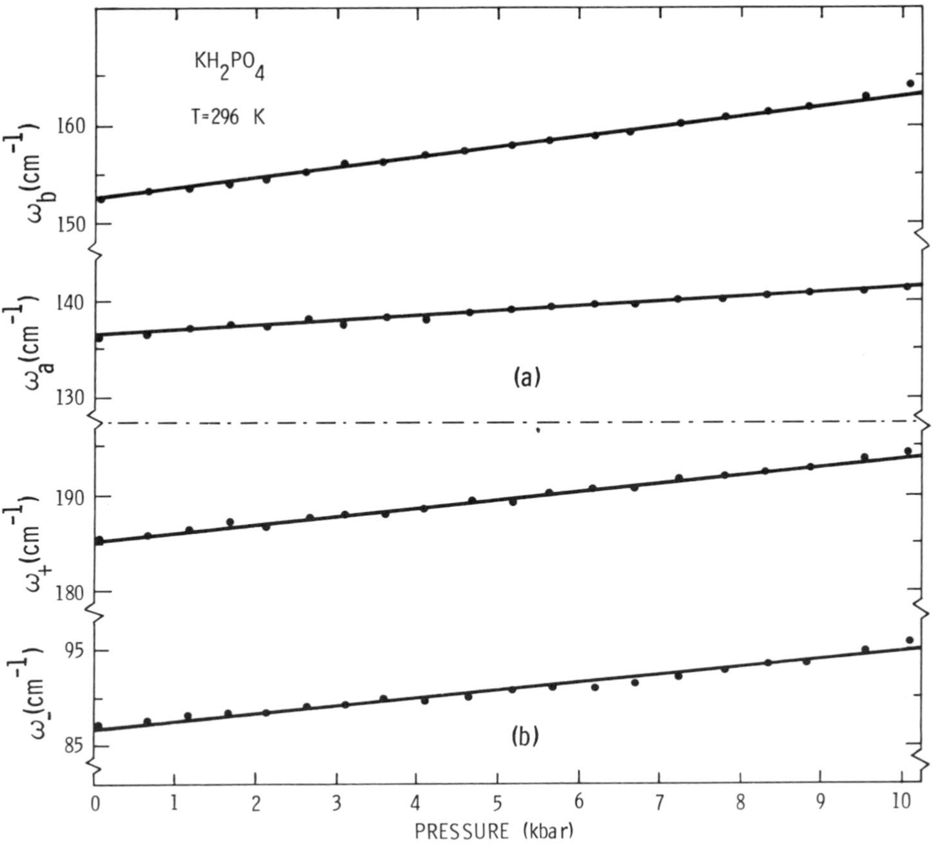


FIG. 23. Pressure dependences of (a) the decoupled modes $\omega_a$ and $\omega_b$ and (b) the coupled responses $\omega_+$ and $\omega_-$ in $KH_2PO_4$ at 296 K. [After P. S. Peercy, *Phys. Rev. B: Solid State* [3] **12,** 2725 (1975).]

free and clamped transition temperatures[38,41]—i.e., it is produced by the interaction of $\omega_-$ with an acoustic mode—although it may contain small contributions from the anharmonic self-energy and the fact that the transition is slightly first order.[146,170] Insofar as the transition is second order and the relaxational self-energy contribution is negligible, the transition temperature $T_c$ is therefore the temperature at which an acoustic mode coupled to $\omega_-$ vanishes.

Both $\omega_\pm$ display frequency increases with pressure $(\partial \ln \omega_-/\partial p)_T =$

[170] See, e.g., J. W. Benepe and W. Reese, *Phys. Rev. B: Solid State* [3] **3,** 3032 (1971).

+0.9%/kbar and $(\partial \ln \omega_+/\partial p)_T = +0.45\%$/kbar at 296 K. The increase in $\omega_-$ is insufficient to produce the dramatic change observed in the spectra with pressure; rather, the spectra become underdamped with pressure because of the large decrease in the damping $\Gamma_-$ with pressure. The pressure dependences $\Gamma_-$ and the relaxation rate $\tau_-^{-1} = \omega_-^2/\Gamma_-$ are shown in Fig. 24. $\Gamma_-$ displays a nonlinear pressure dependence, whereas $\tau_-^{-1}$ increases linearly with pressure at a rate of $(\partial \ln \tau_-^{-1}/\partial p)_T = +9.2\%$/kbar at 296 K. This large pressure dependence of $\tau_-^{-1}$ is produced primarily by the decrease in $\Gamma_-$ with pressure. In fact, it is not the increase in $\omega_-$ but rather the decrease in $\Gamma_-$ that causes the response to become underdamped with pressure.

As noted above, to compare these measurements with the Kobayashi model, the data are examined by fitting the spectra with a real interaction ($\Gamma_{ab} = 0$). Such a procedure yields mathematically equivalent forms for the responses $\omega_\pm$ from the Kobayashi model and the fits to the data as discussed above. The temperature dependences of modes $\omega_a$ and $\omega_b$, which are coupled together with the real interaction $\Delta$ to give the measured responses, are shown in Fig. 26. One of the important checks for consistency between the Kobayashi model and the data is the temperature dependence of the "proton" mode $\Omega_0$. Identifying $\omega_a$ as $\Omega_0$, the result of fitting the temperature dependence of $\omega_a$ (at 6.54 kbar) as deduced from the Raman spectra with the tunneling model result, Eq. (82) for $\Omega_0$, is the solid curve in Fig. 26. The values for the tunneling frequency and $T_0$ obtained from these data are $\Omega = 89\ \text{cm}^{-1}$ and $T_0 = 58$ K. The good quality of the fit attests to the usefulness of the model.

ABLE III. COUPLED-MODE MODEL PARAMETERS AND THEIR LOGARITHMIC PRESSURE DERIVATIVES FOR SEVERAL KDP-TYPE CRYSTALS

| Crystal | $\omega_a$ (cm$^{-1}$) | $\frac{d \ln \omega_a}{dP}$ (% kbar$^{-1}$) | $\omega_b$ (cm$^{-1}$) | $\frac{d \ln \omega_b}{dP}$ (% kbar$^{-1}$) | $\Omega$ (cm$^{-1}$) | $\frac{d \ln \Omega}{dP}$ (% kbar$^{-1}$) | $\tilde{J}$ (cm$^{-1}$) | $\frac{d \ln \tilde{J}}{dP}$ (% kbar$^{-1}$) |
|---|---|---|---|---|---|---|---|---|
| $H_2PO_4$[a] | 136.4 | 0.33 | 152.8 | 0.67 | 86.2 | 0.5 | 440 | −1.5 |
| $bH_2PO_4$[b] | 125 | 0.24 | 126 | 0.15 | 96 | 2.5 | 510 | −1.4 |
| $H_2AsO_4$[c] | 146.5 | −0.14 | 150.4 | 1.04 | 99 | 0.10 | | |
| $bH_2AsO_4$[c] | 156.8 | 0.01 | 116.7 | 0.86 | 125 | 0.02 | | |
| $sH_2AsO_4$[c] | 157.5 | 0.51 | 100.1 | 1.00 | 111 | 0.63 | | |

[a] After P. S. Peercy, *Phys. Rev. B: Solid State* [3] **12**, 2725 (1975).
[b] After P. S. Peercy (unpublished).
[c] After Lowndes *et al.* (Refs. 162, 163). *Note:* The spectral parameters in the arsenates were determined in me cases by fits to relatively limited spectral regions.

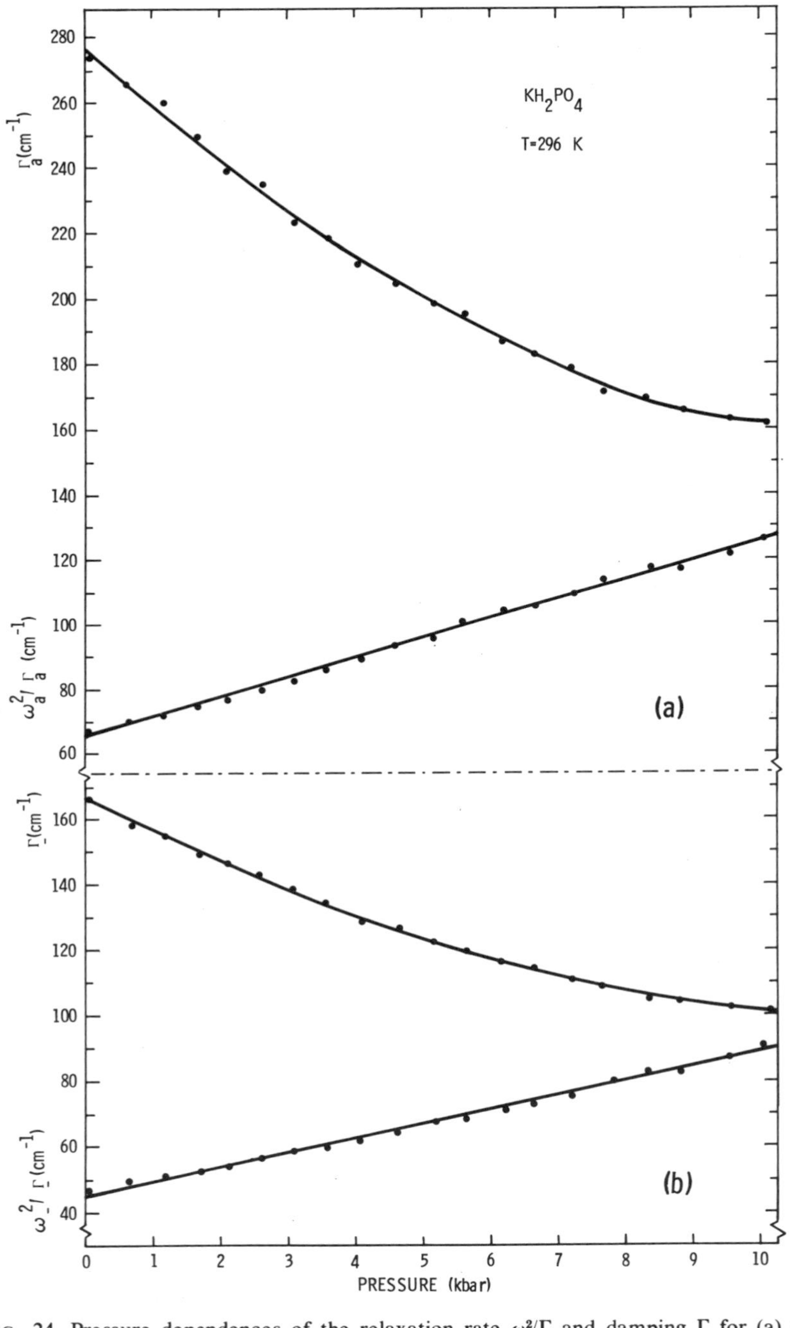


FIG. 24. Pressure dependences of the relaxation rate $\omega^2/\Gamma$ and damping $\Gamma$ for (a) the "proton" mode $\omega_a$ and (b) the soft mode $\omega_-$ in $KH_2PO_4$ at 296 K. [After P. S. Peercy, *Phys. Rev. B: Solid State* [3] **12**, 2725 (1975).]

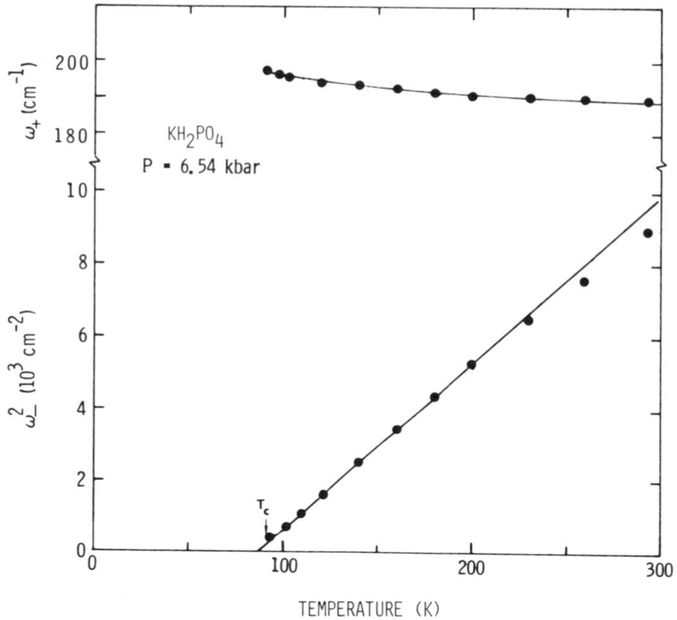


FIG. 25. Temperature dependence of $\omega_\pm$ in the PE phase of $KH_2PO_4$ for $p = 6.54$ kbar. The solid curve in the $\omega_-^2$ plot is $\omega_-^2 = 46.8(T - 86\ \mathrm{K})\ \mathrm{cm}^{-2}$, whereas the solid curve through the $\omega_+$ data is just a guide to the eye. [After P. S. Peercy, *Phys. Rev. B:* **12,** 2725 (1975).]

## 19. Soft Mode in the Ferroelectric Phase

The temperature dependence of the soft-mode Raman spectrum in the FE phase of KDP is illustrated in Fig. 27. For $T \ll T_c$ the spectrum consists of two well-resolved modes with frequencies $\omega_- \simeq 150\ \mathrm{cm}^{-1}$ and $\omega_+ \simeq 215\ \mathrm{cm}^{-1}$ at atmospheric pressure.

Blinc and co-workers[171] investigated the effect of deuteration on this spectrum and observed that the mode labeled $\omega_-$ is absent in fully deuterated KDP. This observation allowed them to associate $\omega_-$ with the proton motion of the system. Measurements of the pressure dependence of these spectra for $T < T_c$ further demonstrated that $\omega_-$ is the soft mode in the FE phase and that the proton motion remains coupled to the optic mode of the lattice $\omega_+$ in this phase.[165]

Although the temperature dependence of the spontaneous polarization should be reflected in the temperature dependence of the soft-mode

[171] R. Blinc, B. Lavrencic, I. Levstek, V. Smolej, and B. Zeks, *Phys. Status Solidi B* **60,** 255 (1973).

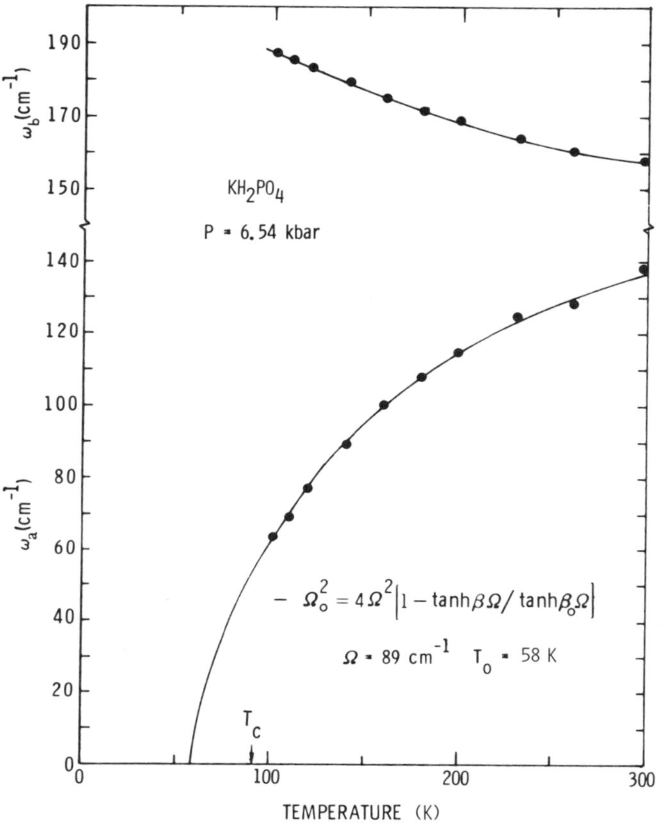


FIG. 26. Temperature dependence of the decoupled frequencies $\omega_a$ and $\omega_b$ in $KH_2PO_4$. The solid curve for $\omega_a$ is calculated from Eq. (82) with the parameters shown, whereas the curve through $\omega_b$ is just a guide to the eye. [After P. S. Peercy, *Phys. Rev. B: Solid State* [3] **12**, 2725 (1975).]

spectrum,[171] direct comparison of these two properties is difficult because the modes become heavily damped near $T_c$. At lower temperatures where the modes are well resolved, neither the mode frequencies nor the spontaneous polarization displays an appreciable temperature dependence, rendering the comparison difficult. Such difficulties can be overcome through pressure measurements, however, since the pressure dependence of the Raman spectra can be measured at low temperatures where the modes are well resolved and the results compared with the pressure dependence of the saturated spontaneous polarization.

The pressure dependences of $\omega_{\pm}$ summarized in Fig. 28 for various

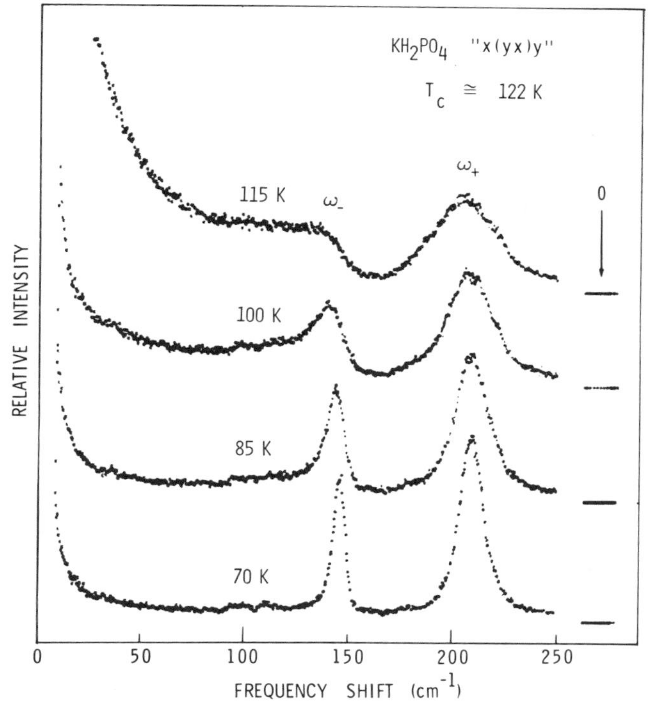


FIG. 27. Raman spectra of $KH_2PO_4$ for various temperatures below $T_c$. [After P. S. Peercy, *Solid State Commun.* **16,** 439 (1975).]

temperatures in the FE phase. Since the transition temperature decreases with pressure ($dT_c/dp = -4.6$ K/kbar for low pressures),[159] increasing the pressure at constant temperature causes the system to approach the transition. As the transition is approached, the soft mode, as well as any mode coupled to it, should decrease in frequency. The data in Fig. 28 illustrate that both $\omega_\pm$ decrease in frequency with increasing pressure, with $\omega_-$ displaying a large, nonlinear pressure dependence very near the transition. The other lattice modes (not shown) on the other hand, exhibit normal frequency increases with increasing pressure. These results thus suggest that $\omega_-$ is the soft mode of the transition in the FE phase and, since $\omega_-$ is determined by the proton motion, that the proton motion is coupled to the optic mode $\omega_+$.

The pressure derivative of $\omega_+$ is temperature independent within experimental uncertainty, $(\partial \ln \omega_+/\partial p)_T = -0.6 \pm 0.1\%$/kbar. For $T \ll T_c$, $\omega_-$ displays a linear pressure dependence over the limited pressure range

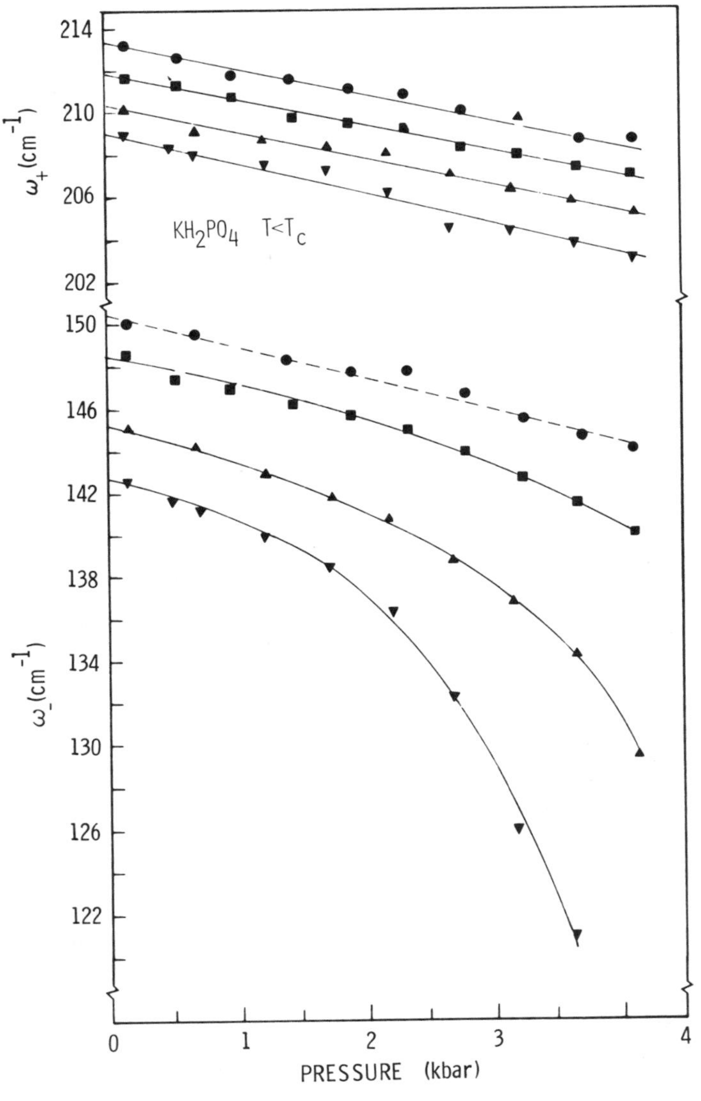


FIG. 28. Pressure dependences of the coupled-mode frequencies at various temperatures: (●) 35 K, (■) 60 K, (▲) 80 K, (▼) 91 K, in the FE phase of $KH_2PO_4$. [After P. S. Peercy, *Solid State Commun.* **16**, 439 (1975).]

examined with $(\partial \ln \omega_- / \partial p)_T = -1.0 \pm 0.2\%$/kbar. For the coupled-mode model, and for the parameters appropriate for KDP, the pressure dependence of the spontaneous polarization is to a first approximation given by[165]

$$(\partial \ln P_s/\partial p)_T \cong (\partial \ln \omega_-/\partial p)_T + (\partial \ln \omega_+/\partial p)_T. \tag{19.1}$$

Quantitative comparison of the measured parameters yields $(\partial \ln P_s/\partial p)_T \cong -(1.6 \pm 0.3)\%$/kbar from the Raman scattering results at 45 K compared to the measured value for the pressure dependence of the saturated spontaneous polarization of $(\partial \ln P_s/\partial p)_T = -(2.1 + 0.2)\%$/kbar,[166] again indicating that this model is consistent with the measurements.

## 20. Pressure Dependence of the Dielectric Properties and the Vanishing of the Ferroelectric State at High Pressure

The effects of temperatures and pressure on the dielectric properties of several crystals of the KDP class have been reported.[156,157,159,161,162,166] The results are qualitatively similar for all the crystals. In the high temperature PE phase the dielectric constant along the polar $c$ axis, $\varepsilon_c$, obeys a Curie–Weiss law [Eq. (17.7)] extremely well. It is also generally observed that the Curie–Weiss temperature is to within 0.1–0.2 K the same as the actual transition temperature $T_c$ for the undeuterated crystals where the transitions are very nearly second order. For the deuterated crystals, on the other hand, $T_c'$ is several degrees lower than $T_c$ reflecting the first-order character of these transitions.[146,166] The main effect of pressure on $\varepsilon_c(T)$ is a relatively large shift of $T_c$ to lower values. The Curie constant $C$ generally decreases by ~0.6–2.0%/kbar.[161,166]

One of the most important results of the pressure studies of KDP-type crystals has been the observation of the vanishing of the ordered phase at high pressure. This has been observed experimentally in KDP[159] and $RbH_2PO_4$,[161] and other crystals of this class are expected to behave similarly. The main features are illustrated for KDP in Fig. 29. The 17.2-kbar isobar shows no evidence for a phase transition, and the FE phase has vanished for this and higher pressures. Figure 30 shows the pressure dependence of $T_c$. At low pressure $T_c$ decreases linearly with pressure, but the rate of decrease increases at high pressure, and the results indicate that $T_c \to 0$ K with infinite slope, i.e., $(dT_c/dp) \to -\infty$ as $T_c \to 0$ K. In Fig. 30 for $p \gtrsim 10$ kbar the data obey the empirical relation $T_c^2 = k(p^* - p)$ with $p^* = 17.1$ kbar and $k = 710 \pm 10$ K²/kbar. The FE phase of KDP thus vanishes for $p \geq 17.1$ kbar.

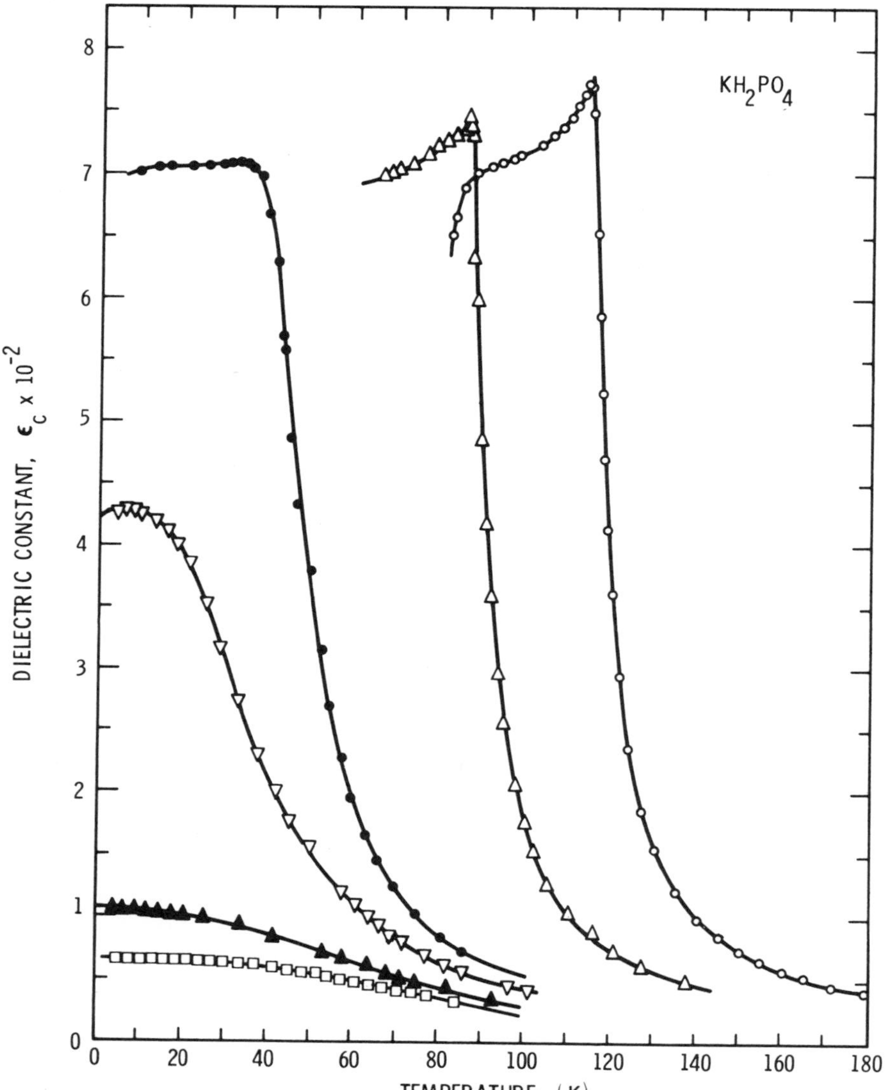


FIG. 29. Temperature dependence of the $c$-axis dielectric constant of $KH_2PO_4$ at various pressures: (○) 1.8, (△) 8.4, (●) 15.4, (▽) 16.9, (▲) 19.3, (□) 21.0 kbar, illustrating the suppression of the FE phase. [After G. A. Samara, *Phys. Rev. Lett.* **27**, 103 (1971).]

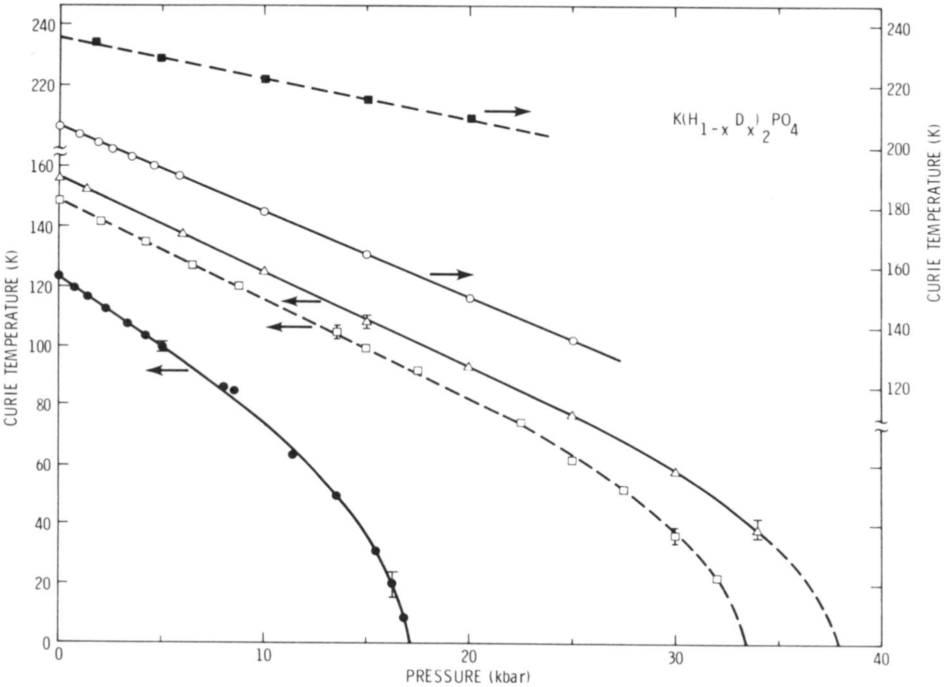


FIG. 30. Pressure dependences of the transition temperatures in $KH_2PO_4$ and $NH_4H_2PO_4$ and their deuterated analogs: (—●—) $x = 0.00$, (—△—) $x = 0.35$, (—○—) $x = 0.82$, (--□--) $NH_4H_2PO_4$, (--■--) $ND_4D_2PO_4$. [After G. A. Samara, *Ferroelectrics* **22**, 925 (1979).]

Figure 30 compares the $T_c(p)$ dependences for KDP and two partially deuterated, $K(H_{1-x}D_x)_2)_2PO_4$, crystals. There are three important features about these results that typify the response of this class of H-bonded FE and AFE crystals. They are (*a*) $T_c$ decreases with pressure [for both FE and AFE transitions (see Section 23)]; (*b*) the magnitude of $dT_c/dp$ decreases markedly with deuteration; and (*c*) for the nondeuterated crystals $T_c$ vanishes with infinite slope, i.e., $dT_c/dp \rightarrow -\infty$ as $T_c \rightarrow 0$ K. A satisfactory theory must account for all of these effects. As we have already seen, at present a physically satisfactory model for the transition in these crystals is the coupled proton tunneling–optic mode model described earlier in this section. We now examine the above results within the framework of this model. In particular, we discuss the macroscopic results summarized above in terms of the microscopic model parameters and their pressure dependences.

## 21. Relation between Macroscopic Properties and Microscopic Parameters of the Coupled Mode Model

In the Kobayashi coupled mode model, $T_c$ is given by Eq. (17.11) (neglecting the small difference between $T_c$ and $T_-$). This equation has solutions, and therefore a finite $T_c$, only if $4\Omega/\tilde{J} < 1$. The FE phase must vanish if $4\Omega/\tilde{J} \geq 1$, for then there is no solution for $T_c$. The physical picture is as follows: $T_c$ is determined by two counterbalancing fields, (*a*) a dipolar field represented by $\tilde{J}$ that is trying to order the protons and induce a spontaneous polarization; and (*b*) a transverse tunneling field represented by $\Omega$ that is trying to disorder the protons. At 1 bar the dipolar field dominates, the ratio $4\Omega/\tilde{J} < 1$, and there is a finite $T_c$. Pressure is the only physical variable that allows one to change the balance between these two fields sufficiently to allow testing of this basic idea.

The decrease of $T_c$ with pressure can be qualitatively understood in terms of an increase in tunneling frequency $\Omega$ and/or a decrease in the dipolar interaction $\tilde{J}$. Both effects are expected on physical grounds[159,161,172] and are observed as confirmed by the analysis of the Raman spectra as discussed earlier (see Table III). The decrease in $\tilde{J}$ with pressure makes the necessary increase in $\Omega$ to achieve a given decrease in $T_c$ less than it would need to be otherwise. The increase in $\Omega$ with pressure is believed to be a consequence of the expectation that as the O–H $\cdots$ O bond length decreases with pressure, the two potential minima for the protons along the bond get closer and the energy barrier between them decreases. Both effects would lead to a higher tunneling frequency. Intuitively one might expect that the O–H $\cdots$ O bond may be the most compressible bond in the KDP lattice and, as a consequence, that the decrease in $T_c$ with pressure might be dominated by the increase in $\Omega$.[173] However, as we have seen for KDP $(\partial \ln \Omega/\partial p)_T = 0.5\%$/kbar and $(\partial \ln \tilde{J}/\partial p) = -1.5\%$/kbar, so that the decrease in $T_c$ is dominated more by the decrease in $\tilde{J}$ than by the increase in $\Omega$. We recall that $\tilde{J} = J + 4G^*$, and examination of the results summarized in Table III shows that the sign and magnitude of $(\partial \ln \tilde{J}/\partial p)_T$ are determined by the proton lattice coupling term $G^*$. In fact the proton–proton dipolar interaction $J$ increases with pressure. This is a natural consequence of the tunneling model when $kT_0 \ll \Omega$, a condition that obtains for KDP-type crystals (see Table III). For

[172] R. Blinc and B. Žeks, *Helv. Phys. Acta* **41**, 700 (1968).

[173] The effects of pressure on the structural parameters of tetragonal $KH_2PO_4$ are discussed in a forthcoming paper by G. M. Meyer, R. J. Nelmes, and C. Vettier.

this condition Eq. (17.4) yields $\tanh(\Omega/kT) \approx 1$, so that $J \simeq 4\Omega$ and $(\partial \ln J/\partial p)_T \simeq (\partial \ln \Omega/\partial p)_T$; i.e., $J$ increases with pressure.

In the Kobayashi model it is also seen that at sufficiently high pressure the condition $4\Omega/\tilde{J} \geq 1$ should obtain and the FE state should vanish. Another important prediction of the model is that Eq. (17.12) yields $dT_c/dp \to -\infty$ as $T_c \to 0$ K.[159,161,174] Both of these model predictions are in agreement with the experimental observations (Fig. 30).

The above discussion of the $T_c(p)$ behavior appears qualitatively satisfactory, and it must be noted that the concept of tunneling is attractive not only from the standpoint of providing an explanation for the effect of deuteration on $T_c$, but also for accounting for the isotope effect on the initial pressure derivative $dT_c/dp$. The larger value of this derivative for the undeuterated crystal results, in part, from the effect of pressure on the tunneling frequency $\Omega$.[79,172] For the fully deuterated crystals it is generally believed that there is negligible tunneling, i.e., $\Omega \approx 0$, so that Eq. (17.12) reduces to the Ising model result $4kT_c = \tilde{J}$. This suggests that in this case $T_c$ should vanish with a finite slope $dT_c/dP$ as opposed to the infinite slope in the case of tunneling. Unfortunately, there are no direct experimental data to test this prediction. The pressures needed to do so for $KD_2PO_4$ or any of its deuterated isomorphs are beyond the range of apparatus used for such measurements thus far (see Fig. 30).

An important way to measure the validity and internal consistency of the Kobayashi model is to examine if the model parameters and their pressure derivatives determined from the Raman data can predict the results of the macroscopic dielectric measurements. Toward this end the pressure dependence of $T_c$ and the temperature and pressure dependences of $\varepsilon_c$ and $P_s$ have been examined for both KDP and $RbH_2PO_4$.[166,175] It is found that the model provides an essentially quantitative account of these macroscopic properties.

On the basis of the above, it can thus be concluded that, although there is some uncertainty about the physical significance of the microscopic parameters determined from the Raman scattering results (as noted in Section 18 and 19), the good quantitative agreement obtained in all cases emphasize the usefulness of the Kobayashi model for treating KDP-type ferroelectrics. The simple tunneling model, which does not explicitly consider proton–lattice interaction, cannot adequately describe the observed temperature dependence of the soft mode frequency $\omega_-^2 = A(T - T_-)$, nor does it lead to a satisfactory comparison and agreement between

[174] R. Blinc, S. Svetina, and B. Žeks, *Solid State Commun.* **10**, 387 (1972).
[175] P. S. Peercy, *Phys. Rev. B: Solid State* [3] **9**, 4868 (1974).

Raman results and dielectric properties. Additional support for the Kobayashi model comes from its ability to account satisfactorily for the effects of deuteration and pressure on both the static[146,166] and dynamic properties of $K(H_{1-x}D_x)_2PO_4$. We shall now briefly consider these effects on the dynamic properties.

## 22. Effects of Deuteration and Pressure on the Dynamic Properties of $KH_2PO_4$

As noted earlier in this section, substitution of deuterons for protons in the KDP crystal class has very strong effects on the transition temperature and the FE properties, as well as their pressure dependences. Comparison of the Raman spectra in deuterated materials with those of the non-deuterated materials has proven useful in investigating the nature of the soft-mode response, and measurements of the pressure dependence of the coupled-mode response as a function of deuteration has provided important information that could be used to examine the applicability of various microscopic models for the phase transition.

The Raman spectrum for the coupled-mode system changes markedly with deuteration in both the PE and FE phases. Although the soft mode can be made underdamped in the PE phase in KDP and its isomorphs with the application of modest pressure,[160,163] this response could not be made underdamped in $KD_2PO_4$ with the application of pressures up to 10 kbar.[175] Attempts to fit the spectrum in $KD_2PO_4$ with damped harmonic oscillator responses similar to those used for KDP yielded such large ratios of $\Gamma$ to $\omega$ that the mode could best be described by a relaxational excitation even at the highest pressures investigated. This conclusion is consistent with the results of Reese *et al.*[41] from atmospheric pressure measurements. However, Lowndes *et al.*[163] were able to fit the coupled mode response using the damped oscillator formalism for some members of this crystal class. The results of their fit to the damped oscillator response yielded qualitatively similar behavior for the temperature dependences of the relaxation rates, $\tau = \omega^2/\Gamma$, for deuterated and nondeuterated materials; however, the value extracted for the deuteron tunneling frequency was surprisingly large for some of the materials. These results raise questions concerning the applicability of the damped oscillator response for the deuterated members of this family.

Raman scattering techniques have been much more successful in investigating the FE phase of these materials. The spectrum of non-deuterated crystals consists of two well-resolved modes $\omega_\pm$ for $T \ll T_c$ (see Fig. 27). the pressure measurements identified $\omega_-$ as the soft mode of nondeuterated KDP,[164] whereas comparison of the spectrum for KDP with that of $KD_2PO_4$ demonstrated that $\omega_-$ was the "proton-like" mode.[171] The manner in which the spectra change with deuteration is shown in Fig. 31 for $T \ll T_c$. Although both $\omega_\pm$ remain underdamped as the deuteron concentration increases, the intensity of the proton branch $\omega_-$ decreases monotonically with decreasing proton concentration.

For $T \ll T_c$, the frequencies of both $\omega_\pm$ increase with increasing deuteration. The concentration dependences of $\omega_\pm$ are shown in Fig. 32.

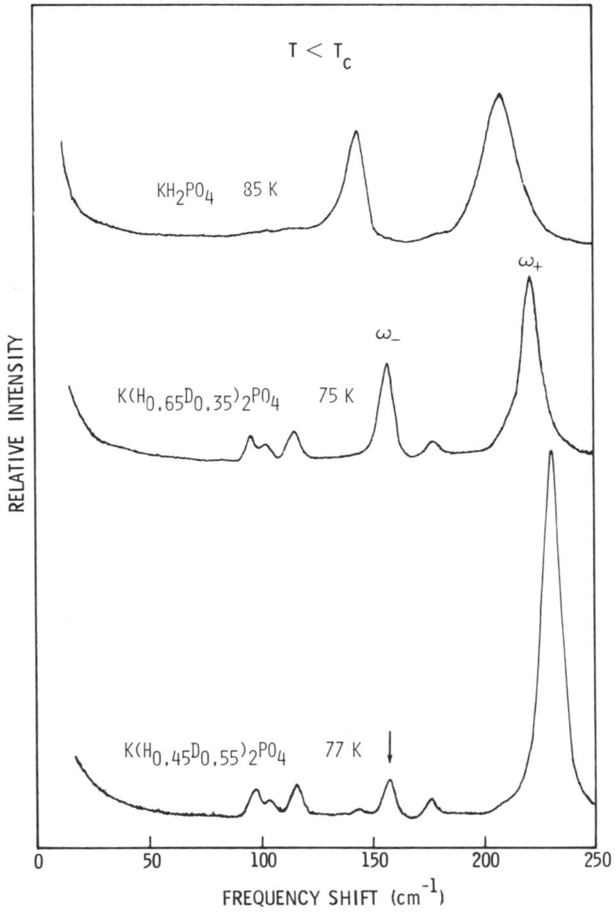


FIG. 31. Raman spectra of the coupled modes in the FE phase of $K(H_{1-x}D_x)_2PO_4$ for various deuterium concentrations at 1 bar. [After P. S. Peercy, *Phys. Rev. B: Solid State* [3] **13,** 3945 (1976).]

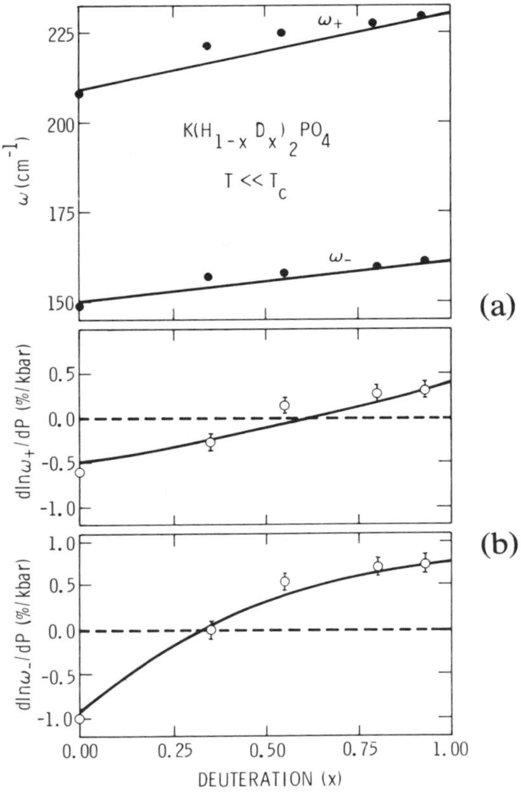


FIG. 32. (a) Calculated deuteration dependences (solid curves) of $\omega_{\pm}$ compared to the measured frequencies in the FE phase of $K(H_{1-x}D_x)_2PO_4$. (b) Comparison between the measured and calculated (solid curves) pressure dependences of $\omega_{\pm}$ in the FE phase of $K(H_{1-x}D_x)_2PO_4$ versus deuteration $x$. [After R. Blinc, B. Pirc, and B. Zeks, *Phys. Rev. B: Solid State* [3] **13,** 2943 (1976).]

However, the pressure dependences for $\omega_{\pm}$ are qualitatively different for KDP and $KD_2PO_4$ as is also illustrated in Fig. 32. For KDP, $d\omega_{\pm}/dp < 0$, which demonstrates that these modes are associated with the phase transition (see Section 19), whereas for $KD_2PO_4$, $d\omega_{\pm}/dp > 0$. In fact, in $KD_2PO_4$ $d \ln \omega_-/dp \cong 0.75\%$/kbar and $d \ln \omega_+/dp \cong 0.4\%$/kbar. These pressure derivatives are comparable to those of the normal lattice modes of KDP, which range from $d \ln \omega_i/dp = +0.1\%$/kbar to $+0.8\%$/kbar.[161,165]

To examine the effects of deuteration on the coupled modes Blinc *et al.*[176] extended the coupled-mode model to examine the dynamics of

[176] R. Blinc, B. Pirc, and B. Zeks, *Phys. Rev. B: Solid State* [3] **13,** 2943 (1976).

partially deuterated crystals. A similar extension of the tunneling model to examine the PE phase of partially deuterated materials was performed by Stinchcombe and Lage.[177] Because of the decoupling procedure used by Blinc *et al.*,[176] this treatment was not valid near the transition temperature. Pirc and Prelovšek[178] have very recently extended these results into this region by the use of a more general decoupling procedure.

The essence of these treatments is to include the deuteron–proton interaction $J_{pd}$ in addition to the proton–proton interaction $J_{pp}$ and deuteron–deuteron interaction $J_{dd}$ considered for KDP and $KD_2PO_4$, respectively. The contribution to the Hamiltonian [Eq. (17.1)] from the protons and deuterons thus becomes

$$\mathcal{H}' = -2\sum_{i,\alpha}\Omega_\alpha C_i^\alpha S_{i\alpha}^x - \sum_{\substack{i,j\\ \alpha,\beta}} J_{ij}^{\alpha\beta} C_i^\alpha C_j^\beta S_{i\alpha}^z S_{j\beta}^z, \qquad (22.1)$$

where $i(j)$ designates the $i(j)$th site on an O–H $\cdots$ O bond; $\alpha,\beta$ = p (proton) or d (deuteron); and $C_i^\alpha$ is the proton (deuteron) concentration. For example, if a proton occupies the $i$th site, $C_i^{\rm p} = 1$ and $C_i^{\rm d} = 0$. The interaction $J_{ij}^{\alpha\beta}$ now can assume three values, $J_{ij}^{\rm pp}$, $J_{ij}^{\rm dd}$, and $J_{ij}^{\rm pd} = J_{ij}^{\rm dp}$ for proton–proton, deuteron–deuteron, and proton–deuteron interactions, respectively.

Solving the contribution $\mathcal{H}'$ in the usual molecular field and random phase approximations yields the susceptibility and response frequencies for the normal modes of the mixed proton–deuteron system. In the approximation $\Omega_{\rm d} \simeq 0$, a diffusive mode $\omega_1$ is obtained for the "deuteron-like" mode, and the "proton-like" mode $\omega_2$ is given by

$$\omega_2^2 = \Omega_{\rm p}^2\left[1 - \left(\frac{\tanh \beta\Omega_{\rm p}}{\tanh \beta_{\rm c0}\Omega_{\rm p}}\right)(1 - C^{\rm d})\right], \qquad T > T_{\rm c} \qquad (22.2)$$

where $\beta_{\rm c0} \equiv (k_B T_{\rm c0})^{-1}$ is the transition temperature for $C^{\rm d} = 0$ and is given by

$$\tanh(\beta_{\rm c0}\Omega_{\rm p}) = 4\Omega_{\rm p}/J_0^{\rm pp}. \qquad (22.3)$$

It can be seen by comparing Eqs. (22.2) and (22.3) with the results of Section 17 that the present results for the mixed proton–deuteron system are quite similar to those for nondeuterated KDP. In fact, Eq. (22.2) for

[177] R. B. Stinchcombe and E. J. S. Lage, *in* "Light Scattering in Solids" (M. Balkanski, R. C. C. Leitz, and S. P. S. Porto, eds.), p. 834. Wiley, New York, 1976; *J. Phys. C* [3] **9**, 3295 (1976).

[178] R. Pirc and P. Prelovšek, *Phys. Rev. B: Solid State* [3] **15**, 4303 (1977).

the proton mode is recovered from the earlier results if the deuteron concentration $C^d = 0$.

To obtain the coupled-mode response, $\mathcal{H}'$ must be added to the lattice contribution and the proton–lattice and deuteron–lattice contributions must be included in the resulting full Hamiltonian [Eq. (17.1)]. Although treatment of the full Hamiltonian is mathematically complex, the calculated concentration dependences of the coupled modes can be described by this treatment for $T \ll T_c$.[176] In particular, the fitted concentration dependences for $\omega_\pm$ are compared with the measured dependences in Fig. 32, and the agreement is quite good.

Because of the large number of parameters involved in determining $\omega_\pm$ versus $C^d$, a much more stringent test of the coupled-mode model is obtained from the pressure dependences of $\omega_\pm$. The calculated pressure dependences of $\omega_\pm$ versus $C^d$ are compared with the measured values in Fig. 32. Again the agreement is very good, providing support for this model for the KDP class of FEs. The study of the pressure dependence of the Raman spectra of mixed crystals has thus been very valuable in investigating this important class of FEs.

## 23. The Case of the Antiferroelectric Crystals

As mentioned earlier, the ammonium dihydrogen phosphates and arsenates, which are isomorphous with KDP at high temperature (tetragonal space group $I\bar{4}2d$), transform on cooling to AFE phases (orthorhombic space group $P2_12_12_1$). These AFE transitions are also triggered by the ordering of the protons along the O–H $\cdots$ O bonds. However, whereas the ordering in the FE crystals is such that in a single domain all protons order near "upper" or "lower" oxygens of the $PO_4$ groups (viewed along the $c$ axis), depending on the polarity, in the AFE case the ordering is such that the protons occupy one "upper" and one "lower" corner of each $PO_4$ tetrahedron. The resulting coupling to the lattice is modified so as to cause antiparallel ionic displacements. The dielectric properties of these crystals are quite different from those of the KDP-type FEs. Relatively small discontinuities in the static dielectric constants are observed at $T_c$, the larger anomaly being along the tetragonal $a$ and not the $c$ axis.[179] The AFE transitions are first order and much less is known about them than about the FE transitions.

On the basis of structural considerations, Cochran[2] suggested that the soft mode which is associated with the transition in ADP should occur at

[179] W. P. Mason, *Phys. Rev.* [2] **69**, 173 (1946); **88**, 477 (1952).

the $Z$ point of the Brillouin zone, i.e., at $(h, 0, l)$ where $h + l$ is odd. This suggestion was confirmed by quasi-elastic neutron scattering results on deuterated ADP by Meister *et al.*[180] The mode is heavily overdamped and softens as the temperature is decreased toward the transition temperature. Apparently no direct measurements of the pressure dependence of this mode have been reported; however, the effects of pressure on the static dielectric constant and $T_c$ have been reported.[159]

Figure 30 shows the $T_c(p)$ results for ADP and $ND_4D_2PO_4$. Three features should be noted: (*a*) there is a large hydrogen isotope effect on $T_c$; (*b*) $T_c$ decreases with pressure; and (*c*) there is a strong deuteration effect on $dT_c/dp$. These features are similar to those observed for the FE crystals of this class. The similarity extends further in that the AFE phase vanishes at sufficiently high pressure in the same manner as does the FE phase, i.e., $dT_c/dp \to -\infty$ as $T_c \to 0$ K. For ADP the AFE phase vanishes at >33.5 kbar. For $ND_4D_2PO_4$, on the other hand, the $T_c(p)$ response is linear over the pressure range covered. Much higher pressure is needed to suppress $T_c$ to 0 K. Note the close similarity of these results to those for KDP and $KD_2PO_4$. It is believed that these similarities emphasize the dominant role of the O–H $\cdots$ O bond in determining the properties of KDP-type FE and AFE crystals.

The decrease of $T_c$ with pressure for the H-bonded AFE crystals is in marked contrast to the behavior of displacive AFE crystals. We recall from Section III and IV that displacive FE and AFE transitions exhibit opposite pressure dependences: $T_c$ for the FE transitions decreases, whereas $T_c$ for the AFE transitions increases with pressure. For KDP-type crystals $T_c$ decreases with pressure for both FE and AFE transitions.

The above results lend important support to the conclusion that even though the low-temperature phase is ferroelectric in KDP and antiferroelectric in ADP (and their related isomorphs), which, as noted earlier, implies differences in the proton lattice coupling and ordering of the protons, it is the O–H $\cdots$ O bond that plays the key role in determining the transitions in these crystals.

## VI. Other Soft-Mode Systems

This section is devoted to a brief discussion of the properties of a few interesting, but generally unrelated soft-mode transitions that have received recent attention. These include the transitions in the *ferroelastic*

[180] H. Meister, J. Skalyo, Jr., B. C. Frazer, and G. Shirane, *Phys. Rev.* [2] **184,** 550 (1969).

crystals paratellurite and the rare earth pentaphosphates and in the $\beta$-tungsten superconductors. Although the microscopic origins of these transitions are not yet fully understood, high-pressure studies have led to important (essential in the case of paratellurite) results.

Ferroelastic crystals are crystals in which the domains can be reoriented by the application of external stress. This domain reorientation is analogous to the domain reorientation that can be achieved in FE crystals by the application of an external electric field. Crystal structures that by symmetry can exhibit ferroelasticity have been discussed by Aizu.[181] Crystals with these structures undergo phase transitions from the paraelastic phase to the ferroelastic phase which are accompanied by soft zone-center phonon modes. Although acoustic modes, which describe the strain in the crystal, must be involved in these transformations, there may be a soft optic mode coupled to the acoustic mode through an internal strain contribution,[182] or the transition may involve only an acoustic mode. Such "pure strain" transitions, in which the soft mode is an acoustic mode, were first discussed theoretically[183] and then observed in $TeO_2$.[184]

## 24. Paratellurite—$TeO_2$

Ever since the introduction of the soft-mode concept there has been much interest in finding and understanding crystals that undergo pressure-induced transitions driven by soft modes. Indeed there are now many examples of materials in which one or more optic or acoustic modes soften with pressure and which undergo pressure-induced transitions.[185] Unfortunately, however, in most cases detailed understanding of the transition has not been possible. The one unique and important exception is paratellurite, $TeO_2$. Not only is it the first crystal observed to undergo a pressure-induced, soft phonon-driven transition for which the mechanism of the transition has been fully established, but, in addition, it is the first known example of a type of pure strain-induced transition that was first discussed theoretically by Anderson and Blount.[183]

Paratellurite (tetragonal $P4_12_12\text{-}D_4^4$) is one of three crystalline modifications of $TeO_2$. Peercy and Fritz[184] discovered that this crystal undergoes an apparently ideal second-order transition at ~9 kbar at 300

[181] K. Aizu, *J. Phys. Soc. Jpn.* **27**, 387 (1969).
[182] P. B. Miller and J. D. Axe, *Phys. Rev.* [2] **163**, 924 (1967).
[183] P. W. Anderson and E. I. Blount, *Phys. Rev. Lett.* **14**, 217 (1965).
[184] P. S. Peercy and I. J. Fritz, *Phys. Rev. Lett.* **32**, 466 (1974).
[185] G. A. Samara, *Proc. Int. Conf. High Pressure, 4th, 1974* p. 247 (1975).

K. The transition has since been studied in more detail.[186–190] It is found that the transition is purely strain-induced, driven by a soft shear acoustic mode propagating along a $\langle 110 \rangle$ crystal direction and polarized along a $\langle 1\bar{1}0 \rangle$ direction. The velocity of this mode is determined by the combination of elastic constants $c' = \frac{1}{2}(c_{11} - c_{12})$. The mode has the low velocity $v_s = 0.61 \times 10^5$ cm/sec at atmospheric pressure and temperature due to near cancellation of $c_{11}$ by $c_{12}$. The mode softens slightly with decreasing temperature, but the transition cannot be induced by temperature alone. The symmetry of the soft mode is $B_1$, and its eigenvector suggested to Peercy and Fritz[184] that the structure of the high-pressure phase should be orthorhombic ($D_2$). This has since been confirmed by neutron[189] and x-ray[190] diffraction at high pressure.

Figure 33 shows the pressure dependence of $c'$ ($c' = \rho v_s^2$, where $\rho$ is the density). Note that $c'$ goes smoothly and nearly linearly to zero as the transition pressure (8.86 kbar) is approached from either the high- or low-pressure side. The inset in Fig. 33 shows the behavior very near the transition. Note that both responses are linear and extrapolate to zero at the same pressure, so that, within experimental resolution the transition is an ideal second-order transition. The linearity also indicates that the transition can be described by classical mean-field theory. An important feature of the data in Fig. 33 is that the ratio of the magnitudes of the slopes of $c'$ in the two phases is ~3 : 1. At first glance this appears to contradict the usual mean field behavior which predicts a slope ratio of 2 : 1. However, a classical Landau-type theory (see Section 2) of the transition[187] predicts this higher ratio. This theory, which employs a primary and a secondary order parameter to describe the lattice distortion, has been successful in explaining the major static and dynamic properties of the transition.

The transition in $TeO_2$ has also been studied by Brillouin scattering,[184,186] inelastic[188] and elastic[189] neutron scattering, x-ray diffraction,[190] and dielectric techniques.[186] The Brillouin measurements examined the soft acoustic mode up to ~10 kbar and agree quite well with the ultrasonic results in Fig. 33.

The inelastic neutron scattering measurements examined the pressure dependence of the dispersion relation for the soft acoustic mode across

[186] P. S. Peercy, I. J. Fritz, and G. A. Samara, *J. Phys. Chem. Solids* **36,** 1105 (1975).

[187] I. J. Fritz and P. S. Peercy, *Solid State Commun.* **16,** 1197 (1975).

[188] D. B. McWhan, R. J. Birgeneau, W. A. Bonner, H. Taub, and J. D. Axe, *J. Phys. C* [3] **8,** L81 (1975).

[189] T. G. Worlton and R. A. Beyerlein, *Phys. Rev. B: Solid State* [3] **12,** 1899 (1975).

[190] E. F. Skelton, J. L. Feldman, C. Y. Liu, and I. L. Spain, *Phys. Rev. B: Solid State* [3] **13,** 2605 (1976).

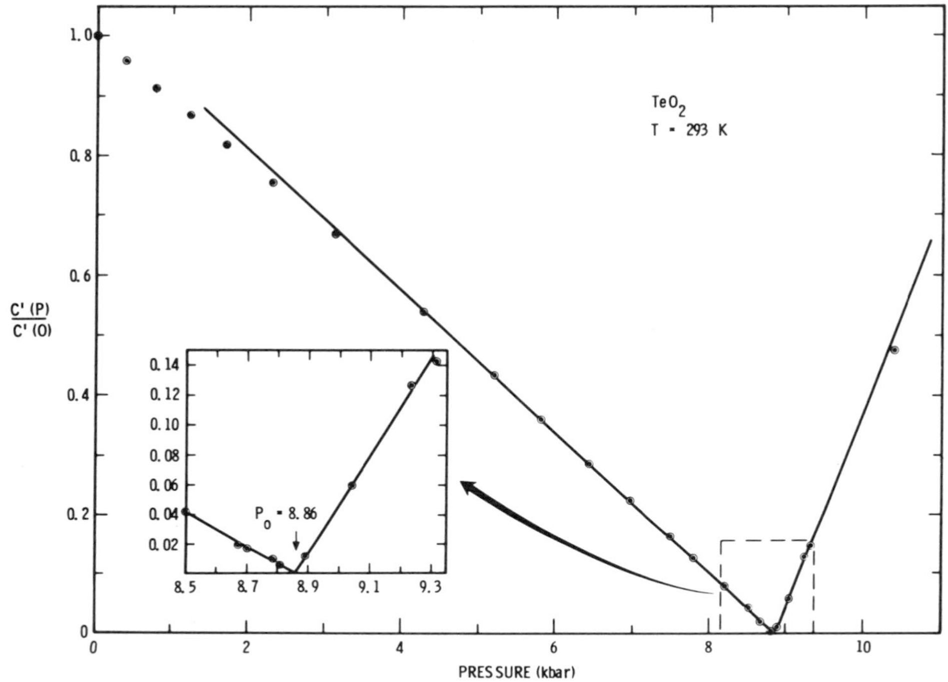


FIG. 33. Pressure dependence of the elastic constant $c' = \frac{1}{2}(c_{11} - c_{12})$ in $TeO_2$ illustrating the second-order nature of the transition in this soft acoustic mode system. [After P. S. Peercy, I. J. Fritz, and G. A. Samara, *J. Phys. Chem. Solids* **36,** 1105 (1975).]

the Brillouin zone up to 18 kbar, i.e., twice the transition pressure $p_c$. The results at 1 bar, 9 kbar, and 18 kbar are shown in Fig. 34. The inset is an expanded view of the long wavelength region near $\mathbf{q} = 0$ showing how the neutron data extrapolate well onto the initial slope determined from the elastic constants. The important feature to note is that at finite **q**, pressure causes a relatively small and nearly uniform shift of the dispersion curve. At pressures up to $p_c$ there is a softening of the curve by ~0.1 meV (0.8 $cm^{-1}$); this is followed by a *stiffening* of the curve by ~0.25 meV (2.0 $cm^{-1}$) between $p_c$ and $2p_c$. It is thus clear from these results and those in Fig. 33 that the dramatic softening of the mode occurs only at very long wavelengths, i.e., $\mathbf{q} = 0$.

The crystal structure of the high-pressure phase and the pressure dependences of the lattice parameters were first determined from neutron diffraction studies on powder samples by Worlton and Beyerlein.[189] The high-pressure phase is indeed orthorhombic ($P2_12_12_1$-$D_2^4$) confirming

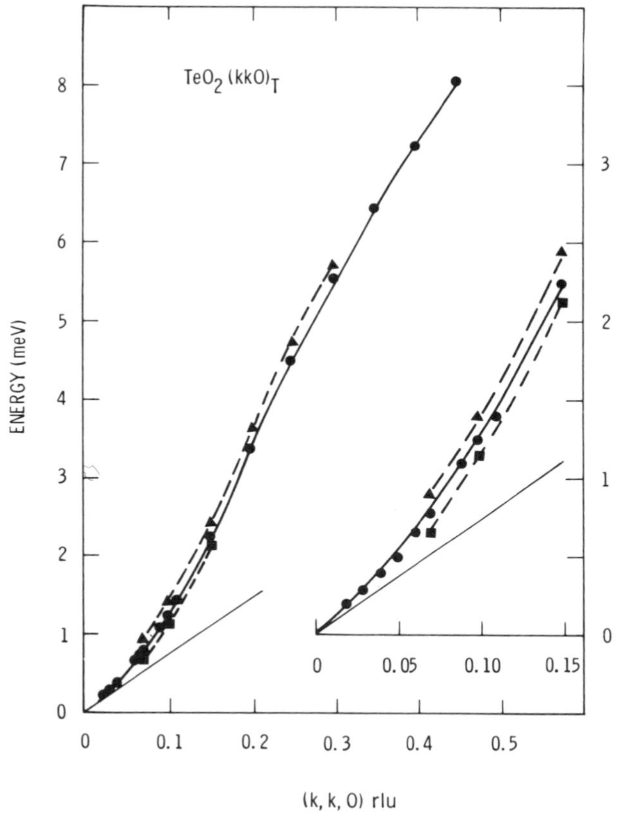


FIG. 34. Dispersion curves for the soft acoustic mode in $TeO_2$ measured at various pressures (after McWhan *et al.*[188]). (●) 1 atm, (■) $p_c$ = 9 kbar, (▲) $2p_c$.

Peercy and Fritz' earlier prediction.[184] Figure 35 shows the pressure dependence of the lattice parameters. Note that the original tetragonal *a* axis ($a_0$) splits into the (unequal) *a* and *b* axes of the orthorhombic cell. Two important features of the results should be noted: (*a*) the distortion $(b - a)$ obeys the relation $(b - a)^2 \propto (p - p_c)$ near the transition indicating mean field–like behavior; and (*b*) there is a small secondary anomaly in the quantity $\frac{1}{2}(a + b)$ at the transition. This anomaly is characterized by a small deviation of this quantity from the background tetragonal value $a_0$, and this deviation is approximately proportional to $(p - p_c)$. It is these two features that dictated that two order parameters are needed in the phenomenological theory of the transition.[187]

The main features of Worlton and Beyerlein's results have been

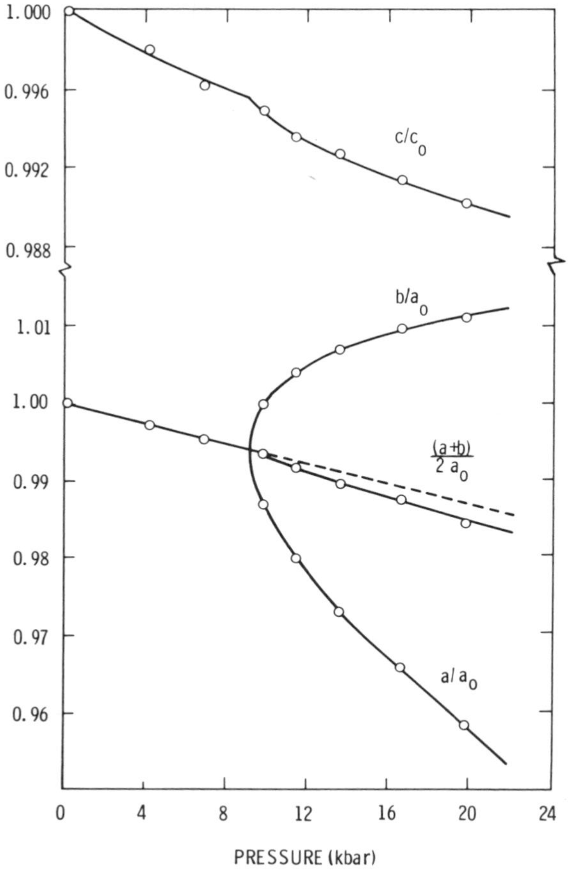


FIG. 35. Pressure dependences of the lattice parameters near the pressure-induced phase transition in $TeO_2$ measured by elastic neutron scattering. [After T. G. Worlton and R. A. Beyerlein, *Phys. Rev. B: Solid State* [3] **12,** 1899 (1975).]

confirmed by single crystal x-ray[190] and neutron[188] diffraction results. The neutron results, however, did not indicate an anomaly in the quantity $\frac{1}{2}(a + b)$, most likely due to relatively large uncertainties and scatter in the measured values of $a$ and $b$.

The temperature dependence of the transition pressure $p_c$ has been measured from 75 to ~400 K.[186] Over this range, $p_c$ increases linearly with temperature with slope $dp_c/dT = 4.8$ bar/K. The physical explanation for this increase is not known and poses an interesting question that deserves attention. We recall here that qualitative physical interpretations can be given for the pressure (temperature) dependence of $T_c(p_c)$ for transitions involving soft FE and short wavelength *optic* modes (see Section III and IV). In the case of $TeO_2$ one is dealing with a soft zone-center *acoustic* mode. The nature of the balance of forces leading to the vanishing of $c'$ is not known at present.

Finally we note that the transition in $TeO_2$ is driven purely by $c'$. There is no evidence of any coupling of this mode to any other acoustic or optic mode. This was established by a detailed study of the elastic and static dielectric constants and Raman-active modes.[186] In this crystal the zone-center optic modes are all either Raman or infrared active, so that Raman scattering and dielectric constant measurements would detect any soft optic modes. Small anomalies in some of these properties are observed at the transition, but these anomalies appear to be a consequence of the lattice strain accompanying the transition and are not directly involved in the mechanism of the transition. The result that only one mode is involved is consistent with the transition being second order.

## 25. Rare Earth Pentaphosphates

In contrast to the case for $TeO_2$ the ferroelastic transition in the rare earth pentaphosphates is accompanied by a soft zone-center optic mode. These materials have the chemical formula $REP_5O_{14}$, where RE indicates a rare earth ion, and crystallize in a variety of structures[191,192] depending on the size of the rare earth ion. Transformations between different structures can be induced with pressure in some cases and with temperature in other cases.[193,194] Rare earth pentaphosphates containing Nd are of particular interest because laser action has been demonstrated in these materials at very low pump powers.[195]

Rare earth pentaphosphates formed from Nd, La, and Tb, as well as mixed crystals containing these ions, are also of significant interest from a lattice dynamical viewpoint. These materials crystallize in an orthorhombic structure with point group symmetry $D_{2h}$ at high temperature and transform on cooling to a monoclinic structure with point group symmetry

[191] A. Durif, *Bull. Soc. Fr. Mineral. Crystallogr.* **94,** 314 (1971); K.-R. Albrand, R. Attig, J. Fenner, J. P. Jeser, and J. P. Mootz, *Mater. Res. Bull.* **9,** 129 (1974).

[192] H. Y.-P. Hong, *Acta Crystallogr. 330,* 468 (1974); H. Y.-P. Hong and J. W. Pierce, *Mater. Res. Bull.* **9,** 179 (1974).

[193] H. Schulz, K.-H. Thiemann, and J. Fenner, *Mater. Res. Bull.* **9,** 1525 (1974).

[194] H. P. Weber, B. C. Tofield, and P. F. Liao, *Phys. Rev. B: Solid State* [3] **11,** 1152 (1974).

[195] See, e.g., H. P. Weber, T. C. Damen, H. G. Danielmeyer, and B. C. Tofield, *Appl. Phys. Lett.* **22,** 534 (1973).

$C_{2h}$. The monoclinic phase is ferroelastic and the phase transition is thought to be an ideal, second-order displacive structural phase transition. The structure can be viewed as being composed of $P_{10}O_{28}$ rings which are in turn composed of $PO_3$ tetrahedra. In the orthorhombic phase the $PO_3$ tetrahedra are aligned on the mirror plane *m*. The change that occurs at the transition consists of small rotations of the $PO_3$ tetrahedra, and the order parameter of the transition is the rotation angle. The order parameter exhibits mean-field behavior near $T_c$.[194,196]

Measurements of the Raman spectra of these compounds[197] demonstrated that the transition is accompanied by a soft zone-center optic mode which is Raman active and underdamped in both phases, and that the soft mode interacts with other optic modes in the ferroelastic phase. Peercy *et al.*[198,199] investigated the pressure and temperature dependences of the Raman spectra. Two remarkable feasures are observed for the ferroelastic transition in these materials: (*a*) the transition temperature increases with pressure[198]; and (*b*) in addition to the soft optic–optic mode couplings, there is a strong soft optic–acoustic mode interaction[199] near $T_c$. The soft-mode frequency increases rapidly with pressure in the ferroelastic phase as shown in Fig. 36 where the temperature dependence of the soft-mode frequency is shown for various pressures. These data yield $dT_c/dp \cong 21$ K/kbar for $Nd_{0.5}La_{0.5}P_5O_{14}$—a remarkably large pressure dependence. Similar results were obtained for $TbP_5O_{14}$ with $dT_c/dp \cong 20$ K/kbar.

The increase in $T_c$ with pressure for the $REP_5O_{14}$ compounds deserves a special comment. In Sections III and IV attention was drawn to the apparent generalizations that (*a*) $T_c$ decreases with pressure for those *ferroelectric* transitions accompanied by soft zone-center optic modes; and (*b*) $T_c$ increases with pressure for those displacive transitions accompanied by soft zone-boundary optic modes. The transition in the $REP_5O_{14}$ compounds is accompanied by a soft zone-center optic mode and the increase of $T_c$ with pressure may appear to contradict generalization (*a*). However, it should be emphasized that the soft mode in the RE pentaphosphates is nonpolar, unlike the case of the soft mode in FEs which involves strong electric dipole forces. The increase of $T_c$ with pressure for the RE pentaphosphates, which is not at present understood

[196] J. P. Budin, A. Milatos-Roufos, N. D. Chinh, and G. Le Roux, *J. Appl. Phys.* **46,** 2867 (1975).

[197] D. L. Fox, J. F. Scott, and P. M. Bridenbaugh, *Solid State Commun.* **18,** 111 (1976).

[198] P. S. Peercy, J. F. Scott, and P. M. Bridenbaugh, *Bull. Am. Phys. Soc.* **21,** 337 (1976).

[199] P. S. Peercy, *in* "Proceedings of the 5th Raman Conference" (E. D. Schmid, J. Brandmüller, W. Kiefer, B. Schrader, and H. W. Schrötter, eds.), p. 571. Hans Ferdinand Schulz Verlag, Freiburg, 1976.

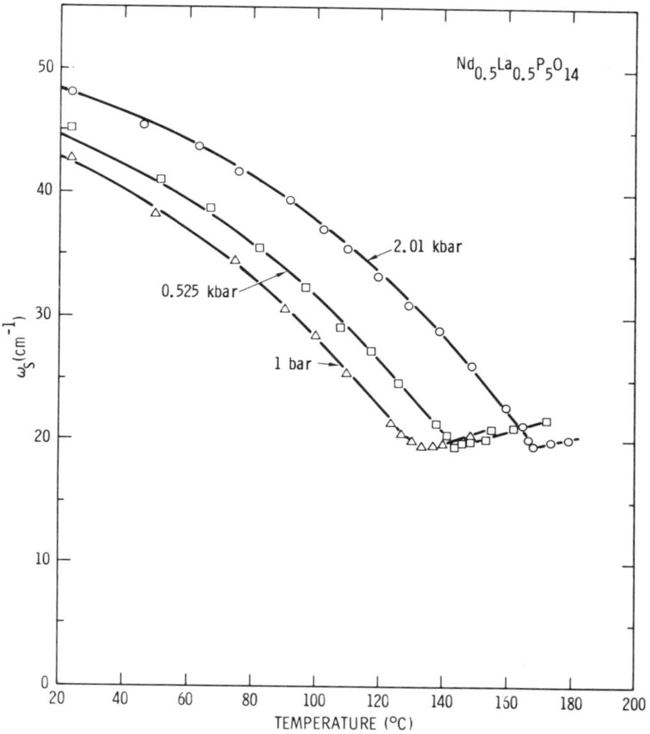


FIG. 36. Temperature dependence of the soft-mode frequency in $Nd_{0.5}La_{0.5}P_5O_{14}$ at various pressures: (△) 1, (□) 0.525, (○) 2.010 bars, indicating the increase in $T_c$ with pressure for this ferroelastic transition. [After P. S. Peercy, *in* "Proceedings of the 5th Raman Conference" (E. D. Schmid *et al.*, eds.), p. 571. Hans Ferdinand Schulz Verlag, Freiberg, 1976.]

from a lattice dynamical viewpoint, may be related to the ferroelastic nature of the transition and the optic–acoustic mode coupling which occurs in these compounds near $T_c$ (see Section VII).

## 26. β-Tungsten Superconductors

The binary intermetallic compounds crystallizing in the cubic β-tungsten (or A-15) structure are best known for their superconductivity. Also important is the fact that several members of this family exhibit, on cooling, a structural phase transition from cubic to tetragonal symmetry at a temperature $T_L$ somewhat higher than the superconducting transition temperature $T_c$. The transition is accompanied by a large elastic instability.[200] Specifically, the shear mode with wave vector **q** ∥ [110] and polarization vector **e** ∥ [1$\bar{1}$0], which is described by the elastic constant $c' = \frac{1}{2}(c_{11} - c_{12})$, softens on cooling. (This is also the mode that softens in $TeO_2$.) The acoustic velocity associated with this mode decreases from a normal value at 300 K to near zero at $\sim T_L$. It is of considerable interest that there is some evidence linking the lattice instability associated with the phase transition with the high superconducting transition temperature of these compounds.[201,202]

The effects of hydrostatic pressure on the elastic constants, $T_c$ and $T_L$, have been reported for some A-15 compounds. These studies have yielded some important results, especially on $V_3Si$ and $Nb_3Sn$, the two most studied members of this group.[203–208] Among the interesting aspects of these results are the findings the pressure dependences of $T_c$ and $T_L$ are different in $V_3Si$ and $Nb_3Sn$: in $V_3Si$, $T_c$ increases and $T_L$ decreases with pressure, whereas the opposite is true for $Nb_3Sn$.

Unfortunately, space limitations do not allow us to go into the details of the results and their interpretation here. The reader is referred to the recent review by Chu[208] for details. It sufficies to note here that both the experimental results and their interpretation on these materials are complex; and although no unique physical picture has emerged thus far, the pressure results provide considerable support for the Weger–Labbé–Friedel electron model.[209,210] In this model, which has proven useful for explaining many of the atmospheric pressure properties of A-15 superconductors, the energy band structure exhibits sharp, narrow peaks in the density of states at the d subband edges. The number $Q$ of d electrons in the subband determines the position of the Fermi level relative to the band edge, and this in turn determines $T_c$, $T_L$, and $c'$. The softening in $c'$ with decreasing $T$ presumably results from a direct electron–acoustic

[200] L. R. Testardi, W. A. Reed, T. B. Bateman, and V. G. Chirba, *Phys. Rev. Lett.* **15,** 250 (1965).

[201] L. R. Testardi, *Comments Solid State Phys.* **6,** 131 (1975).

[202] B. W. Batterman and C. S. Barrett, *Phys. Rev. Lett.* **13,** 390 (1964); *Phys. Rev.* **149,** 296 (1966).

[203] P. F. Carcia, G. R. Barsch, and L. R. Testardi, *Phys. Rev. Lett.* **27,** 944 (1971).

[204] R. E. Larsen and A. L. Ruoff, *J. Appl. Phys.* **44,** 1021 (1973).

[205] C. W. Chu and L. R. Testardi, *Phys. Rev. Lett.* **32,** 766 (1974), and references therein.

[206] P. F. Carcia and G. R. Barsch (to be published).

[207] C. W. Chu, *Phys. Rev. Lett.* **33,** 1283 (1974), and references therein.

[208] C. W. Chu, *in* "High-Pressure and Low-Temperature Physics" (C. W. Chu and J. A. Woollam, eds.), p. 359. Plenum, New York, 1978.

[209] For a review, see M. Weger and I. B. Goldberg, *Solid State Phys.* **28,** 1 (1974).

[210] J. Labbé, S. Barisić, and J. Friedel, *Phys. Rev. Lett.* **19,** 1039 (1967); J. Labbé, *Phys. Rev.* [2] **172,** 451 (1968).

phonon coupling process. The pressure dependences of $T_c$, $T_L$, and $c'$ are thought to arise from a pressure-induced redistribution of electrons leading to an enhancement in $Q$.[207,211,212] This enhancement presumably results from electron transfer from the s to the d bands.

## VII. Effects of Pressure on Coupled Phonon Modes

In several sections of this review we indicated that the soft optic mode associated with a structural phase transition may interact with other phonon modes of the lattice. Several examples of such mode coupling have been investigated to date in experiments that probe the soft mode response as a function of temperature and/or pressure. Of special interest to us here are three distinct types of (zone-center) coupled-mode systems. These are (*a*) soft optic–acoustic coupling in piezoelectric crystals (e.g., $BaTiO_3$ and $KH_2PO_4$); (*b*) soft optic–acoustic coupling in nonpiezoelectric ferroelastics (e.g., $NdP_5O_{14}$); and (*c*) soft optic–optic phonon coupling (e.g., SbSI). Results on these coupled mode systems will be discussed with emphasis on pressure effects. The pressure results have generally led to a more detailed understanding of the interactions involved.

### 27. Optic–Acoustic Mode Coupling: Piezoelectric Crystals

#### *a. $BaTiO_3$*

It was first noted by Fleury and Lazay[40] in measurements of the Brillouin spectra that the soft-optic mode of $BaTiO_3$ interacts with an acoustic mode as the system approaches the tetragonal–orthorhombic phase transition at 280 K. The modes are coupled through a piezoelectric coefficient in an interaction similar to the optic–acoustic coupling in $KH_2PO_4$ previously examined by Brody and Cummins[38] and later in $KD_2PO_4$ by Reese, Fritz, and Cummins.[41] The physical origin of such an interaction is easily visualized: as the atoms undergo the motion involved in the soft TO mode, the long-range dipole moment associated with this motion induces an electric field in the crystal. This induced electric field then produces strains in the crystal through the piezoelectric effect. The atomic motions of the optic and acoustic modes are thus coupled through a piezoelectric coefficient. The strength of the interaction increases as the temperature approaches $T_c$ and the frequency of the optic mode decreases

[211] C. S. Ting and A. K. Ganguly, *Phys. Rev: Solid State B* [3] **9**, 2781 (1974).
[212] G. R. Barsch and D. H. Rogowski, *Mater. Res. Bull.* **8**, 1459 (1973).

toward that of the acoustic mode. Because the frequency of the soft-optic mode is strongly pressure dependent, especially near $T_c$ (see Section 11), pressure can be used to fine tune the strength of the optic–acoustic mode interaction.

The piezoelectric optic–acoustic coupling can be treated using a generalized Green's function formalism as discussed in Section 7. However, more physical insight into the origin of the coupling is obtained using a phenomenological approach. Following Dvorák,[39] the equations of motion, Eqs. (7.4) and (7.5), can be rewritten as

$$(\omega_o^2 - \omega^2 + 2i\Gamma_o\omega)P + aq(B/4\pi)X = F_P \tag{27.1}$$

$$(\omega_a^2 - \omega^2 + 2i\Gamma_a\omega)X + (aq/\rho)P = F_X \tag{27.2}$$

where damped harmonic oscillator responses are assumed for the optic and acoustic modes; $\omega_o(\omega_a)$ and $\Gamma_o(\Gamma_a)$ are the optic (acoustic) frequency and damping, $P$, $X$, $q$, and $\rho$ are the polarization, strain, wave vector, and density, respectively. $a \equiv a_{15}$ is the relevant piezoelectric constant for tetragonal $BaTiO_3$ which couples the optic and acoustic phonons, and $F_P(F_X)$ are the normalized driving forces due to polarization (strains). The parameter $B$ relates the temperature-dependent soft-mode frequency $\omega_o(T)$ to the clamped dielectric constant $\varepsilon_a'(T)$ through the LST relation,

$$\omega_o^2(T) = B/\varepsilon_a'(T). \tag{27.3}$$

Assuming that the coupled system in $BaTiO_3$ is driven by polarization fluctuations so that $F_P \gg F_X$, solution of Eqs. (27.1) and (27.2) yields the coupled susceptibility $\chi(\omega)$ as

$$\chi(\omega) = \frac{\chi_o^{(0)}(\omega)}{1 - A^2\chi_o^{(0)}(\omega)\chi_a^{(0)}(\omega)}, \tag{27.4}$$

where $\chi_j^{(0)}(\omega) = (\omega_j^2 - \omega^2 + 2i\omega^{(0)}/\Gamma_j)^{-1}$ are the decoupled susceptibilities and $A^2 = a^2q^2B/4\pi\rho$ is the coupling constant. All the parameters that comprise $A^2$ are known from independent measurements at atmospheric pressure. In combined Brillouin–Raman scattering measurements as a function of temperature at atmospheric pressure it was shown by Fleury and Lazay[40] that the spectral response predicted for the coupled susceptibility of Eq. (27.4) was in excellent agreement with the measured response. The combined Raman–Brillouin spectrum measured at 296 K is shown in Fig. 37. The effect of the mode coupling is evident from the asymmetric lineshape for the Brillouin components and the background which extends to $\omega = 0$.

The expected effect of hydrostatic pressure on the coupled system can be illustrated by examining the coupled response for $\omega \approx \omega_a \ll \omega_o$.[213] In

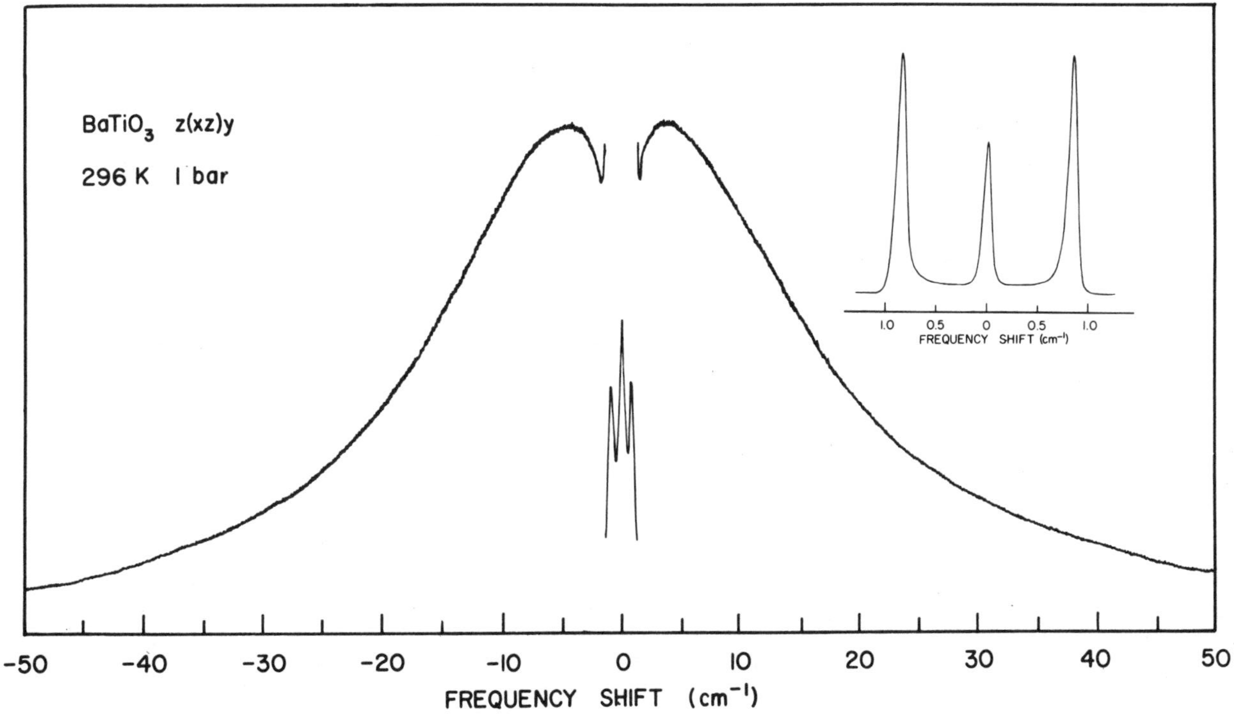


FIG. 37. Combined Raman–Brillouin spectrum in $BaTiO_3$ at 296 K illustrating the optic–acoustic interaction. [After P. S. Peercy and G. A. Samara, *Phys. Rev. B: Solid State* [3] **6**, 2748 (1972).]

this limit the coupled frequency $\omega_a(c)$ and damping $\Gamma_a(c)$ for the low-frequency branch of the system are

$$\omega_a^2(c) \cong \omega_a^2 - A^2/\omega_o^2, \tag{27.5}$$

$$\Gamma_a(c) \cong \Gamma_a + \Gamma_o A^2/\omega_o^4. \tag{27.6}$$

The primary effect of hydrostatic pressure is expected to be to increase the optic soft-mode frequency $\omega_o$ which, as seen from Eqs. (27.5) and (27.6), will increase $\omega_s(c)$ and decrease $\Gamma_a(c)$. In addition to the increase in $\omega_o$ with pressure, both $\Gamma_o$ and $A^2$ are expected to be pressure dependent.

The pressure dependences of $\omega_a(c)$ and $\Gamma_a(c)$ are illustrated in Fig. 38 and show the expected behavior noted above. Whereas these data provide qualitative support for the model, more quantitative support is obtained from the pressure dependence of the other parameters. In particular, detailed analysis of the pressure dependence of $A^2$ reveals that

$$\left(\frac{\partial \ln A^2}{\partial p}\right)_T \cong 2\left(\frac{\partial \ln a_{15}}{\partial p}\right)_T \cong 2\left(\frac{\partial \ln P_s}{\partial p}\right)_T \tag{27.7}$$

if the small pressure dependences in $q$, $\rho$, and the electrostrictive coefficient $Q_{44}$ are neglected.[213] Here $P_s$ is the spontaneous polarization. From the light scattering measurements the pressure dependence of the coupling constant was determined to be $(\partial \ln A^2/\partial p)_T = -3.2\%/\text{kbar}$.[213] Thus, these measurements yield $(\partial \ln P_s/\partial p) \cong -1.6\%/\text{kbar}$ in reasonable agreement with the value of $(\partial \ln P_s/\partial p) = -1.4\%/\text{kbar}$ obtained from static measurements.[9,213]

From the preceding discussion it can be seen that the hydrostatic pressure measurements can be used not only to examine models for the coupled interaction, but also to extract macroscopic dielectric parameters and their pressure dependences in piezoelectric crystals.

### b. $KH_2PO_4$

It was briefly mentioned in Section V that in KDP and its isomorphs there is strong piezoelectric coupling in the PE phase between the soft optic mode $\omega_-$ and an acoustic mode. Specifically, the coupling is to the *xy* shear acoustic mode[214,215] whose velocity is determined by the effective elastic constant $c_{66}$. As $\omega_-$ decreases in frequency on approaching $T_c$ from above, the interaction increases and a level repulsion effect forces the *xy*

[213] P. S. Peercy and G. A. Samara, *Phys. Rev. B Solid State* [3] **6**, 2748 (1972).
[214] W. P. Mason, *Phys. Rev.* [2] **69**, 173 (1946).
[215] C. W. Garland and D. B. Novotny, *Phys. Rev.* [2] **177**, 971 (1969); E. Litov and C. W. Garland, *Phys. Rev. B Solid State* [3] **2**, 4597 (1970).

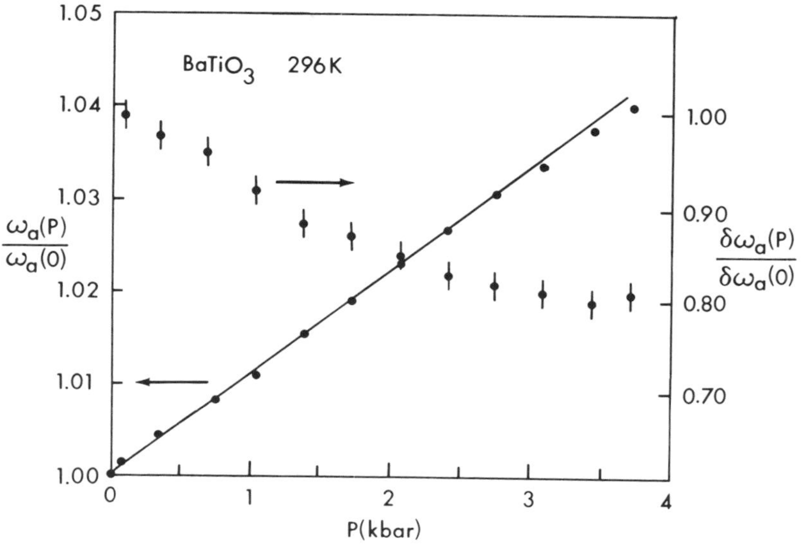


FIG. 38. Increase in the frequency and decrease in the damping with pressure of the coupled acoustic mode in $BaTiO_3$. [After P. S. Peercy and G. A. Samara, *Phys. Rev. B: Solid State* [3] **6**, 2748 (1972).]

shear mode velocity to decrease until it ultimately reaches zero, thereby precipitating the FE transition at $T_c$. In KDP this temperature $T_c$ is ~4 K higher than that at which $\omega_-$ would become zero.[215]

The mode coupling in KDP and its isomorphs is governed by the theoretical considerations outlined above and in Section 7 and is qualitatively similar to that observed in $BaTiO_3$. A detailed study of this coupling in $KD_2PO_4$ has been reported by Reese *et al.*,[41] and an extensive literature exists on the acoustic mode anomaly in KDP and its isomorphs. Both ultrasonic[215] and Brillouin scattering[38,41] techniques have been used to study this mode and its temperature dependence at 1 bar. Fritz[216] has extended the ultrasonic measurements on KDP to high pressure and thereby determined the pressure dependence of the piezoelectric coupling coefficient.

In KDP-type crystals the piezoelectric effect couples the *xy* shear displacements (strain component $x_6$) to the *z*- or *c*-axis polarization component $P_3$. Consequently those normal modes that involve *xy* shears are actually mixed strain–polarization modes and exhibit frequency dispersion in their velocities. For KDP itself there appears to be negligible

[216] I. J. Fritz, *Ferroelectrics* **5**, 17 (1973).

frequency dispersion in the measured $xy$ shear velocity for frequencies up to the gigahertz range,[38] and it can then be shown that

$$c_{66}^{E} = c_{66}^{P} - a_{36}^{2}\chi_{3}^{x} \tag{27.8}$$

where $c_{66}^{E}$ is the *measured,* constant-field, real part of the effective elastic constant $c_{66}$; $c_{66}^{P}$ is the "normal," constant-polarization elastic constant; $a_{36}$ is the piezoelectric coefficient; and $\chi_3^x$ is the clamped, static $c$-axis dielectric susceptibility. From Eq. (27.8) it is seen that as $T_c$ is approached from above, the large increase in $\chi_3^x$ caused by the softening of the optic mode produces a decrease in $c_{66}^{E}$. Ultimately $c_{66}^{E}$ vanishes at $T_c$. $c_{66}^{P}$, on the other hand, is nearly temperature independent through the transition.

By combining his ultrasonic data with dielectric susceptibility data,[166] Fritz obtained $d \ln a_{36}/dp = -2.0 \pm 0.3\%$/kbar. It is of interest to note that this change is of the same magnitude as the decrease in $a_{15}$ of $BaTiO_3$ with pressure (see Section 27,a). Fritz' work also yielded the pressure dependence of the "normal" elastic constant $c_{66}^{P}$ for which $d \ln c_{66}^{P}/dp \simeq -0.1\%$/kbar. Thus, the pleasure dependences of all the parameters that govern the soft acoustic mode behavior of KDP are known.

## 28. Optic–Acoustic Mode Coupling: Nonpiezoelectric Crystals

As was noted in Section 25 the soft zone-center optic mode associated with the ferroelastic phase transition in the $REP_5O_{14}$ system interacts with an acoustic mode near $T_c$. As seen from the data of Fig. 36, the optic soft-mode frequency $\omega_s$ does not vanish at $T_c$. In fact, the lowest value observed for $\omega_s$ is 19.5 $cm^{-1}$ at $T = T_c$, with this value being similar for all isomorphs investigated to date.[199] It was first suggested by Peercy[199] that the finite value of $\omega_s$ at $T_c$ may in part be due to optic–acoustic mode interactions.

Although by symmetry the soft optic mode cannot couple piezoelectrically to an acoustic mode of the lattice in the paraelastic phase of these crystals, Miller and Axe[182] have shown that there can be nonnegligible interactions between Raman-active and acoustic modes even in non-piezoelectric crystals. These authors examined the long-wavelength limit of the equations of motion for the optic and acoustic modes of the crystal within the framework of the rigid-ion model. In this approximation, they obtained a renormalized value for the elastic constant $c_{ijkl}$ given by

$$c_{ijkl} = c_{ijkl}(0) - \sum_{m} \frac{1}{\omega_e^2(m)} F_{ij}(m)F_{kl}(m). \tag{28.1}$$

In Eq. (28.1), $c_{ijkl}(0)$ is the decoupled value of the elastic constant, $\omega_e(m)$ is the frequency of the optic mode with eigenvector $e$ on branch $m$, and the $F(m)$ are coupling constants. The $F(m)$ transform as the eigenvector

$\hat{e}(m)$, so that $F(m)$ is nonzero only for Raman-active modes; if the mode is not Raman active, $F(m) = 0$.

Consider the case of a Raman-active mode which softens as $T$ approaches $T_c$. For nonzero $F(m)$, as $\omega_e(m)$ decreases the second term on the right-hand side of Eq. (28.1), which is the internal strain contribution to the elastic constants, diverges so that at some value of $\omega_e(m)$ the renormalized elastic constant vanishes and the crystal becomes unstable with respect to the associated acoustic mode of the lattice. This treatment thus implies that, for the case of interest, a lattice instability will set in as some elastic constant of the form $c_{iijj}$ vanishes before the optic-mode frequency vanishes. In fact, the optic-mode frequency remains finite at $T_c$.

For the $D_{2h}$ symmetry phase of the rare earth pentaphosphates the soft mode has symmetry $B_{2g}$ with nonzero polarizability components $\alpha_{xz} = \alpha_{zx}$. Combining the Raman tensor with the elastic constant matrix for $D_{2h}$ symmetry indicates that the soft optic mode will couple to the elastic constant $c_{55}$ in the paraelastic phase.

Measurement of the Brillouin spectra[199] of $Nd_{0.5}La_{0.5}P_5O_{14}$ indicates that the elastic constant $c_{55}$ is indeed renormalized. The temperature dependence of the associated acoustic-mode frequency, plotted as $\omega^2$ is shown in Fig. 39. The data reflect the strong interaction expected between the optic and acoustic modes as the optic-mode frequency decreases, and the system can be understood on the basis of the renormalization given in Eq. (28.1). The pressure dependence of this interaction has not been investigated so far, primarily because of the difficulty of doing the experiment on the very small crystals that are available ($<0.2$ mm$^3$).

## 29. Optic–Optic Mode Coupling

In addition to optic–acoustic mode couplings, discussed previously, coupling of the soft optic mode with other optic modes of the lattice is also important in some systems that undergo structural phase transitions. Such interactions can be expected to occur as the frequency of the soft mode approaches the frequency of other optic modes of the same symmetry. In particular, one-phonon $\mathbf{q} = 0$ optic-mode couplings have been observed in the FEs $LiNbO_3$,[217] $LiTaO_3$,[217] and SbSI,[90,91] and in the ferroelastic rare earth pentaphosphates.[197] In addition, an electric field–induced $\mathbf{q} = 0$ optic mode coupling has been observed in the incipient ferroelastic $SrTiO_3$.[218] Of these systems, only the optic-mode interactions in SbSI have been investigated in detail at high pressure, although some prelimi-

[217] W. D. Johnston and I. P. Kaminow, *Phys. Rev.* [2] **168,** 1045 (1968).

[218] J. M. Worlock, J. F. Scott, and P. A. Fleury, *in* "Light Scattering Spectra of Solids" (G. B. Wright, ed.), p. 689. Springer-Verlag, Berlin and New York, 1969.

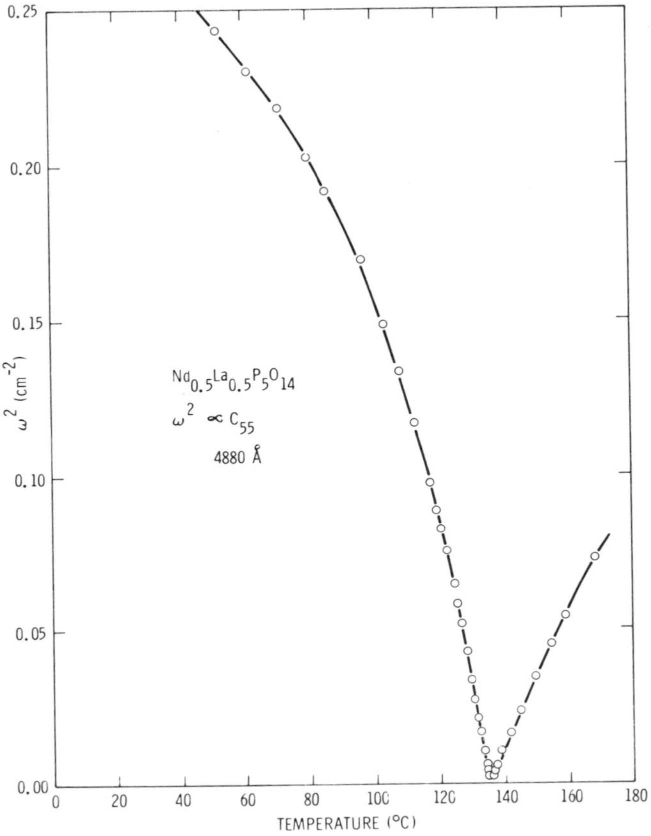


FIG. 39. Temperature dependence of the soft acoustic mode frequency in $Nd_{0.5}La_{0.5}P_5O_{14}$ near the transition temperature. [After P. S. Peercy, *in* "Proceedings of the 5th Raman Conference" (E. D. Schmid *et al.*, eds.), p. 571. Hans Ferdinand Schulz Verlag, Freiburg, 1976.]

nary measurements of the pressure dependence of similar interactions in $LaP_5O_{14}$ and $Nd_{0.5}La_{0.5}P_5O_{14}$ have also been made.[199]

The optic-mode coupling in SbSI was first investigated by Steigmeier *et al.*[90] in measurements of the temperature dependence of the Raman spectra. These authors reported coupling between the soft FE mode and an optic mode near 30 $cm^{-1}$ at atmospheric pressure. The interaction occurred about 20 K below the transition temperature of 292 K. Although they used a somewhat different algebraic formalism than that described in Section 7 to describe the coupling, Scott[35] pointed out that their expressions were numerically equivalent to those derived from the coupled-mode treatment given in Section 7 within experimental uncertainty.

Balkanski and co-workers[97] investigated the pressure dependence of the Raman spectra in SbSI for very low pressures ($\leqslant 0.5$ kbar) at temperatures near the atmospheric pressure transition temperature. These measurements allowed the authors to identify an optic-mode coupling near 40 $cm^{-1}$ in addition to the coupling near 30 $cm^{-1}$ observed in the atmospheric pressure measurements. These optic-mode couplings were also investigated in detail by Peercy.[98] His measurements were performed at lower temperatures and higher pressures than those of Balkanski *et al.*[97] The advantage of using pressure as a complementary parameter to temperature is evident from the results (see Figs. 16 and 17) because, at the higher pressures, the experiments could be performed at much lower temperatures where the modes are less heavily damped. Raman spectra taken at 119 K at pressures such that the modes interact are shown in Fig. 16. The interaction and the accompanying intensity transfer is evident as the soft mode is pressure tuned through the 30 $cm^{-1}$ mode.

The pressure dependences of the optic modes in the low-frequency region of the spectrum are summarized in Fig. 17. Mode couplings are evident at ~1.8 and 3.2 kbar. As seen from these data, the level repulsion is quite weak. Computer fits to the spectra in the region where the 30 $cm^{-1}$ and the soft optic modes interact, assuming the coupled response could be described by two damped harmonic oscillators coupled by a real interaction [Eqs. (7.2) and (7.3)], yielded an interaction of $\Delta \cong 2.0$ $cm^{-1}$. The measured soft mode response is thus very nearly the decoupled response of the system except when the frequency of the soft mode is very close to the other mode frequency.

## VIII. Concluding Remarks

In this review the soft-mode concept and high-pressure studies of soft-mode transitions in solids were discussed. It was emphasized that the decrease of the frequency of the soft mode as the transition is approached is caused by a balance between competing forces, and that this balance can be strongly influenced by hydrostatic pressure. The effects of pressure on soft-mode behavior in different classes of soft-mode transitions were discussed with emphasis on the physical implications of the results. Comparisons of pressure effects among the different classes reveal general trends which have led to a better understanding of the nature of the competing forces involved.

The results presented and the literature cited emphasize the important, often essential, role of high-pressure research in the study of soft-mode transitions. Such research has not only yielded exciting new phenomena that occur at high pressure, but it has also improved our understanding of the response of materials at atmospheric pressure. In general, pressure affords a complementary variable to temperature for investigating anharmonic interactions and soft-mode behavior. With pressure all temperature dependences of the thermal anharmonicities can be removed leaving only the effects of changing the volume. As also discussed in the text, pressure results can be used to evaluate models for soft-mode behavior as well as for coupled-mode interactions. There can be no doubt that high-pressure research will always be important in the study of soft-mode phase transitions and associated phenomena.

Some of the discussion presented has been phenomenological. This is largely due to the lack of adequate detailed models to describe the complex interactions involved. In the cases where model theories are available, these theories are necessarily based on strong assumptions and approximations. The coupled proton–phonon (Kobayashi) model of KDP is a good case in point. However, despite its shortcomings[7,154] it has proved very useful in correlating a variety of experimental data and it provides a great deal of (qualitative) physical insight into what must surely go on in the crystal. Extensive use of this model was made of in Section V. In the displacive transitions, the existence of soft modes is, of course, well established, but nevertheless questions remain such as to the true nature of the temperature dependence of the soft-mode frequency. As discussed earlier in the review, anharmonic interactions are believed to be responsible for this temperature dependence. Undoubtedly this is, at least partly, true and the pressure results provide support for this fact. However, the recent work of Migoni *et al.*[219] draws attention to the importance of the anisotropy of the oxygen polarizability in determining the second-order Raman scattering and temperature dependence of the FE soft mode in $ABO_3$ perovskites. This is an interesting idea that deserves further examination.

Interest in soft-mode behavior and phonon-mode interactions and in the use of pressure in their investigations continues. Looking to the future, we think quite a number of aspects of this general problem are fruitful areas of research. Here we mention a few.

Pressure has already proved to be an important variable in the study of

[219] R. Migoni, H. Bilz, and D. Baüerle, *Phys. Rev. Lett.* **37**, 1155 (1976).

tricritical points in FEs.[98,220,221] This work has been reviewed elsewhere.[222] A tricritical point in a phase diagram is a point at which three lines of second-order transitions (or critical lines) intersect.[223] Three fields are necessary to define a tricritical point. Unfortunately, in several systems where tricritical points are believed to occur, not all three fields are accessible. Examples include the antiferromagnetic transition in $FeCl_2$,[224] the structural transition in $NH_4Cl$,[225] and the two-fluid critical mixing point in $He^3$–$He^4$.[226]

Ferroelectric crystals are potentially uniquely suited for experimental study of tricritical points because they afford three readily accessible fields, namely, temperature, pressure, and the ordering electric field. In the pressure–temperature plane the tricritical point is identified as the point where a line of first-order transitions changes to second order.[98,220,222] This point has also often been referred to as the Curie critical point.[15,227] Thus far the tricritical points in only two FEs (SbSI and $KH_2PO_4$) have been studied in detail.[98,220–222] There is no doubt that this area will grow and expand to include other materials,[222] other multicritical points, and the use of uniaxial stress in addition to hydrostatic pressure.

As noted earlier in this review, much of the work on $KH_2PO_4$ (KDP)-type hydrogen-bonded FE crystals has dealt with KDP itself and a few other isomorphs, and it has generally been believed that all these materials behave similarly. However, recent evidence[228–231] has shown that $CsH_2PO_4$ is different. It crystallizes in a monoclinic ($P2_1/m$) rather than tetragonal structure.[228] A neutron scattering study of the FE transition in

[220] V. H. Schmidt, A. B. Western, and A. G. Baker, *Phys. Rev. Lett.* **37,** 839 (1976); V. H. Schmidt, *in* "High-Pressure and Low-Temperature Physics" (C. W. Chu and J. A. Wollam, eds.), p. 237. Plenum, New York, 1978; P. Bastie, M. Vallade, C. Vettier, and C. M. E. Zeyen, *Phys. Rev. Lett.* **40,** 337 (1978).

[221] G. A. Samara, *Phys. Rev. B: Condens. Matter* [3] **17,** 3020 (1978).

[222] G. A. Samara and P. S. Peercy, *Comments* Solid State Phys. (to be published).

[223] R. B. Griffiths, *Phys. Rev. Lett.* **34,** 715 (1970).

[224] R. J. Birgeneau, G. Shirane, M. Blume, and W. C. Koehler, *Phys. Rev. Lett.* **33,** 1098 (1974).

[225] I. J. Fritz and H. Z. Cummins, *Phys. Rev. Lett.* **28,** 96 (1972).

[226] E. H. Graf, D. M. Lee, and J. D. Reppy, *Phys. Rev. Lett.* **19,** 417 (1967).

[227] V. L. Ginzburg, *Sov. Phys.—Usp.* (*Engl. Transl.*) **5,** 649 (1963).

[228] Y. Vesu and J. Kobayashi, *Phys. Status Solidi A* **34,** 475 (1976).

[229] D. Semmingsen, W. D. Ellenson, B. C. Frazer, and G. Shirane, *Phys. Rev. Lett.* **38,** 1299 (1977).

[230] K. Gesi and K. Ozawa, *Jpn. J. Appl. Phys.* **17,** 435 (1978).

[231] N. Yasuda, M. Okamoto, H. Shimizu, S. Fujimoto, K. Yoshino, and Y. Inuishi, *Phys. Rev. Lett.* **41,** 1311 (1978).

$CsD_2PO_4$ yielded diffuse scattering characteristic of one-dimensional systems with chainlike ordering parallel to the FE $b$ axis.[229] This is unlike the response of tetragonal $KD_2PO_4$ where the scattering is characteristic of three-dimensional dipolar interactions. Also unlike KDP, pressure induces an AFE phase in both $CsH_2PO_4$ and $CsD_2PO_4$.[230,231] It would be extremely desirable to investigate the details of the proton–phonon coupling (as described for KDP in Section V) for these latter crystals with emphasis on features related to the one-dimensional character.

One area of phase transitions that appears not to have been examined in any detail under pressure is that of Jahn–Teller systems. Here pressure will surely have some interesting effects on the interaction of electronic levels with phonons.

In the great majority of phase transitions the transition temperatures ($T_c$) are sufficiently high that there are no experimentally discernible quantum mechanical lattice effects, and consequently, most theoretical models of these transitions have been based on classical mechanics. As $T_c$ decreases and approaches 0K, however, quantum effects can be expected to come into play and influence experimental results. A particularly important case in this regard is the so-called displacive limit. This is defined as that special case where the transition occurs at $T_c = 0$ K. There has been a considerable recent theoretical effort devoted to the study of the displacive limit, especially with regard to the critical behavior.[232,233] Specifically, it is found that zero-point motion and quantum fluctuations cause the critical exponents at the displacive limit to be quite different from those at finite $T_c$.

One method of experimentally studying the behavior at the displacive limit is by changing $T_c$ by chemical substitution, and a very recent study deals with the FE transition in $KTa_{1-x}Nb_xO_3$ where $T_c$ was lowered to 0 K by reducing $x$.[234] However, by changing the composition, i.e., the chemistry, one changes many of the important parameters of the system. We believe that a much "cleaner" way of studying the behavior at the displacive limit and thereby testing theoretical predictions is by the application of pressure. Indeed as seen in Sections III and V there are many structural phase transitions whose $T_c$ decreases with pressure, and in several important cases (e.g., $KH_2PO_4$, $RbH_2PO_4$, $NH_4H_2PO_4$, SbSI, and $K_{1-x}Nb_xTaO_3$) $T_c$ decreases to 0 K at accessible pressures. Examination of the results on these crystals shows the increased influence of

[232] R. Oppermann and H. Thomas. *Z. Phys. B* **22,** 387 (1975).

[233] H. Beck, T. Schneider, and E. Stoll, *Phys. Rev. B: Solid State* [3] **12,** 5198 (1975); **13,** 1123 (1976); see also **16,** 462 (1977).

[234] U. T. Höehli, H. E. Weibel, and L. A. Boatner, *Phys. Rev. Lett.* **39,** 1158 (1977).

quantum effects as $T_c \rightarrow 0$ K. However, the results are generally not sufficiently detailed near $T_c = 0$ K to allow a detailed quantitative comparison with theory, especially with regard to critical exponents. More work should be done in this interesting area.

Closely related to the question of the displacive limit is the crossover region from classical to quantum behavior and the existence of an intrinsic quantum PE state,[235] i.e., a low temperature PE state stabilized by large FE ($\mathbf{q} = 0$) mode fluctuations. This state is characterized by a very large and temperature-independent static dielectric constant. At atmospheric pressure pure $SrTiO_3$[235] and $KTaO_3$ appear to exhibit such a state at helium temperatures, and the results in Sections III and V indicate that several other crystals exhibit such a state immediately past the suppression of $T_c$ to 0 K with pressure. Clearly pressure is an important variable in the study of both the crossover region and the quantum PE state.

Finally, the higher available pressure (up to the 100 kbar range) for light scattering experiments made possible by modifications in the diamond anvil cell[236] opens the door to the investigation of many new systems and pressure-induced transitions. The advent of high-pressure apparatus (in the 20 kbar range)[188] suitable for inelastic neutron scattering affords the opportunity to examine the pressure dependence of phonon frequencies as a function of wave vector. This is especially important in the pressure study of zone-boundary phonon transitions as well as incommensurate phonon transitions. High-intensity synchrotron sources are also becoming available, and they should make it possible to explore (under pressure) detailed structural and lattice dynamical aspects of phase transitions not possible with current x-ray and neutron sources. Because of these recent advances, we expect to see high pressure used more extensively in future studies of structural phase transitions.

## Note Added in Proof

Since this review was completed, significant advances in the study of soft-mode transitions at high pressure have been made. Specifically, some of the issues raised in Section VIII have been addressed. The reader is referred to the following forthcoming papers and references therein for some details.

G. A. Samara, Recent advances in the study of structured phase transitions at high pressure. To be published in *Proc. Int. Symp. Phys. Solids under High Pressure, Bad Honnef, West Germany, Aug. 10–14, 1981.*

[235] H. Burkhard and K. A. Müller, *Helv. Phys. Acta* **49**, 725 (1976), and private communication from K. A. Müller (Feb. 1978).

[236] G. J. Piermarini and S. Block, *Rev. Sci. Instrum.* **46**, 973 (1975).

G. A. Samara, N. E. Massa, and F. G. Ullman, Vanishing of the phase transitions and quantum effects in $K_2SeO_4$ at high pressure. *Proc. 5th Int. Meeting Ferroelectr.* (To be published in *Ferroelectrics.*)
G. A. Samara, Pressure and the study of quantum effects at structural phase transitions. To be published in *Proc. 8th AIRAPT—19th EHPRG Conf. High Pressure Res. Ind., Uppsala, Sweden, Aug. 17–22, 1981.*

These papers deal with incommensurate and cooperative Jahn–Teller transitions as well as with the consequences of quantum effects associated with displacive structural phase transitions.

# Beyond the Binaries—The Chalcopyrite and Related Semiconducting Compounds

A. MILLER*

*Physics Department, North Texas State University, Denton, Texas*

A. MACKINNON

*Physikalisch-Technische Bundesanstalt, Braunschweig, Federal Republic of Germany*

AND

D. WEAIRE†

*Physics Department, Heriot-Watt University, Edinburgh, Scotland*



## I. Introduction

Silicon, one of the simplest of semiconductors, remains by far the most important one in practice. Binary compounds such as GaAs are used for certain special applications. The obvious next step in the pursuit of

* Present address: Royal Signals and Radar Establishment, Great Malvern, Worcestershire, United Kingdom.

† Present address: Department of Experimental Physics, University College, Dublin, Ireland.




ISBN 0-12-607736-3

variety (and lowered symmetry) to meet particular needs leads to the consideration of ternary compounds. These are the "grandchildren" of the Group IV elements,[1] being notionally related to them as shown in Fig. 1. (In some cases, the hexagonal wurtzite structure replaces the zinc blende structure in such a scheme.)

These compounds have attracted widespread interest in the last 20 years. Although they have not yet found extensive commercial applications, they have become established as a topic of continued research. Progress up to 1975 was comprehensively reviewed in a monograph by Shay and Wernick.[2] There is also a series of proceedings of international conferences (the latest of these is in Holah[3]), which may be consulted for further reviews.

The object of the present article is to examine the relationship between structure, vibrational dispersion relations, and band structure for chalcopyrite and related compounds. It is particularly helpful to pursue this question by relating these properties to those of the "parent" or "grandparent" materials. The essential step in this procedure is the "folding" of the dispersion relations [$\omega(\mathbf{k})$ or $E(\mathbf{k})$] for the simplest structure into the smaller Brillouin zone of the more complicated one.

In one-dimensional models such folding is trivial enough and is discussed in solid state textbooks in relation to the nearly free electron approximation, Peierls instability, and so on. In three dimensions it is more complex and therefore merits the detailed and concentrated description that we give here, which we hope may serve as a working guide. (We were tempted to call it "Applied Origami for Beginners.")

In addition to the chalcopyrites, we shall include certain quaternary and "defect" or "vacancy" compounds, such as $HgIn_2Te_4$. The vacancy compounds may be considered to have tetrahedral structures with a sublattice of unfilled sites. In this case, it is obviously questionable whether the compound is to be regarded as derived from its parent structure by a small perturbation, as is often assumed for chalcopyrites.

## II. Ordered Tetrahedral Structures

Compounds that have tetrahedrally coordinated structures generally have an average number of *four* valance electrons per atom. This is the

[1] E. Parthé, "Crystal Chemistry of Tetrahedral Structures." Gordon, Breach, New York, 1964.

[2] J. L. Shay and J. H. Wernick, "Ternary Chalcopyrite Semiconductors: Growth, Electronic Properties and Applications." Pergamon, Oxford, 1975.

[3] G. D. Holah, ed., "Ternary Compounds, 1977," Inst. Phys. Conf. Ser. No. 35. Inst. Phys., London, 1977.

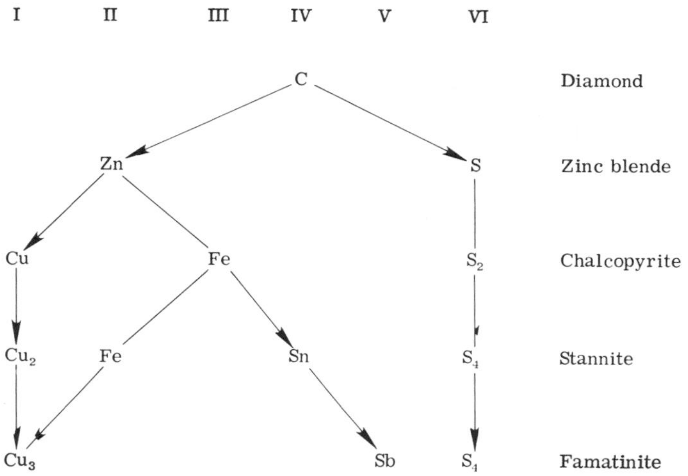


FIG. 1. This family of related compounds (in which the individual atomic species may be replaced by others of similar valence) includes chalcopyrite, one of the main examples discussed in this article. Many such relationships have been analyzed by Parthé (Ref. 1).

*Grimm–Sommerfeld rule.*[4] It holds for the familiar III–V and II–VI compounds, which adopt zinc blende and wurtzite structures, and the more complex I–III–$VI_2$ compounds also obey this rule. These ternaries are considered to be derived from III–V and II–VI compounds by "splitting" (or "cross-substitution") of an atom pair into two different atoms, maintaining the average number of electrons per atom at four, as shown in Fig. 1.

In this case, the splitting sequence gives rise to the minerals ZnS (zinc blende), $CuFeS_2$ (chalcopyrite), $Cu_2FeSnS_4$ (stannite), and $Cu_3SbS_4$ (famatinite). Zinc blende has a cubic structure, space group $F\bar{4}3m$, whereas the others have tetragonal symmetry with space groups $I\bar{4}2d$, $I\bar{4}2m$, and $I\bar{4}2m$, respectively. The names of the minerals have been adopted to describe the crystallographic structure of synthetic compounds with similar atomic arrangements. These structures are shown in Fig. 2.

In *vacancy compounds* some of the tetrahedral sites are vacant, and the average number of valance electrons satisfies the Grimm–Sommerfeld rule *only if these are included* (and, of course, considered to contribute *zero* electrons). This was pointed out by Parthé[1] and Pamplin.[5] It has become conventional to indicate the role of the vacancies in the chemical formula in such a manner: $In_2\square Te_3$, where $\square$ denotes the vacancy.

[4] H. G. Grimm and A. Sommerfeld, *Z. Phys.* **36,** 36 (1926).
[5] B. R. Pamplin, *Nature (London)* **188,** 136 (1960).

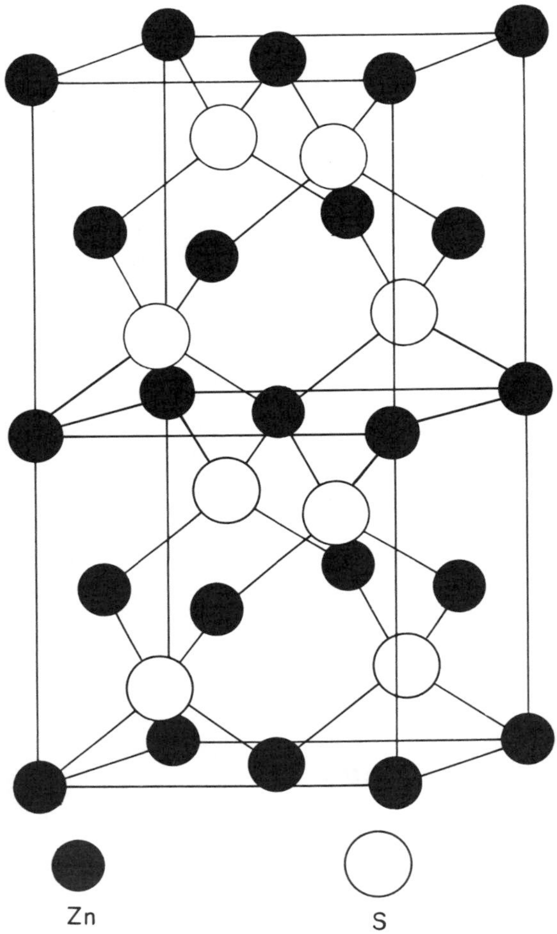


FIG. 2a. Zinc blende structure (e.g., ZnS), $F\bar{4}3m$.

Ternary compounds with the chemical formulas II–$III_2$–□–$VI_4$ and $I_2$–II–□–$VII_4$ have been found to take the two structures shown in Fig. 3a and b. These structures are very similar and can be easily confused. They both contain a body-centered-tetragonal (bct) arrangement of vacant sites and because of the similarities to the normal tetrahedral structures, have been named *defect chalcopyrite* or *thiogallate* (space group $I\bar{4}$) and *defect stannite* (space group $I\bar{4}2m$). Careful comparison of Figs. 2b and 3a shows that the defect chalcopyrite structure is the chalcopyrite structure with half of one of the cations removed. The defect stannite structure, Fig. 3b,

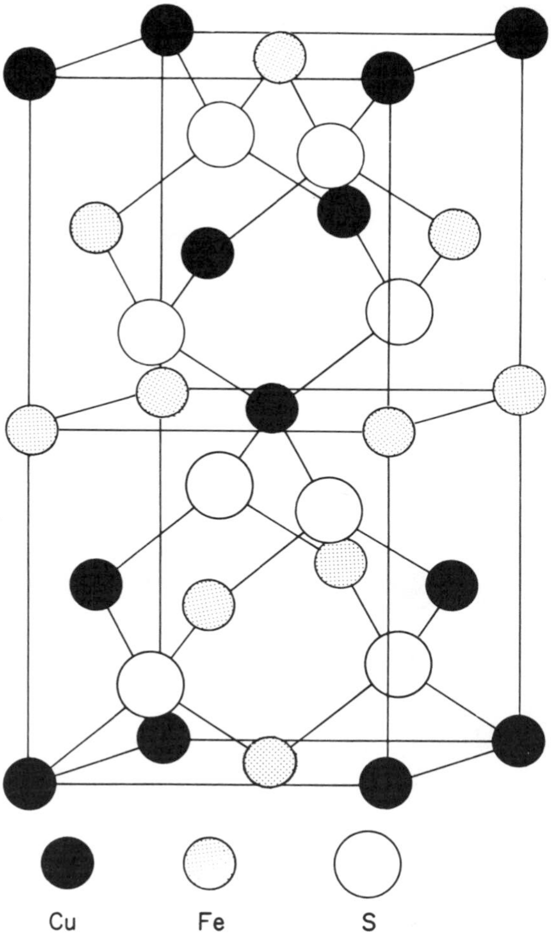


FIG. 2b. Chalcopyrite structure (e.g., $CuFeS_2$), $I\bar{4}2d$.

can be considered as either stannite, Fig. 2c, with the Sn atoms removed, or it may alternatively be viewed as famatinite, Fig. 2d, with one-third of the Cu atoms removed.

In practice, the $II–III_2–VI_4$ compounds adopt several different structures. In addition to the two shown in Fig. 3, these include spinel, phenacite, both zinc blende- and wurtzite-based structures with Group II and III atoms randomly distributed over the four cation sublattice sites, a hexagonal layered structure, and a pseudocubic vacancy structure usually referred to as the $CdIn_2\square Se_4$ structure. No definitive theories have as yet been advanced to explain why any of these structures should be prefer-

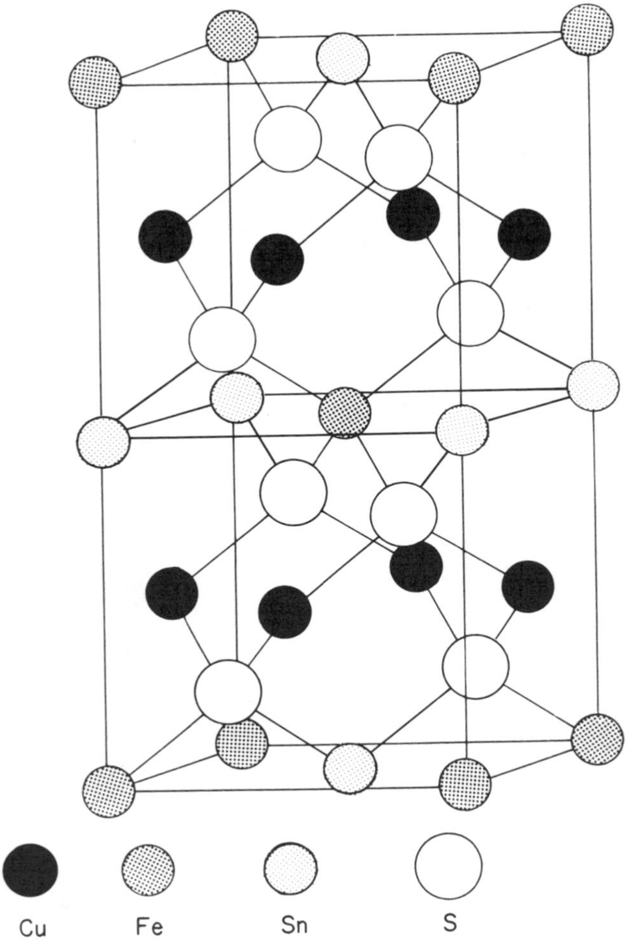


FIG. 2c. Stannite structure (e.g., $CuFeSnS_4$), $I\bar{4}2m$.

red, although von Philipsborn[6] has discussed the occurrence of spinel or ordered vacancy structures in terms of coordination numbers and polarization effects.

We shall compare the properties of the $CdIn_2\square Se_4$ structure with those mentioned previously, as another example of an ordered tetrahedral vacancy compound of recent interest.

[6] H. von Phillipsborn, *Z. Kristallogr., Kristallgeom., Kristallphys., Kristallchem.* **133,** 464 (1971).

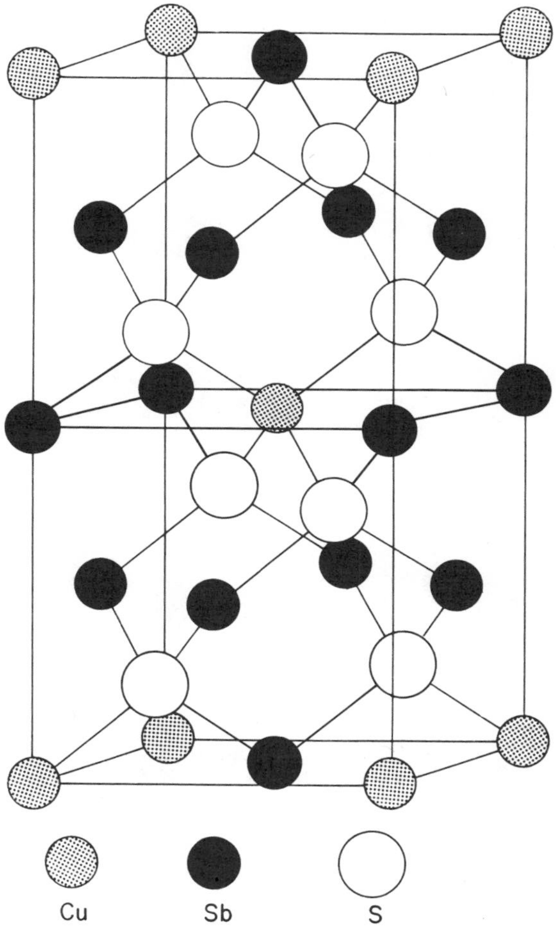


FIG. 2d. Famatinite structure (e.g., $Cu_3SbS_4$), $I\bar{4}2m$.

The $CdIn_2\square Se_4$ structure is shown in Fig. 4b together with what could be considered to be its parent in Fig. 4a. Notice that these structures are also related to zinc blende with different atoms arranged in layers. The structure shown in Fig. 4a has yet to be found in nature but is useful for theoretical comparison.

The vacancy structure in Fig. 4b is obtained by removing half the atoms from one of the cation layers. Both of the structures in Fig. 4 have space group $P\bar{4}2m$ and are often called *pseudocubic*.

The similarity of the structures shown in Figs. 2 to 4 provides an excellent opportunity for analysis of relationships between structure and

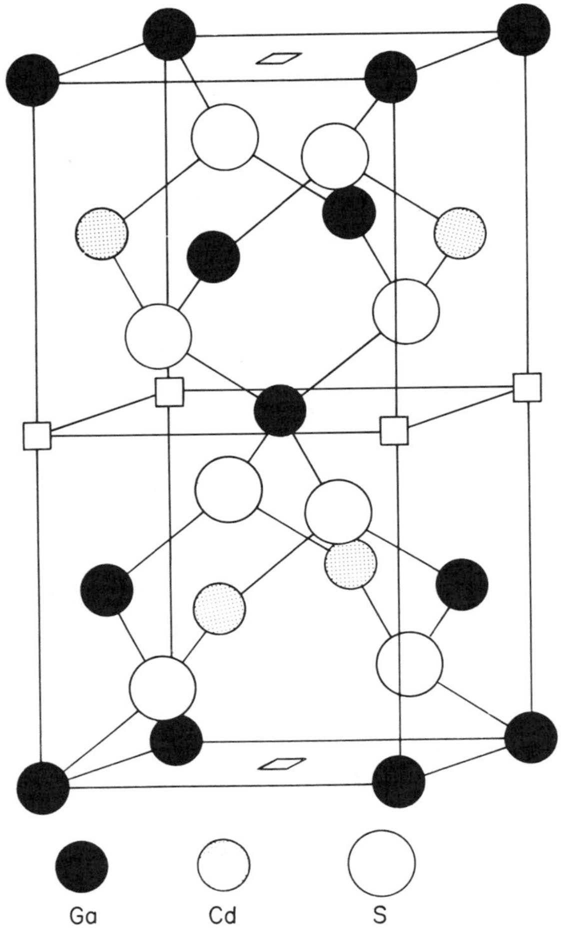


FIG. 3a. Defect chalcopyrite structure (e.g., $CdGa_2\square S_4$), $I\bar{4}$.

properties of tetrahedrally coordinated compounds. In the following section we shall look at these structures in more detail.

### 1. THE CHALCOPYRITE STRUCTURE

Fourteen II–IV–$V_2$ and twenty I–III–$VI_2$ semiconducting compounds are known to have the chalcopyrite structure. They are listed in Table I.[1,7]

This structure may be obtained from the zinc blende structure by

[7] O. Madeling, ed., "Semiconductors and Seminetals," Landolt-Börnstein, Vol. 17a,b, Section 1.5 (to be published).

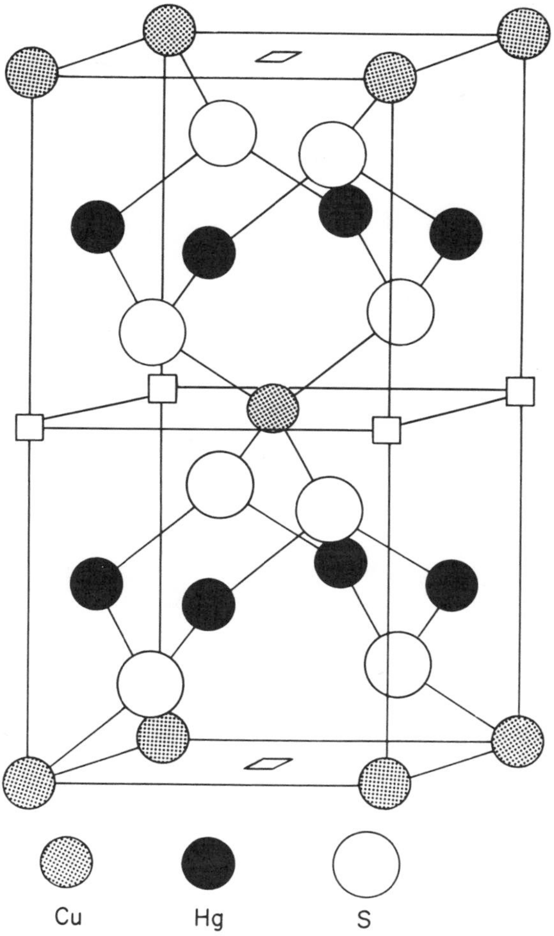


FIG. 3b. Defect stannite structure (e.g., $\beta$-$Cu_2Hg\Box I_4$), $I\bar{4}2m$.

ordering two cations on one sublattice as shown in Fig. 2. The lowering of symmetry gives rise to two distortions of this ideal structure.

First, there is a tetragonal distortion, so that the ratio $c/a$ of the edge lengths of the conventional bct cell departs slightly from the ideal value 2.0. It usually takes a lower value, although it has been found to take the ideal value to within experimental resolution for $ZnSnSb_2$, $CuInSe_2$, $CuInTe_2$, and $CuTlS_2$ and a slightly higher value for $CuInS_2$.[2]

Second, the structure has an internal free parameter $x$, which locates the position of the anion. It is clear from Fig. 2b that this has an asymmetric environment in the ideal structure and may move toward

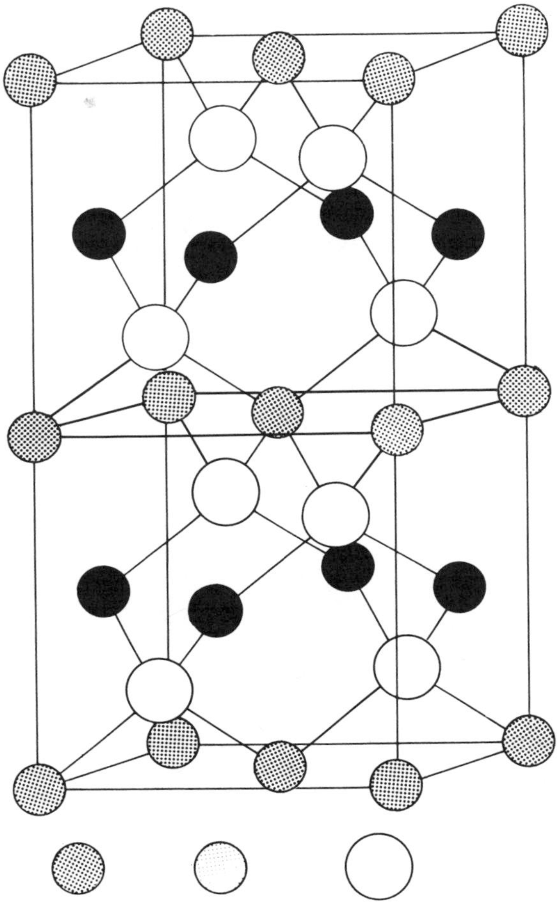

FIG. 4a. Layered structure, $P\bar{4}2m$.

either of the two pairs of cations that surround it. In practice it is found to move toward the Group III/IV atoms. Abrahams and Bernstein[8] showed that for the II–IV–$V_2$ compounds, the Group IV cations generally retain a nearly perfect tetrahedral coordination with their nearest neighbors. This results in a relation between $x$ and $c/a$, which is well obeyed in the case of Si and Ge, less well with Sn, and also less well for the I–III–$VI_2$ compounds.

The variation of tetragonal distortion among these families of compounds

[8] S. C. Abrahams and J. L. Bernstein, *J. Chem. Phys.* **55**, 796 (1971).

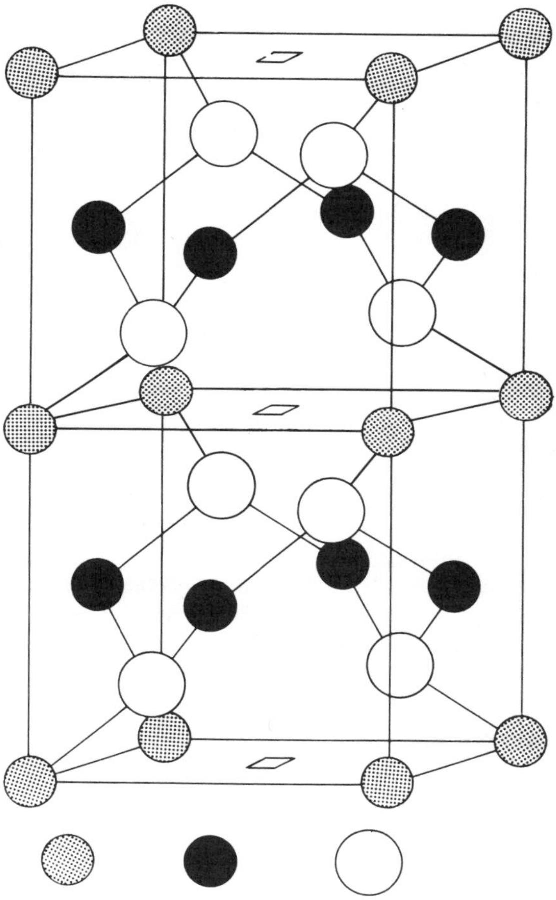


FIG. 4b. Pseudocubic structure (e.g., $CdIn_2\square Se_4$), $P\bar{4}2m$.

presents an obvious challenge to theory, which various authors[9–11] have tried to meet in a semiempirical fashion by looking for correlations with the electronegativities of the constituent atoms. Certainly there is some correlation, roughly expressible as

$$2 - (c/a) \propto (X_A - X_B)^2, \tag{1}$$

[9] J. C. Phillips, *J. Phys. Chem. Solids* **35,** 1205 (1974).
[10] D. Weaire and J. Noolandi, *J. Phys.* (*Orsay, Fr.*) **36,** C3, 27 (1975).
[11] A. Shankat and R. D. Singh, *J. Phys. Chem. Solids* **39,** 1269 (1978).

TABLE I. CHALCOPYRITE COMPOUNDS[a]

| II–IV–$V_2$ | | | I–III–$VI_2$ | |
|---|---|---|---|---|
| $MgSiP_2$ | $ZnSiP_2$ | $CdSiP_2$ | $CuAlS_2$ | $AgAlS_2$ |
| | $ZnSiAs_2$ | $CdSiAs_2$ | $CuAlSe_2$ | $AgAlSe_2$ |
| | $ZnGeP_2$ | $CdGeP_2$ | $CuAlTe_2$ | $AgAlTe_2$ |
| | $ZnGeAs_2$ | $CdGeAs_2$ | $CuGaS_2$ | $AgGaS_2$ |
| | $ZnSnP_2$ | $CdSnP_2$ | $CuGaSe_2$ | $AGGaSe_2$ |
| | $ZnSnAs_2$ | $CdSnAs_2$ | $CuGaTe_2$ | $AgGaTe_2$ |
| | $ZnSnSb_2$ | | $CuInS_2$ | $AgInS_2$ |
| | | | $CuInSe_2$ | $AgInSe_2$ |
| | | | $CuInTe_2$ | $AgInTe_2$ |
| | | | $CuTlS_2$ | |
| | | | $CuTlSe_2$ | |

[a] For details see Refs. 1, 2 (Section 1) and Ref. 7.

where $X_A$ and $X_B$ denote the electronegativities of the two cations, but it is less impressive than some of the other recent correlations of this kind and imperfectly understood.

Thermal expansion has been of some interest, being of crucial importance in crystal growth, and has been measured for several compounds by high temperature X-ray or optical interference techniques. Rather surprisingly, it is found that the axial ratio $c/a$ *decreases* with increasing temperature. Also the rate of decrease of $c/a$ is proportional to $(2.0 - c/a)$—the bigger the distortion, the more strongly it varies.[10]

## 2. ORDERED VACANCY COMPOUNDS

Ternary compounds that take tetrahedral ordered vacancy structures are shown in Table II.[12–18] These include 24 II–$III_2$–□–$VI_4$ compounds and 2 $I_2$–II–□–$VII_4$ compounds. The majority of compounds take either the $I\bar{4}$ or $I\bar{4}2m$ structures shown in Figs. 3a and b; however when the X-

[12] H. Hahn, G. Frank, W. Klinger, A. D. Störger, and G. Störger, *Z. Anorg. Allg. Chem.* **279,** 241 (1955).

[13] H. Hahn, G. Frank, and W. Klinger, *Z. Anorg. Allg. Chem.* **279,** 271 (1955).

[14] F. Lappe, A. Niggli, R. Nitsche, and J. G. White, *Z. Kristallogr., Kristallgeom., Kristallphys., Kristallchem.* **117,** 146 (1962).

[15] K. Range, W. Becker, and A. Weiss, *Z. Naturforsch., B: Anorg. Chem., Org. Chem., Biochem., Biophys., Biol.* **23B,** 1009 (1968).

[16] S. T. Kshirsagar and A. P. B. Sinha, *J. Mater. Sci.* **12,** 2441 (1977).

[17] G. A. Saunders and T. Seddon, *J. Phys. Chem. Solids* **37,** 873 (1976).

[18] K. W. Browall, J. S. Kasper, and H. Weidmeir, *J. Solid State Chem.* **10,** 20, (1974).

TABLE II. ORDERED VACANCY COMPOUNDS AND THEIR STRUCTURES[a]

| Compound | Structure | | Compound | Structure | |
|---|---|---|---|---|---|
| $ZnAl_2S_4$ | $P6_3mc$ | $R3m$ (Ref. 14) | $HgAl_2S_4$ | $I\bar{4}$ | $I\bar{4}$ (Ref. 15) |
| $ZnAl_2Se_4$ | $I\bar{4}$ | $I\bar{4}$ (Ref. 15) | $HgAl_2Se_4$ | $I\bar{4}$ | $I\bar{4}$ (Ref. 15) |
| $ZnAl_2Te_4$ | $I\bar{4}$, $I\bar{4}2m$ $P\bar{4}2m$, $I\bar{4}m2$ | | $HgAl_2Te_4$ | $I\bar{4}$, $I\bar{4}2m$ $P\bar{4}2m$, $I\bar{4}m2$ | |
| $ZnGa_2S_4$ | $I\bar{4}$, $I\bar{4}2m$ | | $HgGa_2S_4$ | $I\bar{4}$ | |
| $ZnGa_2Se_4$ | $I\bar{4}$, $I\bar{4}2m$ | | $HgGa_2Se_4$ | $I\bar{4}$ | |
| $ZnGa_2Te_4$ | $I\bar{4}$, $I\bar{4}2m$ | | $HgGa_2Te_4$ | $I\bar{4}$, $I\bar{4}2m$ $F\bar{4}3m$ | |
| $AnIn_2Se_4$ | $I\bar{4}$ | | $HgIn_2Se_4$ | $I\bar{4}$ | |
| $ZnIn_2Te_4$ | $I\bar{4}$ | | $HgIn_2Te_4$ | $I\bar{4}$ | $I\bar{4}2m$ (Ref. 17) |
| $CdAl_2S_4$ | $I\bar{4}$ | $I\bar{4}$ (Ref. 15) | $\beta$-$Cu_2HgI_4$ | $I\bar{4}2m$ | |
| $CdAl_2Se_4$ | $I\bar{4}$ | $I\bar{4}$ (Ref. 15) | $\beta$-$Ag_2HgI_4$ | $I\bar{4}$ | $I\bar{4}$ (Ref. 18) |
| $CdAl_2Te_4$ | $I\bar{4}$, $I\bar{4}2m$ $P\bar{4}2m$, $I\bar{4}m2$ | | | | |
| $CdGa_2S_4$ | $I\bar{4}$ | $I\bar{4}$ (Ref. 16) | | | |
| $CdGa_2Se_4$ | $I\bar{4}$ | $I\bar{4}$ (Ref. 16) | | | |
| $CdGa_2Te_4$ | $I\bar{4}$ | | | | |
| $CdIn_2Se_4$ | $P\bar{4}2m$ | | | | |
| $CdIn_2Te_4$ | $I\bar{4}$ | | | | |

[a] Structures quoted are from Hahn *et al.*[12,13] unless otherwise indicated.

ray scattering powers of the two cations are similar, as in the case of Zn and Ga, or if the axial ratio $c/a \simeq 2.0$, the exact space group can be difficult to determine, and several possible structures have been listed for some of these compounds. For example, Saunders and Seddon[17] have found $HgIn_2\square Te_4$ to have the $I\bar{4}2m$ space group in contrast to Hahn's earlier result. This seems to be confirmed by Raman scattering measurements.[19] There is a possibility that for this material a small amount of disorder between Hg and In atoms exists,[20] which may be the source of some of the confusion. This may well be the case in other materials of this

[19] A. Miller, D. J. Lockwood, A. MacKinnon, and D. Weaire, *J. Phys. C* **9**, 2997 (1976).
[20] R. Nelmes, private communication.

type also. The exact structures of the tellurides, $ZnAl_2\square Te_4$, $CdAl_2\square Te_4$, $HgAl_2\square Te_4$, and $HgGa_2\square Te_4$ are still in doubt. $ZnAl_2\square S_4$ and $CdIn_2\square Se_4$ are two puzzling exceptions in this list because they both form their own individual layered vacancy structures. $ZnIn_2\square S_4$ has a wurtzite-type structure, whereas $CdIn_2\square Se_4$ is pseudocubic, as mentioned in Section II. No exact determination of the atomic position or bond lengths of II–$III_2$–$\square$–$VI_4$ compounds has been undertaken, although $x$, $y$, and $z$ parameters for the $I\bar{4}$ and $I\bar{4}2m$ structures have been determined. These are the free parameters of the structures which determine the position of the anion, no longer restricted to a displacement in the $x$ direction as in the case of chalcopyrite. Like the chalcopyrite structured compounds, these materials show considerable distortion from the ideal zinc blende atomic positions. The axial ratio varies from 1.8 for $CdGa_2\square S_4$ to 2.04 for $HgIn_2\square Se_4$.

We have shown the $I\bar{4}$ structure in Fig. 3a in a slightly different way to most authors in order to highlight the differences between it, chalcopyrite, and defect stannite, $I\bar{4}2m$. Comparing Figs. 3a and 3b we see that the $I\bar{4}$ and $I\bar{4}2m$ structures both have a body-centered-tetragonal array of vacant tetrahedral sites, the difference being the ordering of the two cations. Defect chalcopyrite has mixed cation layers, whereas defect stannite has a single atom type in each layer.

The two I–$II_2$–$\square$–$VII_4$ compounds make very interesting examples of crystals with these structures. Both $Ag_2Hg\square I_4$ and $Cu_2Hg\square I_4$ have a first-order phase transition[21] at relatively low temperatures to a zinc blende-type structure in which the cations disorder leaving the anions in the same sites.[21–29] Both materials show ionic conduction properties in the higher temperature $\alpha$-phase, and $Ag_2Hg\square I_4$ has been very well studied recently. This compound has been confirmed to have the defect chalcopyrite $I\bar{4}$ structure at low temperature (bond lengths have been determined by Browall and Kapser[29] in single crystals) and a disordered ionic conductive phase above 50°C. Recent evidence suggests that the Hg ions may remain locally ordered in the $\alpha$-phase, whereas the ionic conduction is via mobile

[21] S. M. Girvan and G. D. Mahan, *Solid State Commun.* **23,** 629 (1977).

[22] J. A. A. Ketalaar, *Z. Kristallogr., Kristallgeom., Kristallphys., Kristallchem.* **80,** 190 (1931); **87,** 436 (1934); *Z. Phys. Chem. Abt. B* **26,** 327 (1934); **30,** 53 (1935); *Trans. Faraday Soc.* **34,** 874 (1938).

[23] S. Hoshino, *J. Phys. Soc. Jpn.* **10,** 197 (1955).

[24] L. Suchow and G. R. Pond, *J. Am. Chem. Soc.* **75,** 5242 (1953).

[25] T. J. Neubert and G. M. Nichols, *J. Am. Chem. Soc.* **80,** 2619 (1958).

[26] R. Weil and A. W. Lawson, *J. Chem. Phys.* **41,** 832 (1964).

[27] A. W. Webb, *J. Phys. Chem. Solids* **34,** 501 (1973).

[28] J. S. Kasper and K. W. Browall, *J. Solid State Chem.* **13,** 49 (1975).

[29] K. W. Browall and J. S. Kapser, *J. Solid State Chem.* **15,** 54 (1975).

Ag ions on the Ag and vacancy sublattices.[30] The low-temperature phase of $Cu_2Hg\square I_4$ has the defect stannite, $I\bar{4}2m$ structure. In this case the transition is at 67°C, but no detailed structure analysis has yet been undertaken for this material.

## III. The Folding Method

### 3. Introduction

Since the crystal structures under consideration are closely related to those of zinc blende compounds, we would expect that much useful information can be gained from a perturbation treatment based on the well-established properties of the simpler zinc blende materials. Special consideration must, of course, be given to the vacancy compounds where the effect of the vacancies is too large to be reliably described by perturbation theory.

As a simple example of the folding procedure which is the basis of such a theory, we consider first the problem of a one-dimensional binary system. Suppose we wish to study the vibrational and electronic eigenstates of a chain of atoms such as that in Fig. 5a. First, consider the case of a monatomic chain (i.e., A = B). The unit cell of the lattice has length $a$, so that the smallest reciprocal lattice vector has length $2\pi/a$, and the Brillouin zone occupies the space between $-\pi/a$ and $+\pi/a$. Typical phonon dispersion curves and typical electronic energy bands are shown by the solid lines in Fig. 5b.

For the diatomic chain (i.e., A ≠ B), the unit cell doubles to $2a$, so that the smallest reciprocal lattice vector is now halved to $\pi/a$, and the Brillouin zone is reduced to the range between $-\pi/2a$ and $\pi/2a$. The dashed curves in Fig. 5b are typical phonon dispersion and electronic energy band curves for such a system.

How can these curves be related to the solid curves for the monatomic system? In the diatomic case the region $-\pi/a$ to $-\pi/2a$ and $\pi/2a$ to $\pi/a$ lies outside the Brillouin zone. Every point in this region is therefore equivalent to a point between $-\pi/2a$ and $\pi/2a$ (i.e., inside the Brillouin zone), into which it can be transformed by the addition of a reciprocal lattice vector. When the portion of the solid curve between $-\pi/a$ and $-\pi/2a$ is shifted by $\pi/a$ to the right and then between $\pi/2a$ and $\pi/a$ by $\pi/a$ to the left, this results in the dashed curves. In particular the states from $-\pi/a$ and $+\pi/a$ now lie at the origin.

[30] T. Hibma, H. V. Beyeler, and H. R. Zeller, *J. Phys. C* **9**, 1691 (1976).

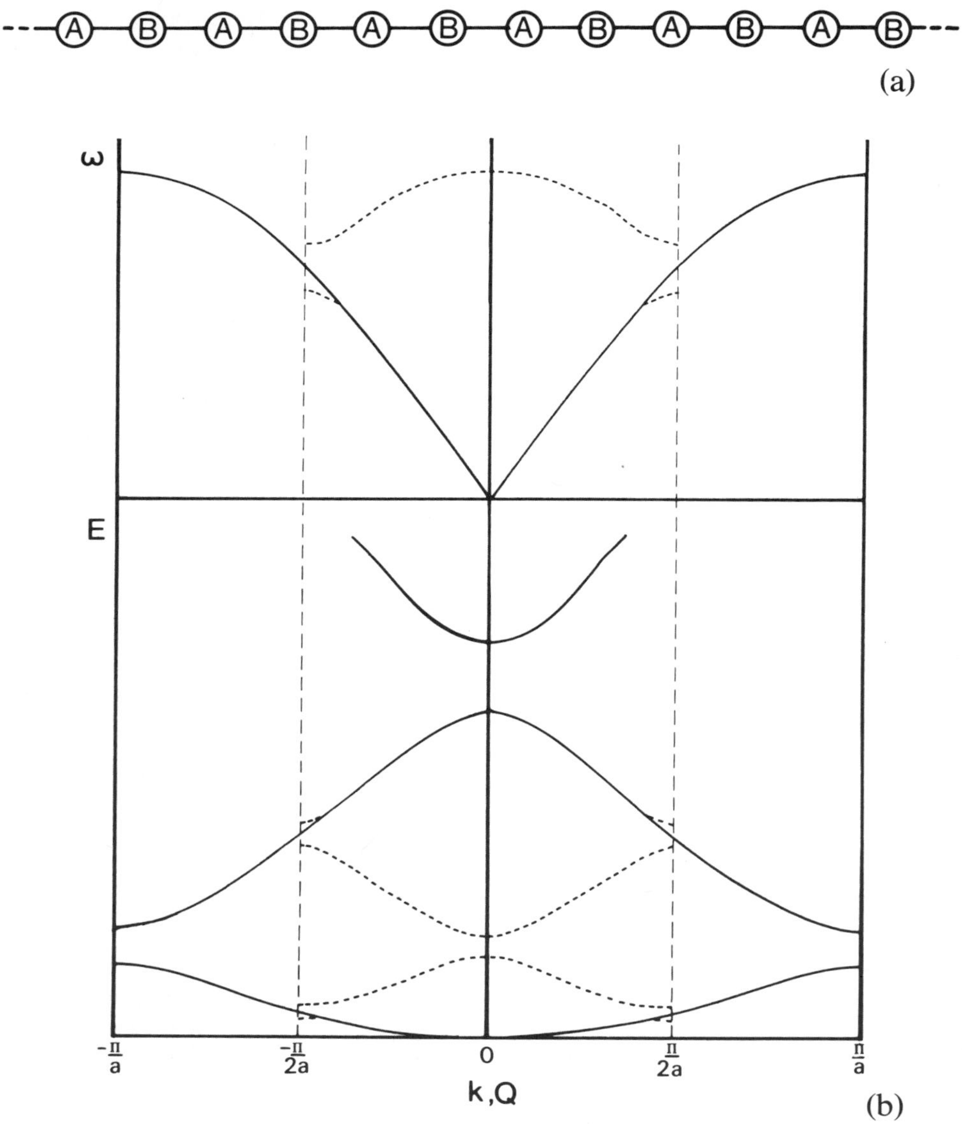


FIG. 5. (a) Diatomic linear chain of atoms. (b) Folding of energy bands $E(\mathbf{k})$ and phonon dispersion relations $\omega(\mathbf{Q})$ to relate properties of a diatomic chain to those of the monatomic chain.

The solid curves now lie over the dashed curves except near the zone edges, $-\pi/2a$ and $\pi/2a$, where gaps have opened. The exact size of these gaps depends on the details of the potentials involved in the system, but there is a simple rule of thumb that gives the qualitative effect: *dispersion curves that have been folded back tend to repel each other*. This repulsion is larger when the energy difference is smaller, so that, in Fig. 5b, the effect is only significant where the states that have been folded back are almost degenerate.

In more complex cases we shall see that symmetry considerations can give us useful information on the relative strength of the repulsion, although the exact results require knowledge of the appropriate matrix elements.

## 4. Folding in Three Dimensions

In considering the application of this folding method to the crystal structures discussed in Section II, we shall concentrate on the Γ point, the zone center of their Brillouin zones. This is by far the most interesting point as far as present theoretical and experimental results are concerned. Figure 6 shows the Brillouin zone for the body-centered-tetragonal (bct) structures (chalcopyrie, stannite, etc.) superimposed on the Brillouin

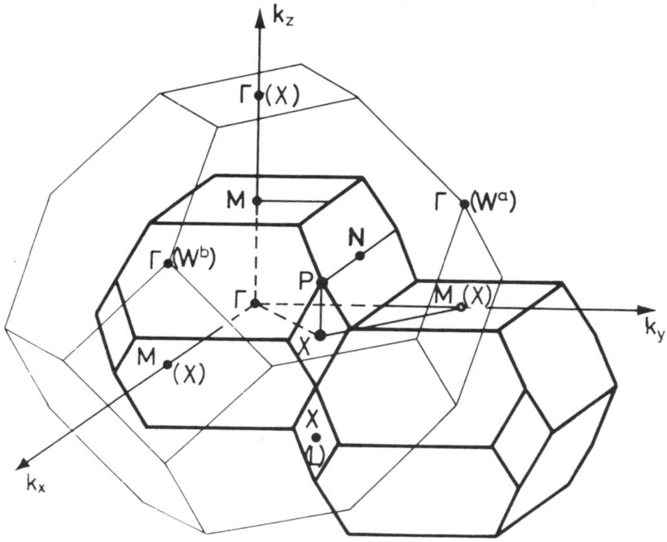


FIG. 6. Brillouin zone of bct lattice (with ideal axial ratio 2.0 appropriate to chalcopyrite) and corresponding fcc Brillouin zone of the parent zinc blende structure.

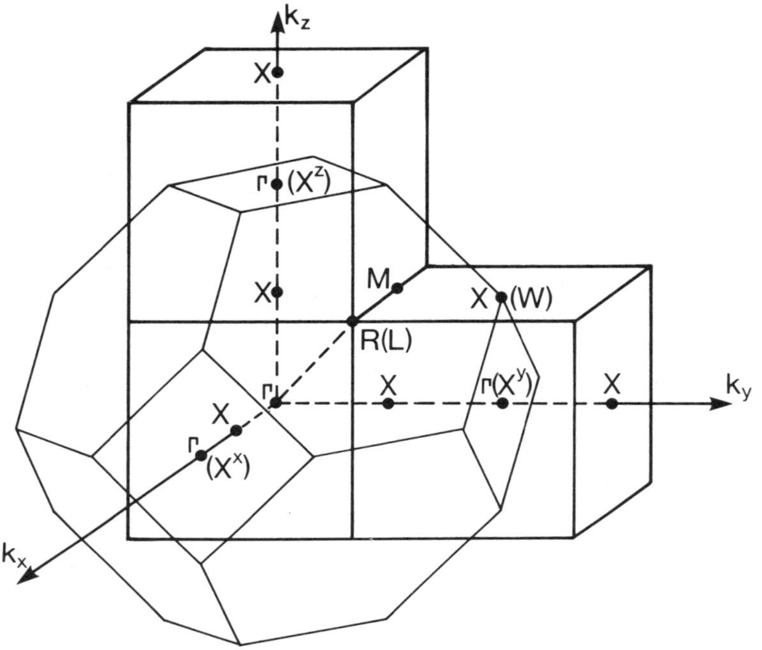


FIG. 7. Brillouin zone of simple tetragonal lattice (with ideal axial ratio 1.0 appropriate to the pseudocubic structure) and corresponding fcc Brillouin zone of the parent zinc blende structure.

zone of zinc blende, whereas Fig. 7 depicts the Brillouin zone of the pseudocubic structure similarly. The labeling of the various symmetry points and lines is as given by Lax.[31] The Γ points of bct map onto Γ, $X^z$, and two distinct $W$ points ($W^a$, $W^b$), whereas in the pseudocubic case the Γ points map onto Γ and three distinct $X$ points ($X^x$, $X^y$, $X^z$). Just as in one dimension, a first approximation to the new dispersion curves or band structure can be found by folding back the dispersion curves or band structure of an appropriate zinc blende compound (see Sections 7 and 9 for examples). The new Γ point states are simply those of Γ, $X$, and $2W$ (Γ and $3X$) superimposed on one another. However, in order to make this information more useful it is necessary to find out to which irreducible representations (IRs) of the new group the folded-back states belong. To avoid confusion between the representations of the various space groups, we shall use the space group notation ($\Gamma_n$, $X_m$, etc) for the zinc blende IRs

[31] M. J. Lax, "Symmetry Principles in Solid State and Molecular Physics." Wiley, New York, 1974.

and the point group notation ($A_n, B_n, E, T$) for the IRs of the $\Gamma$ point of the other systems. The character table of point group $\bar{4}2m$ in the Appendix gives the labels of the IRs in both notations, together with the combinations of $x$, $y$, and $z$ vectors and $R_x$, $R_y$, $R_z$ rotations which transform as each IR.

The IRs of the new group are found by treating the IRs of zinc blende as primitive basis functions and doing exactly the same operations with them as would be done with more conventional basis functions, such as $x$, $y$, $z$ displacement vectors on each atom or $sp^3$ orbitals. We divide the basis functions into sets that transform into each other under operations of the group. Such sets comprise displacements or orbitals associated with the same type of atoms or IRs from zinc blende associated with similar Brillouin zone points (e.g., $W_1^a$, $W_1^b$). For each such set we can determine the character of the set under a particular symmetry operation of the new group by summing, with appropriate coefficients, over those basis functions that transform into themselves or multiples of themselves. That is, we find the trace or character of the transformation matrix.

For example, the IRs $X_2^x$ and $X_2^y$ transform into each other under the operations of the point group $\bar{4}2m$ of the pseudocubic structure, as should be obvious from an examination of Fig. 7. Table III(a) shows how they transform under the various operations and the consequent characters. Forming the scalar product of these characters with the rows of the $\bar{4}2m$ character table and dividing by the number of operations (in this case, 8) gives the IRs of the new group into which the set ($X_2^x$, $X_2^y$) transforms as $A_1 + B_1$. Another example [Table III(b)] is the IR $\Gamma_{15}$ of the point group $\bar{4}3m$. This three-dimensional IR of zinc blende transforms into $B_2 + E$; $\Gamma_{15}$ transforms like ($x$, $y$, $z$) and $B_2$ and $E$ transform as $z$ and ($x$, $y$), respectively.

TABLE III. EXAMPLES OF HOW THE IRREDUCIBLE REPRESENTATION OF ZINC BLENDE TRANSFORM IN THE SPACE GROUPS OF THE STRUCTURES

| | $E$ | $C_{2z}$ | $S_{4z}^+$ | $S_{4z}^-$ | $C_{2x}$ | $C_{2y}$ | $\sigma_{xy}$ | $\sigma_{\bar{x}y}$ |
|---|---|---|---|---|---|---|---|---|
| (a) | $X_2^x$ | $X_2^x$ | $X_2^y$ | $X_2^y$ | $X_2^x$ | $X_2^x$ | $X_2^y$ | $X_2^y$ |
| | $X_2^y$ | $X_2^y$ | $X_2^x$ | $X_2^x$ | $X_2^y$ | $X_2^y$ | $X_2^x$ | $X_2^x$ |
| | 2 | 2 | 0 | 0 | 2 | 2 | 0 | 0 |
| (b) | 3 | 1 | −1 | −1 | −1 | −1 | 1 | 1 |
| (c) | 1 | 1 | −1 | −1 | 1 | 1 | −1 | −1 |
| | 1 | 1 | −1 | −1 | −1 | −1 | 1 | 1 |

A state of the new system can only be related to a set of zinc blende states as defined above and not to a particular zinc blende state. This causes no problems since all states in such sets are degenerate. Normally a set contains IRs of only one type but from different **k**-points (e.g., all $X_1$'s), but it is possible when the IRs have complex characters that this is not the case (e.g., $W_3^a$, $W_4^b$). When this occurs there can be some confusion in the naming of the IRs.

Before proceeding further it is necessary to consider the special case of a nonsymmorphic space group such as $I\bar{4}2d$ (e.g., chalcopyrite). In those symmetry operations ($C_{2x}$, $C_{2y}$, $\sigma_{xy}$, $\sigma_{x\bar{y}}$) that involve a nonprimitive lattice translation, this must be taken into account in carrying out the above procedure. For these operations the character must be multiplied by a factor $e^{i\mathbf{k}\cdot\mathbf{v}}$ ($= -1$ for the $X$ point), where **k** is the coordinate of the Brillouin zone point and **v** the nonprimitive lattice translation.[32,33] Table III(c) shows the characters of $X_2$ under the operations of $\bar{4}2m$ (stannite) and $\bar{4}2m$ (chalcopyrite), respectively. The results of the above calculations for the structures discussed in Section II are tabulated in Table IV. Note the differences in the $X$-IRs for the $I\bar{4}2m$ and $I\bar{4}2d$ cases. The two columns from $X$ given under $P\bar{4}2m$ are for the $X^z$ and the ($X^x$, $X^y$) sets, respectively, required in the treatment of the pseudocubic structure. The monolayer system requires only $X^z$.

We are now in a position to rewrite the "repulsion" rule given in Section 3: *States that when folded back belong to the same irreducible representation repel each other.*

## 5. The Symmetry Properties of Perturbations

In order to carry the analysis one stage further, it is necessary to consider the various perturbations contributing to the mixing of the zinc blende states and hence their mutual repulsion. This is also important in determining various transition probabilities. For the structures discussed above (in the case of either the electronic or vibrational states) the perturbations may be classified as (*a*) cationic asymmetry (the presence of two or more different cations); (*b*) anionic displacement; or (*c*) tetragonal compression. Which of the zinc blende states that are to be mapped into Γ are connected by matrix elements of these perturbations? The answer is given in Table V. With this in mind, we can now add a further rule to those already enunciated: *The largest interactions between folded states* (*and hence mixing and repulsion*) *are in general between states whose parent*

[32] R. Sandrock and J. Treusch, *Z. Naturforsch.*, **19A**, 844 (1964).
[33] G. D. Holah, J. S. Webb, and H. Montgomery, *J. Phys. C* **7**, 3875 (1974).

TABLE IV. IRREDUCIBLE REPRESENTATIONS FROM Γ, $X$, AND $W$ POINTS OF SPACE GROUPS $Fd3m$ (DIAMOND) AND $F\bar{4}3m$ (ZINC BLENDE), TOGETHER WITH THE REPRESENTATIONS INTO WHICH THEY TRANSFORM

| $Fd3m$ | $F\bar{4}3m$ | $I\bar{4}2d$ | $I\bar{4}$ | $I\bar{4}2m$ | $P\bar{4}2m$ | |
|---|---|---|---|---|---|---|
| $\Gamma_1^+ + \Gamma_2^-$ | $\Gamma_1$ | $A_1$ | $A$ | $A_1$ | $A_1$ | |
| $\Gamma_1^- + \Gamma_2^+$ | $\Gamma_2$ | $B_1$ | $B$ | $B_1$ | $B_1$ | |
| $\Gamma_{12}^+ + \Gamma_{12}^-$ | $\Gamma_{12}$ | $A + B_1$ | $A + B$ | $A_1 + B_1$ | $A_1 + B_1$ | |
| $\Gamma_{15}^- + \Gamma_{25}^+$ | $\Gamma_{25}$ | $A_2 + E$ | $A + E$ | $A_2 + E$ | $A_2 + E$ | |
| $\Gamma_{15}^+ + \Gamma_{25}^-$ | $\Gamma_{15}$ | $B_2 + E$ | $B + E$ | $B_2 + E$ | $B_2 + E$ | |
| $X_1$ | $X_1$ | $A_2$ | $A$ | $A_1$ | $A_1$ | $A_1 + B_1$ |
| $X_3$ | $X_2$ | $B_2$ | $B$ | $B_1$ | $B_1$ | $A_1 + B_1$ |
| $X_1$ | $X_3$ | $B_1$ | $B$ | $B_2$ | $B_2$ | $E$ |
| $X_3$ | $X_4$ | $A_1$ | $A$ | $A_2$ | $A_2$ | $E$ |
| $X_2 + X_4$ | $X_5$ | $E$ | $E$ | $E$ | $E$ | $A_2 + B_2 + E$ |
| $W_1$ | $W_1$ | $A_1 + A_2$ | $A$ | $A_1 + A_2$ | | |
| $W_2$ | $W_2$ | $B_1 + B_2$ | $B$ | $B_1 + B_2$ | | |
| $W_1 + W_2$ | $W_3 + W_4$ | $E$ | $E$ | $E$ | | |

TABLE V. FIRST-ORDER COUPLINGS DUE TO PERTURBATIONS WITH Γ, $X$, AND $W$ (Γ, $X^z$, $X^{x,y}$) SYMMETRY, TOGETHER WITH DETAILS OF THE APPLICATION OF THESE TO VARIOUS STRUCTURES AND PERTURBATIONS

| Bravais lattice | Perturbation Γ | Perturbation $X(X^z)$ | Perturbation $W(X^{x,y})$ |
|---|---|---|---|
| bct ($I$) | $\Gamma - \Gamma$ | $\Gamma - X$ | $\Gamma - W$ |
| | $X - X$ | $W - W$ | $X - W$ |
| | $W - W$ | | $W - W$ |
| st ($P$) | $\Gamma - \Gamma$ | $\Gamma - X^z$ | $\Gamma - X^{x,y}$ |
| | $X^z - X^z$ | $X^{x,y} - X^{x,y}$ | $X^z - X^{x,y}$ |
| | $X^{x,y} - X^{x,y}$ | | |

| Structure | Cationic asymmetry | Tetragonal compression | Anionic displacement |
|---|---|---|---|
| $I\bar{4}2d$ | $\Gamma + W$ | Γ | $W$ |
| $I\bar{4}$ and $I\bar{4}2m$ | $\Gamma + X + W$ | Γ | $X + W$ |
| $P\bar{4}2m$ (layer) | $\Gamma + X^z$ | Γ | $X^z$ |
| $P\bar{4}2m$ ($CdIn_2Se_4$) | $\Gamma + X^z + X^{x,y}$ | Γ | $X^z + X^{x,y}$ |

*(zinc blende) states are connected in first order by the perturbations (a)–(c)*. In the above form, the entries in Table V are obvious enough if one considers what **k** vectors would be associated with each perturbation if it were thought of as a Bloch state. More detailed selection rules can be derived for first-order perturbations by finding the space group in the presence of that perturbation alone and using the method of Section 4 to find out which states have the same IR and therefore repel each other. The simplest example is *tetragonal compression*. This perturbation alone changes the point group at $\Gamma$ from $\bar{4}3m$ to $\bar{4}2m$. The previously triply degenerate $\Gamma_{15}$ state becomes a singly degenerate $B_2$ state and a doubly degenerate $E$ state. The top valence band of zinc blende is split in this way, and the $B_2$ state moves to slightly higher energy than $E$. This is an example of so-called *crystal field splitting* as observed in chalcopyrite[2,34–37] The effect is generally small.

Tetragonal compression is not, however, the only perturbation that can cause a crystal field splitting. Where they exist, one would expect the $\Gamma$–$X$ matrix elements to give rise to a much more significant splitting due to the coupling $\Gamma_{15}$–$X_5$. This applies to $HgIn_2\square Te_4$, which has a defect stannite structure ($I\bar{4}2m$) and the term in question has the opposite sign to that due to tetragonal compression.[38] Thus the *crystal field splitting in chalcopyrite* ($I\bar{4}2d$) *has opposite sign to that of the* $I\bar{4}2m$, $I\bar{4}$, *and monolayer* ($P\bar{4}2m$) *structures*. The case of the pseudocubic structure is a little more complex. The $X_5$ modes transform as $A_2 + B_2 + 2E$, where the presence of $B_2$ in addition to $E$ makes the prediction of a crystal field splitting more difficult. Moreover, the valence band maximum of the most studied crystal of this type $CdIn_2\square Se_4$ may not be at $\Gamma$,[39,40] so that the above discussion does not apply in this case. These arguments regarding crystal field splittings are summarized in Fig. 8.

An additional complication in the case of those compounds containing heavier elements (e.g., $HgIn_2\square Te_4$, $ZnSnSb_2$) is the importance of *spin*–

[34] C. Varea de Alvarez, M. L. Cohen, S. E. Kohn, Y. Petroff, and Y. R. Shen, *Phys. Rev. B: Solid State* [3] **10,** 5175 (1974).

[35] L. Pasemann, W. Cordlts, A. Heinrich, and J. Monecke, *Phys. Status Solidi* B **77,** 527 (1976).

[36] Y. I. Polygalov, A. S. Poplavnoi, and A. M. Ratner, *J. Phys. (Orsay, Fr.)* **36,** C3, 129 (1975).

[37] F. Aymerich, G. Mula, A. Baldereschi, and F. Meloni, *Conf. Ser.—Inst. Phys.* **35,** 159 (1977).

[38] A. MacKinnon, A. Miller, and G. Ross, *Conf. Ser.—Inst. Phys.* **35,** 171 (1977).

[39] L. S. Koval, M. M. Markus, S. I. Radautsan, V. V. Sabolev, and A. V. Stanchu, *Phys. Status Solidi A* **9,** K69 (1972).

[40] A. Baldereschi, F. Meloni, F. Aymerich, and G. Mula, *Solid State Commun.* **21,** 113 (1977).

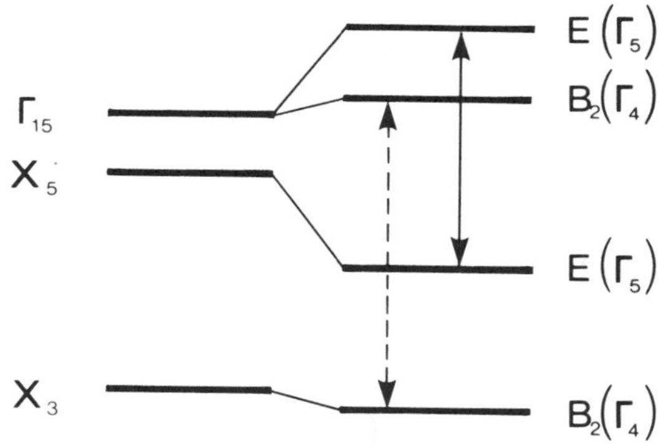


FIG. 8. Hypothetical example of crystal field splitting due to the interaction of folded states.

*orbit coupling*. Since the crystal field and spin–orbit splittings are of similar size, they must both be considered to the same order of perturbation. The simplest way to do this is by applying the above rules to the spin–orbit split zinc blende levels but using the IRs of the unsplit system (see Fig. 8).

We have so far emphasized the splitting of degenerate states by the interaction of various states via the perturbations (a)–(c). Associated with this is the *mixing* of the original zinc blende states, giving rise to allowed processes, forbidden for the original states. Consider, for example, the infrared/Raman activity of $\mathbf{k} = 0$ vibrational modes in chalcopyrite. Only mixing with active modes derived from Γ can render active those from other Brillouin zone points, so they may be expected to be comparatively weak. This is particularly true for those modes associated with $X$; for chalcopyrite there are no Γ–$X$ matrix elements of any of the perturbations. Thus $X$-derived modes would be exceptionally weak,[33,41] wherever this picture based on first-order mixing applies. Again, optical absorption between states derived from zinc blende Γ and $X$ states is very weak in chalcopyrite compounds, since there is very little mixing of Γ and $X$. When the zinc blende crystal has an indirect gap Γ–$X$ and the minimum of the conduction band of the chalcopyrite is derived from $X$, as in $ZnGeP_2$,[34] the optical properties are those of an indirect gap. Such a gap is referred to as *pseudodirect*. (Strictly speaking, *pseudoindirect* would be a much more reasonable term for this, but it is probably too late to change it!) Γ and $X$ are coupled to *second order* in chalcopyrite by a combination of two

[41] M. Bettini, *Phys. Status Solidi B* **69**, 201 (1975).

$W$ perturbations. This accounts for the small amount of optical absorption just above the band edge in pseudodirect materials.

There thus emerges a hierarchy of levels of perturbation, in terms of which transition probabilities can be classified, as follows

*Allowed:*

- *Strong:* Processes that are allowed for the original zinc blende states.
- *Weak:* Processes that become allowed when first-order mixing is considered.

*Very weak:* Processes that become allowed when higher order mixing is considered.

*Forbidden:* Processes that are strictly forbidden.

This hierarchy can be further elaborated in at least two ways. One may add the further effects of spin–orbit splitting, and one may distinguish processes that are allowed in zinc blende but not diamond cubic.

Such a picture is admittedly an oversimplification in some cases, owing to accidental degeneracies, as we shall see, particularly in Section IV. It also needs further qualification in the case of structures with ordered vacancies, as described in the next section.

## 6. The Role of Ordered Vacancies

An example of modification of zinc blende which cannot be treated as a perturbation is the creation of vacancies.

The most obvious correction to the rules of Section 5 occurs in lattice dynamics. Figure 9 illustrates the "defect diamond" model. On the left are the vibrational states of diamond mapped onto the $\Gamma$ point of chalcopyrite, stannite, and so on, as calculated using a Keating model.[42] Moving toward the center of the figure, the mass of one atom in the unit cell is gradually reduced to zero. The most significant effect is that three modes rise to infinity: The removal of an atom implies that the number of degrees of freedom, and therefore of modes, is reduced by three. This reduction was not taken into account in early attempts[43] to apply the folding method to the vacancy compound, $In_2\square Te_3$. Moving further to the right, the force constants associated with the missing atom go to zero. In this case one mode drops significantly in frequency. This is the so-called *breathing mode,* which turns out to have several interesting properties. In

[42] P. N. Keating, *Phys. Rev.* [2] **145,** 637 (1966).

[43] E. Finkman, J. Tauc, R. Kershaw, and A. Wold, *Phys. Rev. B: Solid State* [3] **7,** 3785 (1975).

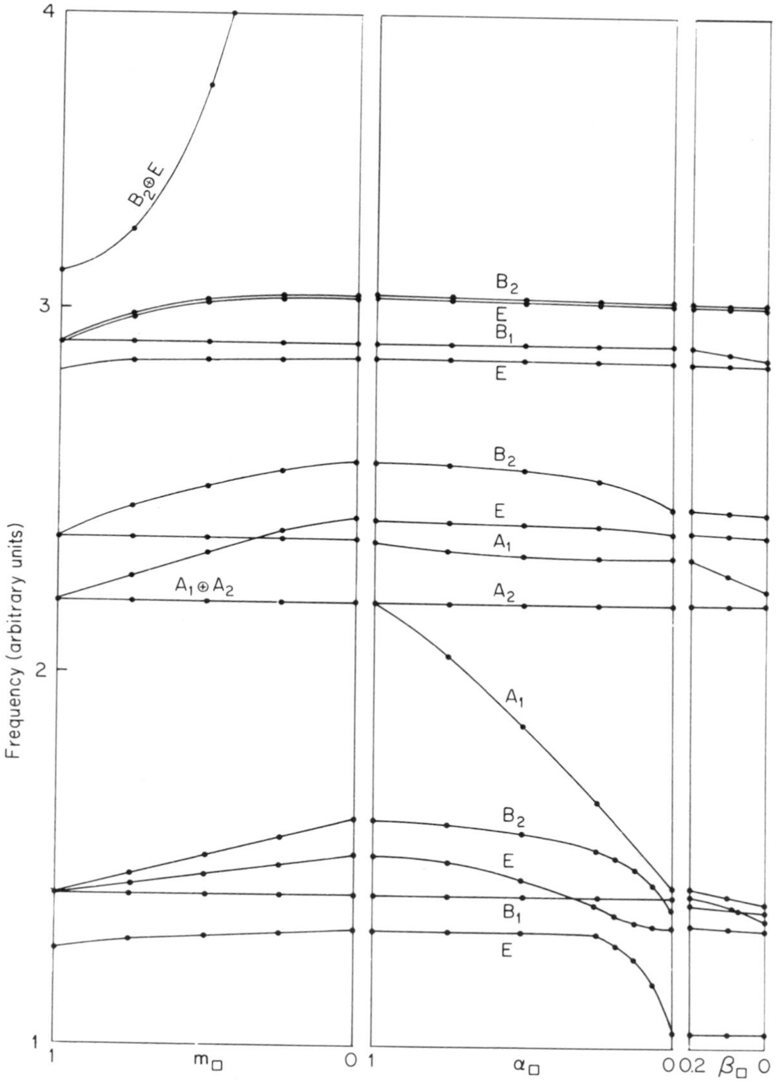


FIG. 9. The calculation of zone-center vibrational modes for a "defect diamond" structure shows the effect of reducing to zero (a) the mass of the vacant site; (b) the central force constant associated with the vacant site; and (c) the noncentral force associated with the vacant site. The final distribution of modes is a useful first approximation to those of $HgIn_2\square Te_4$. For further details see Ref. 19.

(a)

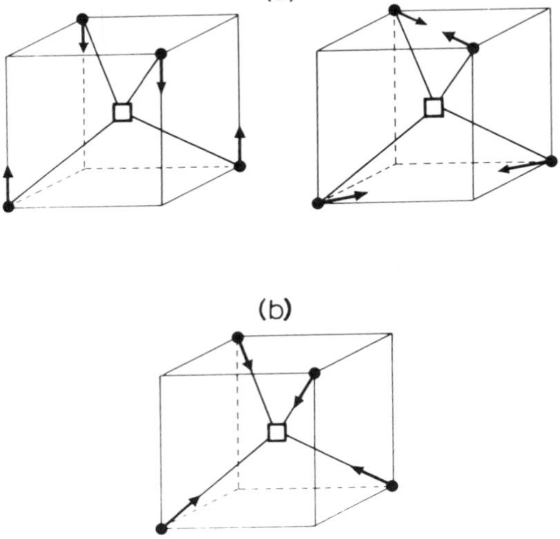

(b)

FIG. 10. Nature of $A_1$ modes indicated in Fig. 9. (a) The folding of zinc blende modes gives two $A_1$ modes with displacements as shown in the neighborhood of a vacancy. The correct $A_1$ modes are linear combinations of these. (b) As the central force constant is reduced, the lower lying $A_1$ mode takes the form shown (breathing mode).

this mode the four neighbors of the vacancy move together along the anion-vacancy bond direction (Fig. 10). In group theory, such a mode may be treated as having the spherical symmetry, like an s orbital on the vacant site.

The IRs of the breathing modes may be found by performing all symmetry operations on such spherical basis functions and forming the scalar products of their characters with the rows of the character table for the appropriate point group.[44] The IRs of the *lost* modes may be found in a similar way using $x$, $y$, and $z$ displacement basis functions or p orbitals. Thus one vacancy per unit cell has the symmetry of $x^2 + y^2 + z^2$ $(A_1)$, and the missing modes have that of $x$, $y$, $z$ $(B_2 + E)$.

The Grimm–Sommerfeld rule ensures that no electronic states are lost after folding back. However, the positive potential energy at the vacant site ensures that in the highest-valence states the electrons are concentrated near the vacancies. The IRs of these states are the same as those of the lost vibrational states. Such bands are very flat because they are made up of weakly interacting orbitals associated with the vacancies and their

[44] A. MacKinnon, A. Miller, D. J. Lockwood, G. Ross, and G. Holah, *Conf. Ser.—Inst. Phys.* **35**, 179 (1977).

energies are extremely sensitive to the breathing modes. This, in turn, implies a sensitivity of the optical dielectric function to the breathing modes, just the requirement for strong Raman scattering. Miller *et al.*[19] observed that the scattered intensity from this mode is about an order of magnitude greater than from the other phonon modes of $HgIn_2\square Te_4$.

Experimentally, however, the frequency of the breathing mode is not observed to be significantly lower than the simple "folded back" frequency.[19] This is due to the Coulomb forces which were neglected in the model of Fig. 9. The absence of an ion on the vacant site makes the Coulomb field seen by the anions not only significant but also extremely asymmetric, especially as seen by the breathing mode. We should therefore expect to observe nonlinear effects.

If we consider only the $\Gamma$ point, where the asymmetry is strongest, two-phonon generation should occur for those combinations of breathing phonons whose product IRs contain a representation having $x$, $y$, or $z$ character.[44] Unfortunately this cannot occur for crystals with only one vacancy per unit cell.

## IV. Lattice Vibrations

The similarity of the II–IV–$V_2$ and I–III–$VI_2$ chalcopyrite compounds to III–V and II–VI zinc blende materials has been the basis for interpretation of the phonon spectra of these compounds, along the lines of Section III.

Much less information is available for the phonon spectra of ordered vacancy structures.

### 7. Chalcopyrite Structure

Chalcopyrite has two formula units per primitive cell, yielding 24 degrees of freedom. The character table is given in Table A.2 in the Appendix. Of the 21 optic modes only 3 $B_2$ and 6 (doubly degenerate) $E$ modes are infrared active. These vibrations are also Raman active together with the $A_1$ and three $B_1$ modes. The remaining two $A_2$ modes cannot be observed in first-order Raman scattering. As a first approximation, these zone-center frequencies will be the same as those for the isoelectronic binary analog at $\Gamma, X$, and $W$ as explained in Section 3. Holah *et al.*[33] have given the displacement coordinates for the atomic vibrations in this approximation from an analysis of the normal modes of the zinc blende structure. Only the $A_1$ coordinate corresponds exactly to a normal mode because of its unique symmetry. All other modes are likely to have a degree of mixing particularly where the frequencies are closely spaced. This can be understood by a consideration of the phonon dispersion curves of GaAs shown in Fig. 11. A comparison of the $X$ and $W$ point modes with Table IV will show that $\Gamma_{15}$, $X_5$, and $W_4$ vibrations of zinc blende all become $E$ modes of chalcopyrite resulting in three vibrations of the same symmetry at the upper end of the phonon spectrum. Thus, although the $X_5$ mode of chalcopyrite in the zinc blende approximation might be expected to have only a small dipole moment, as discussed in Section 3, mixing of nearby states in higher order can result in a dipole oscillator of considerable strength.

*a. II–IV–$V_2$ Compounds*

The results of infrared and Raman studies on II–IV–$V_2$ compounds[45–72] are compiled in Fig. 11. The values and assignments given have been selected from the literature on the basis of consistency and agreement

[45] L. B. Zlatkin, J. F. Makov, A. I. Stekhanov, and M. S. Shur, *Phys. Status Solidi* **30,** 473 (1969).
[46] I. P. Kamino, E. Buehler, and J. H. Wernick, *Phys. Rev. B: Solid State* [3] **2,** 960 (1970).
[47] N. S. Boltovets, B. Kh. Mamedov, and E. O. Osmanov, *Sov. Phys.—Semicond. (Engl. Transl.)* **4,** 499 (1970).
[48] G. F. Karavaev, A. S. Poplavnoi, and V. G. Tyuterev, *Sov. Phys. J. (Engl. Transl.)* **10,** 42 (1970).
[49] L. B. Zlatkin, E. K. Ivanov, and G. P. Startsev, *J. Phys. Chem. Solids* **31,** 567 (1970).
[50] E. E. Alekperova, M. A. Gazalov, N. A. Goryunova, and B. K. Mamedov, *Phys. Status Solidi* **41,** K57 (1970).
[51] S. Isomoura and K. Masumoto, *Phys. Status Solidi* A **6,** K139 (1971).
[52] L. B. Zlatkin, Yu. F. Markov, V. M. Orlov, V. I. Sokolova, and M. S. Shur, *Sov. Phys.—Semicond. (Engl. Transl.)* **4,** 1181 (1971).
[53] G. D. Holah, *J. Phys. C* **5,** 1893 (1972).
[54] G. C. Bhar and R. C. Smith, *Phys. Status Solidi* A **13,** 157 (1972).
[55] V. F. Markov and N. B. Reshetnyak, *Opt. Spectrosc. (Engl. Transl.)* **33,** 280 (1972).
[56] L. B. Zlatkin and Yu. F. Markov, *Opt. Spectrosc. (Engl. Transl.)* **32,** 403 (1972).
[57] M. Bettini and A. Miller, *Phys. Status Solidi* B **66,** 579 (1974).
[58] M. Bettini, W. Baunhofer, M. Cardona, and R. Nitsche, *Phys. Status Solidi* B **63,** 641 (1974).
[59] W. H. Koschel, F. Sorger, and J. Baars, *Solid State Commun.* **15,** 719 (1974).
[60] Yu. F. Markov, V. S. Grigoreva, B. S. Zadokhin, and T. V. Rybakova, *Opt. Spectrosc. (Engl. Transl.)* **36,** 93 (1974).
[61] A. Miller, G. D. Holah, and W. C. Clark, *J. Phys. Chem. Solids* **35,** 685 (1974).
[62] M. Attorresi, A. Pinczuk, and A. Gavini, *Proc. Int. Conf. Phys. Semicond., 12th, 1974* p. 321 (1974).
[63] I. S. Gorban, V. I. Lugovoi, I. I. Tychina and I. Yu. Tkachuk, *Sov. Phys.—Solid State (Engl. Transl.)* **16,** 1794 (1974).
[64] M. Bettini and W. B. Holzapfel, *Solid State Commun.* **16,** 27 (1975).

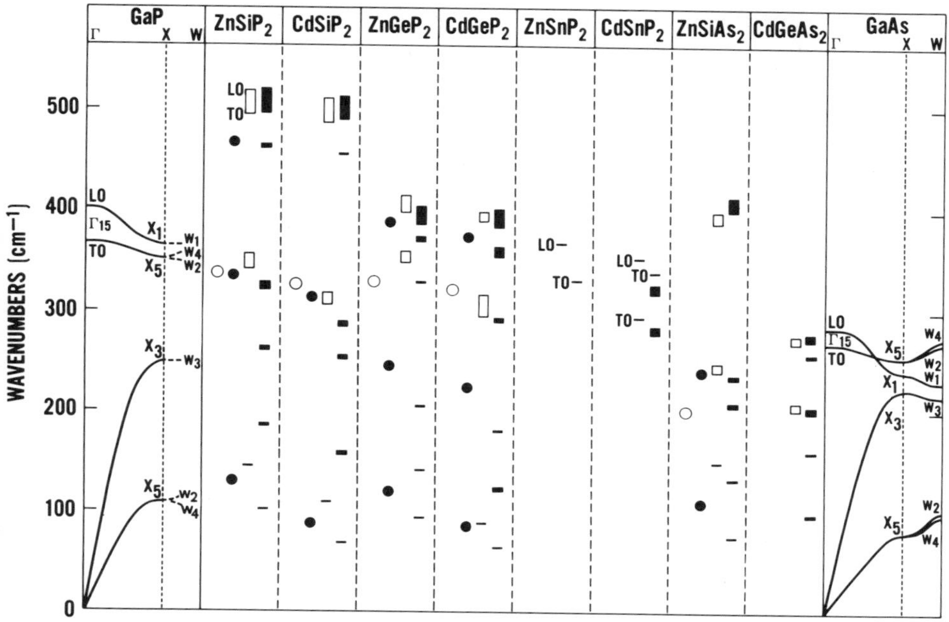


FIG. 11. Observed zone-center phonon frequencies of II–IV–$V_2$ compounds. $A_1$: ○, $B_1$: ●, $B_2$: □, $E$: ■. The lengths of the boxes for modes of $B_2$ and $E$ symmetry give the Coulomb splittings of these modes. The phonon dispersion curves of GaP and GaAs in the Γ–$X$–W directions are shown for comparison.

---

[65] I. S. Gorban, V. A. Gorynya, V. I. Lugovoi, and I. I. Tychina, *Sov. Phys.—Solid State* (*Engl. Transl.*) **17,** 1749 (1975).

[66] I. S. Gorban, V. A. Gorynya, V. I. Seryi, I. I. Tychina, and M. A. Il'in, *Sov. Phys.—Solid State* (*Engl. Transl.*) **17,** 24 (1975).

[67] V. A. Gorynya, V. I. Lugovoi, and I. I. Tychina, *Ukr. Fiz. Zh.* (*Russ. Ed.*) **20,** 1428 (1975).

[68] V. S. Grigoreva, Yu. F. Markov, and T. V. Rybakova, *Sov. Phys.—Solid State* (*Engl. Transl.*) **17,** 1303 (1975).

[69] Yu. F. Markov, T. M. Gromova, Yu. V. Rud, and M. Tashanova, *Sov. Phys.—Solid State* (*Engl. Transl.*) **17,** 796 (1975).

[70] I. S. Gorban, V. A. Gorynya, V. I. Gugovoi, I. I. Tychina, and I. Yu. Tkacuk, *Sov. Phys.—Solid State* (*Engl. Transl.*) **18,** 1036 (1976).

[71] G. D. Holah, A. Miller, W. D. Dunnett, and G. W. Iseler, *Solid State Commun.* **23,** 75 (1977).

[72] R. G. Humphreys, *Conf. Ser.—Inst. Phys.* **35,** 105 (1977).

with existing theory. Bettini[41] has applied a Keating-type model calculation for the zone-center phonons of the phosphides with some success. The model describes the high-energy phonons reasonably well and shows that the spectra are determined predominantly by the masses of the cations and the force constants between nearest neighbors. A rigid-ion model developed by Poplavnoi and Tyuturev[73–76] has also given insight into the lattice dynamics of the chalcopyrite structure, and the phonon dispersion curves for $ZnSiP_2$ have been calculated.[76]

The $\Gamma_{15}$ mode of zinc blende splits into $B_2$ and $E$ modes in chalcopyrite which remain nearly degenerate and show no relative change in frequency because of the lattice compression. The optical $\Gamma_{15}$ mode of zinc blende is a motion of the cation lattice vibration in antiphase with the anion lattice. These remain the strongest modes, but mixing with $X$ and $W$ modes causes considerable distortion.[41] For the compounds shown in Fig. 11 the $\Gamma_{15}$-derived vibration is principally a group IV–group V atomic motion, owing to the larger force constant of this bond. The LO–TO splittings of these highest modes are generally smaller than that of their III–V analog because of the smaller effective charge associated with the IV–V bond. The mixing of $W_4$ and $X_5$ modes with $\Gamma_{15}$ gives considerable coupling of these oscillators to the infrared radiation.

There is a surprisingly large splitting of the $B_1$ and $B_2$ modes, which derive from the same $W_{2\text{op}}$ mode of zinc blende, in several compounds. In these materials, the $B_1$ mode consists mainly of motions of the lighter Group IV atoms and thus moves to higher energies, whereas the $B_2$ mode is influenced more by the motion of the heavier Group II atoms, lowering its energy. The $B_1$ modes have been notoriously difficult to locate by Raman scattering. This is because electron redistribution tends to concentrate the highest valence band electrons on the higher valence cation or on the anion bond associated with this cation.[38] The fact that the Group II atom environment is not substantially changed in this motion gives the reason why Raman scattering is weak for these phonons. Conversely, the $A_1$ mode has always been found to be the dominant Raman mode in chalcopyrites. This mode consists of movement of only the anions in an asymmetric environment because of the charge redistribution and hence strong scattering.

[73] A. S. Poplavoni and V. G. Tyuterev, *J. Phys.* (*Orsay, Fr.*) **36,** C3 169 (1975).

[74] A. S. Poplavnoi and V. G. Tyuterev, *Sov. Phys.—Solid State* (*Engl. Transl.*) **17,** 189 (1975).

[75] A. S. Poplavnoi and V. G. Tyuterev, *Sov. Phys.—Solid State* (*Engl. Transl.*) **17,** 672 (1975).

[76] A. S. Poplavnoi and V. G. Tyuterev, *Izv. Vyssh. Uchebn. Zaved., Fiz.* **6,** 51 (1975).

### *b. I–III–$VI_2$ Compounds*

The results of infrared and Raman measurements on I–III–$VI_2$ compounds are shown in Fig. 12.[33,64,77–96] Several I–III–$VI_2$ alloy ranges have also been investigated.[87,92,93] The phonon spectra of these compounds are very similar to those of the II–IV–$V_2$ crystals. It is interesting to compare the frequencies for compounds with similar masses, such as $CuAlS_2$/$ZnSiP_2$, $CuGaS_2$/$ZnGeP_2$, $AgGaS_2$/$CdGeP_2$, and $AgGaSe_2$/$CdGeAs_2$. Differences in the phonon frequencies that cannot be accounted for by the small mass change derive from the greater ionicity of the I–III–$VI_2$ family and the resulting changes in force constants and effective charges.

In this case, the high-frequency modes derived from the $\Gamma_{15}$ of zinc blende are found at higher energies than the zone-center phonons of II–VI analogs. In fact, the frequencies of these modes correspond more closely to those found in III–V analogs. This is again because of the difference in the force constants of the two bonds of the chalcopyrite lattice. This has been analyzed by Koschel and Bettini[85] who have made a rigid-ion model calculation for the sulfides similar to that previously carried out for the II–IV–$V_2$ phosphides. Although this model is not so suitable for the sulfides,

[77] G. D. Holah, J. S. Webb, *Proc. Int. Conf. Semicond., 11th, 1972* p. 1161.

[78] G. D. Holah, *Opt. Commun.* **5,** 10 (1972).

[79] W. H. Koschel, V. Hohler, A. Rauber, and J. Baars, *Solid State Commun.* **13,** 1011 (1973).

[80] J. P. Van Der Ziel, A. E. Meinner, H. M. Kasper, and J. A. Ditzenberger, *Phys. Rev. B: Solid State* [3] **9,** 4286 (1974).

[81] G. D. Holah, H. Montgomery, and J. S. Webb, *Proc. Int. Conf. Phys. Semicond., 12th, 1974* p. 316 (1974).

[82] D. J. Lockwood and H. Montgomery, *J. Phys. C* [3] **7,** 3241 (1974).

[83] D. J. Lockwood and H. Montgomery, *J. Phys. (Orsay, Fr.)* **36,** C3 183 (1975).

[84] W. H. Koschel, F. Sorger, and J. Baars, *J. Phys. (Orsay, Fr.)* **36,** C3, 173 (1975).

[85] W. H. Koschel and M. Bettini, *Phys. Status Solidi* B **72,** 729 (1975).

[86] A. Miller, G. D. Holah, W. D. Dunnett, and G. W. Iseler, *Phys. Status Solidi* B **78,** 569 (1976).

[87] J. N. Gan, J. Tauc, V. G. Lambrecht, and M. Robbins, *Phys. Rev. B: Solid State* **13,** 3610 (1976).

[88] D. J. Lockwood, *Conf. Ser.—Inst. Phys.* **35,** 97 (1977).

[89] G. Kanellis and K. Kampas, *J. Phys. (Orsay, Fr.)* **38,** 833 (1977).

[90] G. Kanellis, *in* "Lattice Dynamics" (M. Balkanski, ed.), p. 99. Flammarion Sciences, Paris, 1977.

[91] G. Kanellis and K. Kampas, *Mater. Res. Bull.* **13,** 9 (1978).

[92] I. V. Bodnar, A. G. Karoza, and G. F. Smirnova, *Phys. Status Solidi* B **84,** K65 (1977).

[93] I. V. Bodnar, A. G. Karoza, and G. F. Smirnova, *Phys. Status Solidi* B **86,** K.171 (1978).

[94] V. Riede, H. Sobotta, H. Neumann, Hang Xuan Nguyen, W. Möller, and G. Kühn, *Solid State Commun.* **28,** 449 (1978).

[95] Z. Petrović, P. M. Nikolić, and S. S. Vujatović, *Teh. Fiz.* **19,** 31 (1978).

[96] N. V. Joshi, *J. Phys. Chem. Solids* **40,** 93 (1979).

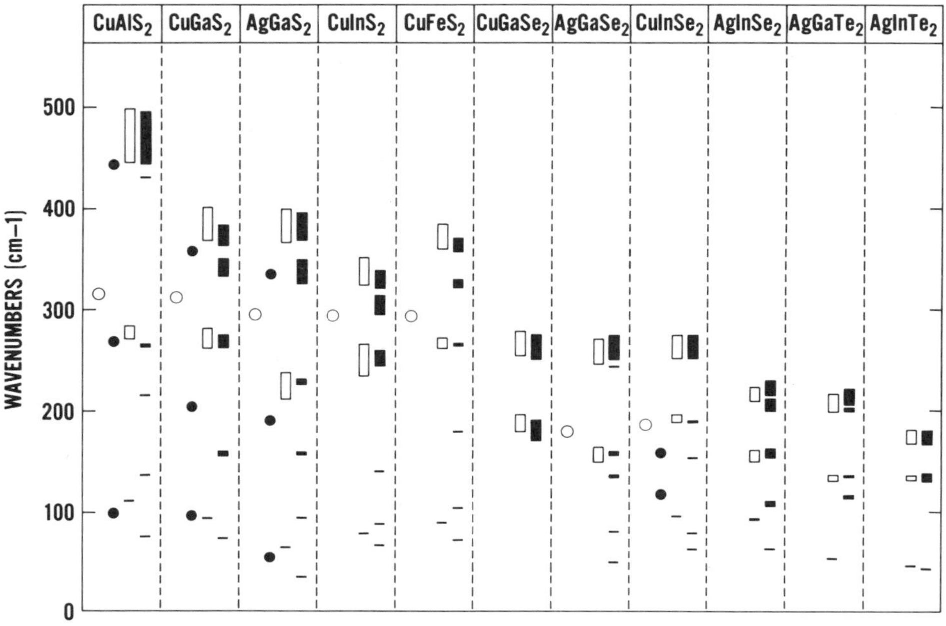


FIG. 12. Observed zone-center phonon frequencies of I–III–$VI_2$ compounds. $A_1$: ○, $B_1$: ●, $B_2$: ▯, $E$: ▮. The lengths of the boxes for modes of $B_2$ and $E$ symmetry give the Coulomb splittings of these modes.

the main qualitative features are represented. Koschel and Bettini found that the bond stretching force constant of the III–VI bond can be twice that of the heteropolar I–VI bond containing the noble metal. This results in the high energy modes being governed principally by the III–VI force constants, and the low energy modes being influenced more by the I–VI force constants.

The effect of the lower I–VI force constant on the acoustic phonons has been emphasized by a Brillouin scattering study of $AgGaS_2$.[97,98] There are six elastic constants that characterize the chalcopyrite structure, and the values that were obtained for $AgGaS_2$ are shown in Table VI. These results agree well with the averaged elastic moduli of ZnS and CdS.

The $A_1$ vibration is again the dominant mode observed by Raman scattering in I–III–$VI_2$ crystals; however, a significant decrease in the scattered intensity has been observed in $AgGaS_2$ and $CuGaS_2$ for this mode when the excitation laser frequency is close to the band gap

[97] M. H. Grimsditch and G. D. Holah, *Phys. Rev. B: Solid State* [3] **12,** 4377 (1975).
[98] G. D. Holah and M. H. Grimsditch, *J. Phys.* (*Orsay, Fr.*) **36,** C3, 185 (1975).

TABLE VI. ELASTIC MODULI OF $AgGaS_2$ (REF. 97) AND CORRESPONDING CALCULATED VALUES IN THE ZINC BLENDE APPROXIMATION (REF. 85)[a]

| | $C_{11}$ | $C_{33}$ | $C_{44}$ | $C_{66}$ | $C_{12}$ | $C_{13}$ | |
|---|---|---|---|---|---|---|---|
| | 8.79 | 7.58 | 2.41 | 3.08 | 5.84 | 5.92 | Experiment |
| | 8.03 | | 3.27 | | 4.92 | | Calculation |
| ½(ZnS + CdS) | 9.11 | | 3.51 | | 5.90 | | Experiment |

[a] The experimental values for the averaged isoelectronic analog are also shown. (Units: $10^{11}$ dynes/cm$^2$.)

energy.[80] This was attirbuted to a resonant interference between a nonresonant and a resonant term in the Raman scattering tensor.

The dependences of the most prominant zone-center phonons of $CdSiP_2$, $CuAlS_2$, and $CuGaS_2$ on hydrostatic pressure were measured by Bettini and Holzapfel[64] and comparison made with $\Gamma$, $X$, and $W$ modes in binary analogs. It was found that some $X$- and $W$-derived modes could be distinguished by this technique. The high-energy modes with $\Gamma_{15}$ character had Grüneisen parameters comparable with those of III–V compounds, and the lowest-energy $X_{5ac}$ modes of $CdSiP_2$ and $CuGaS_2$ have negative constants, consistent with the results for transverse acoustic modes at the zone boundary of zinc blende-structured compounds.

## 8. ORDERED VACANCY STRUCTURES

In Section 6, we discussed the effect of having an ordered array of vacant sites on the lattice vibrational properties of tetrahedral structures. The hypothetical "defect diamond" structure was considered. The bct array of vacant sites used for defect diamond is identical to that found in both $I\bar{4}$ defect chalcopyrite and $I\bar{4}2m$ defect stannite, so that some of the features of Fig. 9 deduced from the Keating model calculation and the group theoretical analysis should be apparent in these structures. Only three compounds of this type have been studied in any detail. $HgIn_2\square Te_4$, with $I\bar{4}2m$ structure has been the subject of ultrasonic,[17] infrared, and Raman[19,99] measurements, and two $I\bar{4}$ structured compounds, $Ag_2Hg\square I_4$[100,101] and $CdGa_2\square Se_4$,[102] have been studied by Raman scatter-

[99] A. Miller, A. MacKinnon, D. Weaire, C. R. Pidgeon, D. J. Lockwood, and G. A. Saunders, *Phys. Semicond., Proc. Int. Conf., 13th, 1976* p. 509 (1976).

[100] D. Greig, D. F. Shriver, and J. R. Ferraro, *J. Chem. Phys.* **66,** 5248 (1977).

[101] D. Greig, G. C. Joy, III, and D. F. Shriver, *J. Chem. Phys.* **67,** 3189 (1977).

[102] R. Bacewicz, P. P. Lottici, and C. Razzetti, *J. Phys. C* [3] **12,** 3603 (1979).

ing with particular emphasis on the "superionic" phase transition in the case of $Ag_2Hg\square I_4$. Some results of two-phonon absorption exist for $I\bar{4}$ $CdGa_2\square S_4$ and $CdGa_2\square Se_2$ crystals.[103]

*a. $HgIn_2\square Te_4$*

The defect stannite compound, $HgIn_2\square Te_4$, is one of a series intermediate between HgTe, in which every tetrahedral site is filled, and $In_2\square Te_3$, in which one in every six sites is unfilled. This series was chosen by Saunders and Seddon[17,104] for an investigation of the influence of the vacancies on elastic constants. $HgIn_2\square Te_4$ is tetragonal, whereas the other intermediate members of the series, $Hg_5In_2\square Te_8$ and $Hg_3In_2\square Te_6$, are cubic with ordered zinc blende superlattices. (Measurements have also been made on most of the equivalent gallium series, but crystals of tetragonal $HgGa_2\square Te_4$ have yet to be grown to sufficient size and quality.) These studies were carried out at 77 K and room temperature using the technique of pulse superposition of ultrasonic waves at 10 MHz. The resulting elastic constants obtained for $HgIn_2\square Te_4$ are shown in Table VII. It is interesting to compare these values with the results of Brillouin scattering studies[97] in $AgGaS_2$ which were given in Table VI. The elastic stiffness constants for $HgIn_2\square Te_4$ are smaller and are much closer to those of a cubic compound ($C_{11} = C_{33}$, $C_{12} = C_{13}$, $C_{44} = C_{66}$) than to those of $AgGaS_2$. It has been suggested that this may be due to the large tetragonal compression in $AgGaS_2$ ($c/a = 1.789$) compared to $HgIn_2\square Te_4$ ($c/a = 2.0$); however, in the light of the calculations for chalcopyrite structured materials, this is more likely to derive from differences between nearest-neighbor force constants. Saunders and Seddon[17] have shown the marked dependence of elastic properties on vacancy concentration by plotting the reduced bulk modulus for a number of ordered vacancy compounds against the percentage of vacancies (Fig. 13). Similar results for indium and gallium compounds with the same concentration of vacancies reinforce the notion that the ordered array of vacant sites determines the elastic behavior.

Further intriguing properties of the vacancies have been obtained by infrared reflectivity and Raman scattering measurements.[19] To be clear on how the optical phonon modes in defect stannite relate to those of filled structures, first recall that chalcopyrite and stannite both have the same point group $\bar{4}2m$, they both have eight atoms per unit cell, but the space group of stannite is symmorphic, i.e., $C_2'$ and $\sigma_d$ operations do not have a

[103] S. I. Radautsan, N. N. Syrbu, I. I. Nebola, V. G. Tyrziu, and D. M. Bercha, *Sov. Phys.—Semicond. (Engl. Transl.)* **11**, 38 (1977).

[104] G. A. Saunders and T. Seddon, *J. Phys. Chem. Solids* **31**, 2495 (1920).

TABLE VII. ELASTIC STIFFNESS CONSTANT ($C$) AND BULK MODULUS ($B$) FOR $HgIn_2\square Te_4$[a]

| $C_{11}$ | $C_{33}$ | $C_{44}$ | $C_{66}$ | $C_{12}$ | $C_{13}$ | $B$ | |
|---|---|---|---|---|---|---|---|
| 4.31 | 4.47 | 2.14 | 2.41 | 2.54 | 2.18 | 2.99 | Experiment (Ref. 17) |
| 4.20 | 4.4 | 1.18 | 1.21 | 2.88 | 2.9 | 3.35 | Calculation (Ref. 19) |

[a] Units: $10^{11}$ dynes/cm$^2$.

nonprimitive lattice translation associated with them as in chalcopyrite. Thus, $X_1$, $X_3$, and $X_5$ zinc blende phonons have $A_1$, $B_2$, and $E$ irreducible representations, respectively (Table IV), and the zone-center phonons comprise $2A_1 + 1A_2 + 2B_1 + 4B_2 + 6E$. By withdrawing one atom per unit cell to give defect stannite, one $B_2$ and a doubly degenerate $E$ mode, which correspond to the $Z$ and $(X, Y)$ components of the atomic displacement, are lost, as was observed for defect diamond in Fig. 9.

The experimental results for $HgIn_2\square Te_4$ are shown in Table VIII, and the distribution of these zone-center phonons is very similar to that shown in Fig. 9 for defect diamond except that no low-energy $A_1$ phonon was observed. As for chalcopyrite, one $A_1$ mode at 100 cm$^{-1}$ dominates the Raman spectrum, but now a second weaker mode of $A_1$ symmetry could

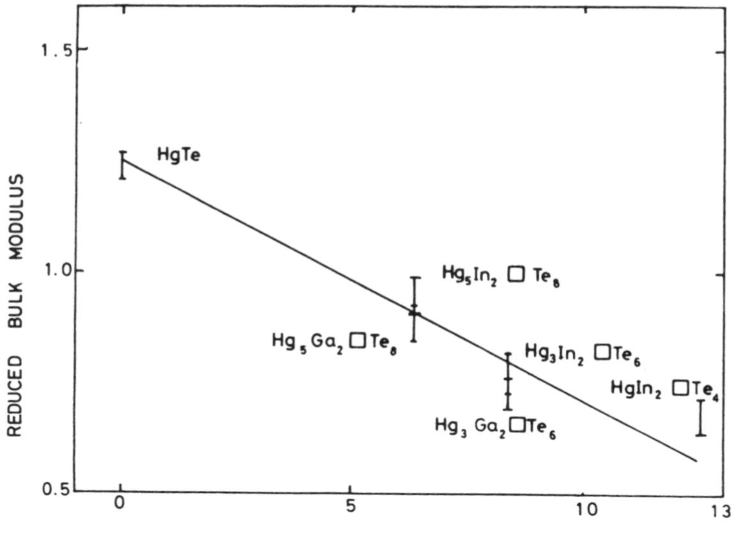


FIG. 13. The reduced bulk modulus plotted against the percentage of sited vacancies for two families of vacancy compounds (Ref. 17).

TABLE VIII. EXPERIMENTAL AND CALCULATED VIBRATIONAL FREQUENCIES FOR $HgIn_2\square Te_4$[a,b]

| Irreducible representation | Infrared | | Raman | | Calculated |
|---|---|---|---|---|---|
| | | | $\nu(cm^{-1})$ | | |
| $A_1(\Gamma_1)$ | Inactive | | 100 | | 99 |
| | | | 132 | | 140 |
| $A_2(\Gamma_2)$ | Inactive | | Inactive | | 114 |
| $B_1(\Gamma_3)$ | Inactive | | ~50 | | 57 |
| | | | 155 | | 158 |
| | $\nu_{TO}$ | $\nu_{LO}$ | $\nu_{TO}$ | $\nu_{LO}$ | $\nu_{TO}$ |
| $B_2(\Gamma_4)$ | | | | 74 | 58 |
| | 147 | 153 | — | 152 | 139 |
| | 180 | 184 | 184 | 189 | 181 |
| | | | | 42 | 50 |
| | | | | 61 | 57 |
| $E(\Gamma_5)$ | 118 | 127 | 122 | — | 132 |
| | 160 | 169 | 162 | — | 160 |
| | — | — | 181 | 188 | 179 |

[a] Taken from A. Miller, D. J. Lockwood, A. MacKinnon, D. Weaire, *J. Phys. C* **9**, 2997 (1976).
[b] Units: $cm^{-1}$.

be observed at 132 $cm^{-1}$. A clue to the origin of these modes was obtained from a Keating model calculation similar to that used for chalcopyrites by Bettini.[41] Surprisingly, it was necessary to incorporate a □–Te force constant in order to reproduce the phonon spectrum adequately (Table VIII), particularly for the $A_1$ mode frequencies. Thus, as pointed out in Section 6, this force constant represents the Coulomb forces neglected in the model, so that the vacancy must be considered not as an empty space in the lattice but as a positive potential attracting the valence electrons of the surrounding anions. The strong $A_1$ mode corresponds approximately to the breathing mode of its tellurium atoms about the vacant site as shown in Fig. 10. The force constants found for the best fit to the optical phonons were used to calculate the elastic stiffness constants and are compared with the experimental results in Table VII. The results are reasonable considering the simplicity of the model, apart from the $C_{44}$ and $C_{66}$ values.

An unexplained feature of the optical phonon results for $HgIn_2\square Te_4$ is the small dipole moment of the highest $E$ mode. This mode was not observed in reflectivity measurements, whereas the 160- and 118-$cm^{-1}$ phonons have strong infrared coupling. This is a clear departure from the vibrational properties of chalcopyrite compounds.

*b.* $Ag_2Hg\square I_4$

The temperature dependence of the Raman spectrum of $Ag_2Hg\square I_4$ has been investigated by Greig *et al.*[101] between 8 and 470 K. The structure is defect chalcopyrite $I\bar{4}$ below 323 K, and above this temperature there is a partial disorder of the cation lattice. Single-crystal, polarized Raman scattering measurements were carried out at room temperature to determine the symmetry of the observed modes, whereas at the other temperatures, the orientation was not specified. Large enough crystals have not been grown for a full Raman scattering analysis.

The character table of the point group $I\bar{4}$ is given in Table A.1 in the Appendix. The 21 zone-center phonons may be associated with the 3 irreducible representations in the combination $3A + 6B + 6E$, where $E$ is doubly degenerate. One $B$ and one $E$ comprise the acoustic phonons, leaving 13 Raman-active optical phonons to be observed. Although the vacancies in this structure again have the same bct distribution as defect diamond, the reduced symmetry in this case allows mixing of the three $A$ modes derived from the $X_1$ and $W_1$ zinc blende symmetries. $B_1$ and $B_2$ become modes of $B$ symmetry in $I\bar{4}$.

Table IX reproduces the Raman scattering results[101] for this compound. As for the other ternary structures we have considered, a dominant Raman mode of $A$ symmetry was observed at 122 $cm^{-1}$ (295 K). Two other weak modes were tentatively assigned $A$ symmetry at 81 and 106 $cm^{-1}$. The four low-frequency modes were labeled, $B$: 30 and 42 $cm^{-1}$, and $E$: 25 and 35 $cm^{-1}$. We again note the similarity of this distribution to the low-lying modes of those calculated for defect diamond (Fig. 9), but we see that none of the $A$ modes have dropped to the energy of the lowest $B$ and $E$ modes. We would thus conclude that the very strong $A$ mode observed in $Ag_2Hg\square I_4$ is similar to the breathing mode described for $HgIn_2\square Te_4$.

TABLE IX. PEAK POSITIONS FOR POLYCRYSTALLINE $Ag_2Hg\square I_4$ AT 8K[a,b]

| | | |
|---|---|---|
| 26.1 | 42.8 | 108.6 |
| 27 | 82.4 | 119(?) |
| 32.1 | 99.5 | 124.5 |
| 36.7 | 105.5 | 130 |
| 38.0 | 106.7 | 136 |

[a] Taken from D. Greig, G. C. Joy, III, and D. F. Shriver, *J. Chem. Phys.* **67**, 3189 (1977).

[b] Units: $cm^{-1}$.

Iodine lies next to tellurium in the periodic table, so the difference in frequency of these dominant modes 122/100 in the two compounds is due to the difference in the bond stretching force constants associated with them. Further work and a model calculation are required before a full interpretation can be made for this compound.

*c.* $CdGa_2\square Se_4$

Bacewicz *et al.*[102] investigated the Raman spectrum of single crystals of $CdGa_2\square Se_4$ in backreflection. A polarized angular analysis of the spectra gave the symmetry assignment for the results shown in Table X. The structure is believed to be $I\bar{4}$ for the compound giving 21 zone-center phonons as discussed for $Ag_2Hg\square I_4$; however, it is interesting to compare the results for $CdGa_2\square Se_4$ and $HgIn_2\square Te_4$, since the ratios of the masses of the constituent atoms are very similar—each atom of the $HgIn_2\square Te_4$ being one row lower than $CdGa_2\square Se_4$ in the periodic table. In particular, the 141 and 188 $cm^{-1}$ $A$ modes of $CdGa_2\square Se_4$, which comprise motions of only the selenium atoms, compared to the 100 and 132 $cm^{-1}$ modes in the tellurium compound are accounted for to a first approximation simply by the differences in mass of selenium and tellurium atoms. The 141 $cm^{-1}$ mode of $CdGa_2\square Se_4$ dominates the Raman spectrum and is likely to be the breathing mode previously described. Doubt exists as to the assignment of $A$ symmetry to the 277 $cm^{-1}$ mode.[105] If we associate the two $A$ modes discussed above with the $A_1$ modes of $HgIn_2\square Te_4$, then the third $A$ mode should correspond to the $A_2$ mode of $I\bar{4}2m$, which derives from the same $W$ point as the breathing mode in the folding picture (Fig. 9). From the discussion of Section 6, we would expect this mode to have an energy in the same region as the other two $A$ modes. MacKinnon[105] points out that the 277-$cm^{-1}$ peak could be due to scattering from two "breathing" phonons, since strong multiphonon effects are expected because of the large asymmetry in the Coulomb forces with respect to the breathing mode (Section 6).

## V. Band Structure

The search for new semiconducting materials for applications in light-emitting diodes, infrared detectors, and solar cells has resulted in a great deal of activity in the investigation of the energy band structures of

[105] A. MacKinnon, *J. Phys. C* [3] **12,** L655 (1979).

TABLE X. PEAK POSITIONS OF RAMAN FREQUENCIES IN $CdGa_2\square Se_4$[a,b]

| | | | |
|---|---|---|---|
| 47 | *E* | 196 | *B* (TO) |
| 53 | *E* | 202 | *B* (LO) |
| 70 | *E* | 210 | *B* (TO) |
| 78 | *B* (LO + TO) | 222 | *B* (TO) |
| 107 | *E* | 237 | *B* (LO) |
| 125 | *B* (TO) | 240 | *E* |
| 127 | *B* (LO) | | |
| 141 | *A* | 248 | *E* |
| 178 | *E* | 261 | *E* ? |
| 182 | *E* | 277 | *A* ? |
| 188 | *A* | 279 | *B* (LO) |

[a] Taken from R. Bacewicz, P. P. Lottici, and C. Razzetti, *J. Phys. C* **12,** 3603 (1979).
[b] Units: $cm^{-1}$.

ternary compounds during the last decade. They offer a wide selection of energy gaps throughout the infrared and visible spectrum.

Modulation techniques have now identified the lowest interband transitions in most of the chalcopyrite compounds. The method of folding into the zone center has proved successful in identifying the main features of these energy band gap transitions.

The electronic band structures of II–IV–$V_2$ and I–III–$VI_2$ chalcopyrite semiconductors have been reviewed several times.[2,106–110] We shall present only the essential features and recent results here to give the current understanding of the principal interband transitions in the light of the group theoretical discussion of Section 5. Results for II–$III_2$–$\square$–$VI_4$ compounds are discussed in terms of the effect of their different lattice symmetries on band properties.

## 9. CHALCOPYRITE COMPOUNDS

The relative splittings and polarizations of the lowest interband transitions of II–IV–$V_2$ compounds have been explained by folding the energy band structure of a zinc blende analog—such as GaP (Fig. 14)—and

[106] N. A. Goryunova, A. S. Poplavnoi, Yu. I. Polygalov, and V. A. Chaldyshev, *Phys. Status Solidi* **39,** 9 (1970).
[107] A. Shileika, *Surf. Sci.* **37,** 730 (1973).
[108] J. L. Shay and B. Tell, *Surf. Sci.* **37,** 748 (1973).
[109] V. A. Chaldeshev, "Poluprovodniki, $A^2B^4C_2^5$," p. 212. Soveskoe Radio, Moskva, 1974.
[110] A. Shileika, *Conf. Ser.—Inst. Phys.* **35,** 129 (1977).

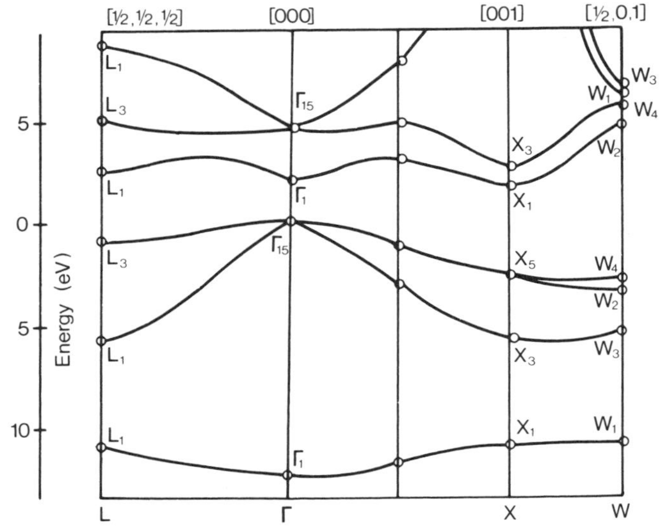


FIG. 14. Energy band structure of GaP neglecting spin–orbit splitting.

considering the effects of crystal field splitting. Hopfield[111] has given a useful formula for the splitting of the $\Gamma_{15}$ level in the presence of both crystal field and spin–orbit terms. This is

$$E_{\Gamma_7} = E_{\Gamma_6} - \tfrac{1}{2}(\Delta_{so} + \Delta_{cf}) \pm \tfrac{1}{2}[(\Delta_{so} + \Delta_{cf})^2 - \tfrac{8}{3}\Delta_{so}\Delta_{cf}]^{1/2}.$$

We shall not pursue the precise definition of $\Delta_{so}$ and $\Delta_{cf}$, which have generally been used as semiempirical fitting parameters. In Section 5 we saw that in principle, there are various contributions to the crystal field, but that in the case of the chalcopyrite structure, the $\Gamma$–$X$ mixing terms vanished in first order, leaving the tetragonal compression as the dominant factor.

*a. II–IV–V$_2$ Compounds*

The zone-center triply degenerate $\Gamma_{15}$ valence band of zinc blende compounds is crystal field split into $\Gamma_4$ and $\Gamma_5$ bands because of the lower symmetry of the chalcopyrite structure (Fig. 15). Inclusion of spin-orbit splitting results in three zone-center bands. Mapping of $X$ and $W$ points of zinc blende to the zone center of chalcopyrite gives eight extra valence bands, all at energies between the $\Gamma_{15}$- and $\Gamma_1$-derived bands. The negative crystal field splitting (Table XI) has been found to be correlated with

[111] J. J. Hopfield, *J. Phys. Chem. Solids* **15**, 97 (1960).

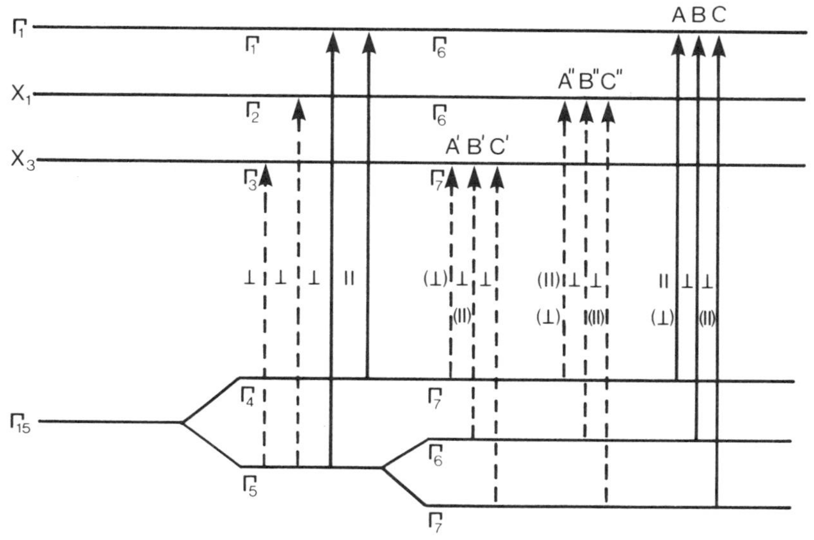


FIG. 15. Chalcopyrite zone-center energy bands and selection rules derived from zinc blende Γ and $X$ points by including the crystal field and spin–orbit splittings. The transitions labeled $A$, $B$, $C$ are the direct transitions associated with $\Gamma_{15} \rightarrow \Gamma_1$ transitions of zinc blende. $A'$, $B'$, $C'$ and $A''$, $B''$, $C''$ are "pseudodirect" transitions in which the final state derives from the $X$ point of the zinc blende Brillouin zone.

TABLE XI. ENERGIES OF INTERBAND TRANSITIONS IN SOME II–IV–$V_2$ COMPOUNDS AND CORRESPONDING CRYSTAL FIELD PARAMETERS, TOGETHER WITH RELEVANT DATA FOR BINARY ANALOGS[a]

| Compound | | $A$ | $B$ | $C$ | $-\Delta_{cf}$ | $\Delta_{so}$ | Binary analogs | $E_g$ | $\Delta_{so}$ |
|---|---|---|---|---|---|---|---|---|---|
| $ZnSiP_2$ | P[b] | 2.07 | 2.16 | 2.21 | 0.13 | 0.07 | $Ga_{0.5}Al_{0.5}P$ | 2.35 | 0.08 |
| $ZnSiAs_2$ | P | 1.74 | 1.83 | — | 0.13 | 0.28 | $Ga_{0.5}A_{0.5}As$ | 2.01 | 0.31 |
| $ZnGeP_2$ | P | 1.99 | 2.05 | 2.12 | 0.08 | 0.07 | GaP | 2.26 | 0.10 |
| $ZnGeAs_2$ | | 1.15 | 1.19 | 1.48 | 0.06 | 0.31 | GaAs | 1.44 | 0.34 |
| $ZnSnP_2$ | | 1.66 | 1.66 | 1.75 | 0 | 0.09 | $In_{0.5}Ga_{0.5}P$ | 1.81 | 0.10 |
| $ZnSnAs_2$ | | 0.73 | 0.73 | 1.07 | 0 | 0.34 | $In_{0.5}Ga_{0.5}As$ | 0.82 | 0.38 |
| $CdSiP_2$ | P | 2.10[c] | 2.30[c] | 2.30[c] | — | — | $In_{0.5}Al_{0.5}P$ | 2.25 | 0.09 |
| $CdSiAs_2$ | | 1.55 | 1.74 | 1.99 | 0.24 | 0.29 | $In_{0.5}Al_{0.5}As$ | 1.60 | 0.24 |
| $CdGeP_2$ | | 1.72 | 1.90 | 1.99 | 0.20 | 0.11 | $In_{0.5}Ga_{0.5}P$ | 1.87 | 0.10 |
| $CdGeAs_2$ | | 0.57 | 0.73 | 1.02 | 0.21 | 0.33 | $In_{0.5}Ga_{0.5}As$ | 0.82 | 0.10 |
| $CdSnP_2$ | | 1.17 | 1.25 | 1.33 | 0.10 | 0.10 | InP | 1.35 | 0.11 |
| $CdSnAs_2$ | | 0.26 | 0.30 | 0.77 | 0.06 | 0.48 | InAs | 0.36 | 0.43 |

[a] Units: electron volts. [b] P = pseudodirect. [c] At 77 K.

tetragonal compression and approximately equal to that expected for a zinc blende analog strained along the ⟨001⟩ direction by an amount corresponding to the tetragonal compression[112,113] (Fig. 16). However, the deviation from a linear dependence for the II–IV–$V_2$ materials with large distortion in Fig. 16 shows that the crystal field splitting is caused partly by other contributions in chalcopyrite. The effect of the anion shift from the "ideal" zinc blende position in $ZnSiAs_2$, $ZnSnAs_2$[36] and $ZnSiP_2$[35] has been studied by pseudopotential band structure calculations, and it was found that this could cause a reverse crystal field splitting with the $\Gamma_5$ above the $T_4$ band at large enough values of the $x$ parameter. Possible evidence of this has been found in $CdSiP_2$ which has the largest II–IV–$V_2$ anion displacement. The polarization dependences of the thermoreflectance spectra of the $\Gamma_4$, $\Gamma_5$ to $\Gamma_1$ transitions for this material seem to indicate a reverse ordering,[114] although this conflicts with earlier results for the $\Gamma_4$, $\Gamma_5$ to $\Gamma_3$ transitions.[115] A similar dilemma exists for $ZnSiP_2$ in which a reverse ordering of valence bands seems to be apparent from the polarization dependence of electroreflectance.[116] ($ZnSiP_2$ has a much smaller anion distortion than $CdSiP_2$.) It was tentatively suggested that this might be because the lowest Γ-derived conduction band in $ZnSiP_2$ is $\Gamma_{15}$ rather than $\Gamma_1$ as is normal for chalcopyrites, although this seems unlikely by a comparison with the corresponding band positions of zinc blende.[116]

The spin–orbit splittings $\Delta_{so}$, observed in II–IV–$V_2$ compounds, are close to III–V analogs. The phosphides have values around 0.1 eV, whereas the arsenides have larger values, around 0.3 eV.

The lowest conduction band of II–IV–$V_2$ compounds can take one of two forms. One type derives from zinc blende compounds with direct band gaps, such as GaAs, and has a lowest conduction band of $\Gamma_1$ symmetry equivalent to $\Gamma_1$ of zinc blende. The energy bands of GaAs embedded in the chalcopyrite Brillouin zone are shown in Fig. 17.[107] The second type derives from indirect band gap materials, such as GaP (Fig. 14), and has a lowest conduction band of $\Gamma_3$ symmetry that originates in the $X_1$ or $X_3$ bands of zinc blende. (*Note:* There is an inconsistency in the published literature between the designation of $\Gamma_2$ and $\Gamma_3$ modes derived from $X_1$ and $X_3$ in both vibrational and electronic spectra. This appears to be caused by a change in origin for zinc blende band structure calcula-

[112] J. L. Shay, E. Buehler, and J. H. Wernick, *Phys. Rev. B: Solid State* [3] **2**, 4104 (1970).

[113] J. L. Rowe and J. L. Shay, *Phys. Rev. B: Solid State* [3] **3**, 451 (1971).

[114] G. Ambrazevicius, G. Babonas, and A. Shileika, *Phys. Status Solidi* B **82**, K45 (1971).

[115] G. Babonas, G. Ambrazevicius, V. S. Grigoreva, V. Neviera, and A. Shileika, *Phys. Status Solidi* B **62**, 327 (1974).

[116] J. L. Shay, B. Tell, E. Buehler, and J. H. Wernick, *Phys. Rev. Lett.* **30**, 983 (1973).

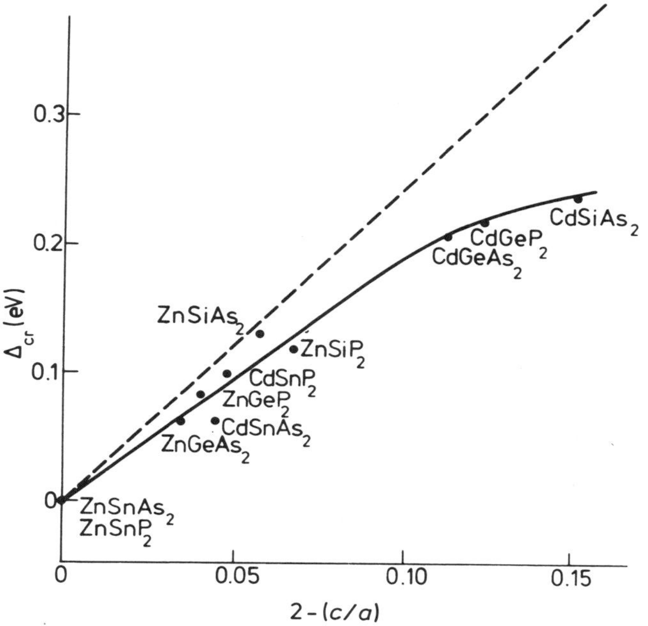


FIG. 16. The crystal field splitting parameter plotted against the tetragonal compression for several II–IV–$V_2$ compounds (Ref. 110).

tions, since the $X_1$ and $X_3$ designations will interchange on shifting the origin from the cation to anion sublattice.) The energy bands of GaP embedded in the chalcopyrite Brillouin zone are shown in Fig. 18.[107] Although the three transitions *A*, *B*, and *C* between the valence bands and the $\Gamma_1$ conduction band have been located easily by reflectance modulation techniques for crystals of the first type, *A'*, *B'*, and *C'* pseudodirect transitions of the second type of crystal have been very difficult to locate. In Section 5 we explained how these transitions are very weak for symmetry reasons, even if the pseudopotentials of the two cations are quite different. Unlike the $X_5$ phonon in infrared reflectivity measurements discussed in Section 7, there are no Γ- or *W*-derived bands of the same symmetry available in this energy range to give any strong mode mixing. Four materials have pseudodirect energy band gaps: $ZnSiP_2$, $CdSiP_2$, $ZnGeP_2$, and $ZnSiAs_2$. These compounds were first identified by the pressure dependence of the absorption edge.[117] The energies of the three transitions *A'*, *B'*, and *C'* have been determined for these compounds by

[117] R. Bendorius, V. D. Prochukan, and A. Shileika, *Phys. Status Solidi* B **53**, 745 (1972).

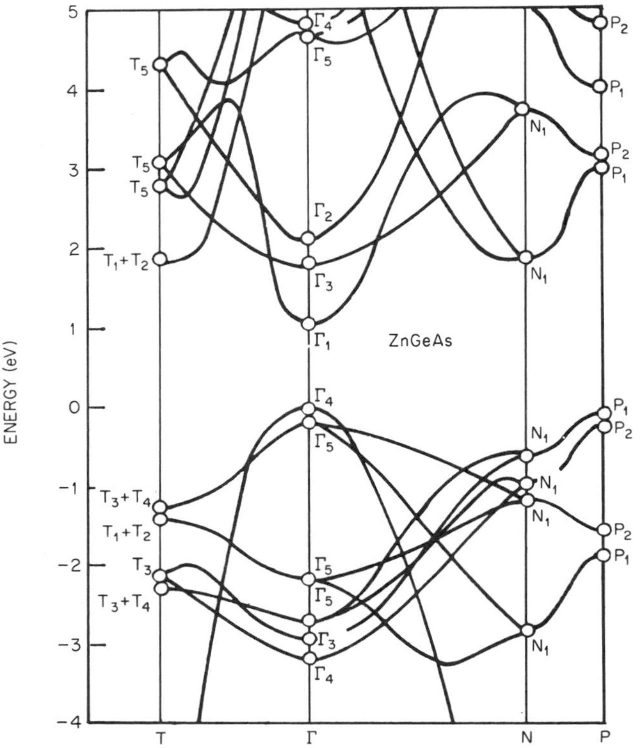


FIG. 17. Approximate band structure of a direct gap II–IV–$V_2$ compound obtained by embedding the band structure of GaAs into the chalcopyrite Brillouin zone (Ref. 107).

wavelength-modulated absorption[114] (see Table XI). The polarization dependences were consistent with transitions to the conduction band of $\Gamma_3$ symmetry (Fig. 18).

Transitions between the $\Gamma_3$ and $\Gamma_2$ conduction bands have also been observed in *n*-type $ZnSiP_2$ at 0.76 eV by photomodulated absorption.[118] This is in reasonable agreement with the calculated separation.[35]

Pseudopotential calculations[34–37,119–122] have been carried out for all of

[118] R. G. Humphreys, *J. Phys. C* **9**, 4491 (1976).

[119] A. S. Poplavnoi, Ju. I. Polygalov, and V. A. Chaldyshev, *Izv. Vyssh. Uchebn. Zaved., Fiz.* **11**, 58 (1969).

[120] A. S. Poplavnoi, Ju. I. Polygalov, and V. A. Chaldyshev, *Izv. Vyssh. Uchebn. Zaved., Fiz.* **6**, 95 (1970).

[121] A. S. Poplavnoi, Ju. I. Polygalov, and V. A. Chaldyshev, *Izv. Vyssh. Uchebn. Zaved., Fiz.*, **7**, 17 (1970).

[122] J. F. Alward, C. Y. Fong, and F. Wooten, *Phys. Rev. B: Condens. Matter* [3] **19**, 6337 (1979).

the II–IV–$V_2$ compounds listed in Table XI, and the most recent of these have emphasized the importance of taking into account the anion displacement in chalcopyrite.[35–37] All calculations except those of Varea de Alvarez *et al.*,[34] which did not include the anion displacement, give the $\Gamma_4$ band above the $\Gamma_5$ band. Aymerich *et al.*[37] find that the chemical difference between cations tends to compensate the effect of tetragonal distortion. A possible explanation for the apparently anomalous transitions in the 2.9–3.2 eV range in $ZnSiP_2$ is given by Pasemann *et al.*[35] from the results of calculations. These workers proposed that direct transitions at *X* and *M* points of the chalcopyrite Brillouin zone (Fig. 6) will be close to the *A*, *B*, and *C* transitions in energy and will therefore confuse the reflectance spectrum. A very similar band structure is calculated for $CdSiP_2$ by Aymerich *et al.*,[37] and so the same zone-boundary transitions may well be responsible for some of the structure in $CdSiP_2$ in the 2.7–3.0 eV range.[144] Aymerich *et al.*[37] have also deduced valence electron charge densities along II–V and IV–V bonds for the 12 compounds in Table XI from pseudopotential calculations. They find that their calculated effec-

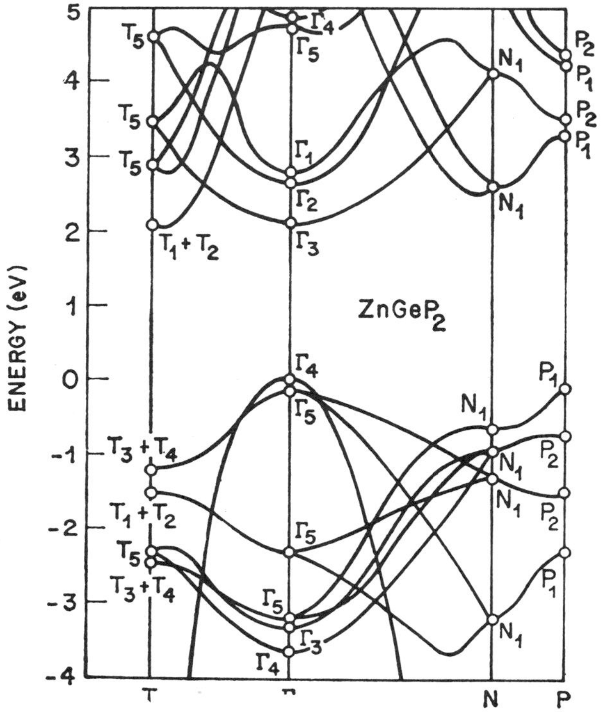


FIG. 18. Approximate band structure of a "pseudodirect" II–IV–$V_2$ compound obtained by embedding the band structure of GaP into the chalcopyrite Brillouin zone (Ref. 107).

tive valences for the constituent elements are approximately proportional to electronegativities as deduced by Phillips.[123]

We can thus conclude that the quasi-cubic model for II–IV–$V_2$ compounds gives a good description of the lowest interband electronic transitions. Tetragonal compression is the principal distortion that accounts for the crystal field splitting.

### b. *I–III–$VI_2$ Compounds*

The energy gaps observed in 14 I–III–$VI_2$ compounds are listed in Table XII.[2,124,125] These materials all have direct band gaps, and like the II–IV–$V_2$ compounds are characterized by three transitions $A$, $B$, and $C$ because of the crystal field and spin–orbit splitting of the $\Gamma_{15}$ valence band of zinc blende. The quasi-cubic model is again fairly successful in describing the crystal field splitting in terms of the tetragonal compression as the principal contribution for I–III–$VI_2$ compounds. The observed crystal field parameters are plotted as a function of lattice compression in Fig. 19. The deduced values of deformation potential are what would be expected from the properties of II–VI compounds.

It is when we compare the energy gaps and spin–orbit splitting with isoelectronic II–VI compounds that we realize the zinc blende approximation breaks down for the electronic structures of I–III–$VI_2$ compounds. The energy gaps are between 0.4 and 2.4 eV lower and the spin–orbit splittings are considerably smaller (negative for $CuInSe_2$[126] and $CuGaS_2$[127]) than those of II–VI analogs. These effects have been attributed to the proximity and consequent hybridization of the nobel metal d-levels with the p-bands of the other atoms, as shown in Fig. 20. The fivefold degenerate d-levels split into a threefold $\Gamma_{15}$ and a twofold $\Gamma_{12}$ in a tetrahedral field. The spin–orbit contribution splits the p-like $\Gamma_{15}$ into $\Gamma_8$ and $\Gamma_7$ levels, whereas the d-like $\Gamma_{15}$ gives $\Gamma_7$ and $\Gamma_8$ levels in reverse order. The repulsion rule of Section 4 explains why the band gap is reduced and smaller spin–orbit splittings result. The $\Gamma_8$ band derived from the d-like $\Gamma_{15}$ band will force the p-like $\Gamma_8$ upward, reducing the band gap energy. Likewise, the close proximity of the two $\Gamma_7$ bands will reduce the spin–orbit splitting of the p-levels in chalcopyrite. A mixing of the wave

[123] J. C. Phillips, *Rev. Mod. Phys.* **42**, 317 (1970).

[124] B. Tell, J. L. Shay, and H. M. Kasper, *Phys. Rev. B: Solid State* [3] **9**, 5203 (1974).

[125] M. J. Thwaites, R. D. Tomlinson, and M. J. Hampshire, *Conf. Ser.—Inst. Phys.* **35**, 237 (1977).

[126] J. L. Shay, B. Tell, H. M. Kasper, and L. M. Shiavane, *Phys. Rev. B: Solid State* [3] **7**, 4485 (1973).

[127] B. Tell and P. M. Bridenbaugh, *Phys. Rev. B: Solid State* [3] **12**, 3330 (1975).

TABLE XII. ENERGIES OF INTERBAND TRANSITIONS IN SOME I–III–$VI_2$ COMPOUNDS AND CORRESPONDING CRYSTAL FIELD PARAMETERS, TOGETHER WITH RELEVANT DATA FOR BINARY ANALOGS

| Compound | $A$ | $B$ | $C$ | $-\Delta_{cf}$ | $\Delta_{so}$ | Binary analogs | $E_g$ | $\Delta_{so}$ |
|---|---|---|---|---|---|---|---|---|
| $CuAlS_2$ | 3.49 | 3.62 | 3.62 | 0.13 | 0 | $Mg_{0.5}Zn_{0.5}S$ | 5.9 | — |
| $CuGaS_2$ | 2.43 | 2.55 | 2.55 | 0.12 | 0 | ZnS | 3.8 | 0.07 |
| $CuGaSe_2$ | 1.68 | 1.75 | 1.96 | 0.09 | 0.23 | ZnSe | 2.68 | 0.43 |
| $CuGaTe_2$ | 1.24 | 1.27 | 1.85 | 0.04 | 0.6 | ZnTe | 2.56 | 0.93 |
| $CuInS_2$ | 1.53 | 1.53 | 1.53 | <0.005 | −0.02 | $Zn_{0.5}Cd_{0.5}S$ | 3.17 | 0.07 |
| $CuInSe_2^b$ | 1.04 | 1.04 | 1.27 | −0.006 | 0.23 | $Zn_{0.5}Cd_{0.5}Se$ | 2.33 | 0.43 |
| $CuInTe_2$ | 1.06 | 1.06 | 1.67 | 0.0 | 0.61 | $Zn_{0.5}Cd_{0.5}Te$ | 2.18 | 0.93 |
| $AgAlTe_2$ | 2.27 | 2.38 | — | 0.16 | — | $Mg_{0.5}Cd_{0.5}Te$ | 3.30 | — |
| $AgGaS_2^b$ | 2.73 | 3.01 | 3.01 | 0.28 | 0 | $Zn_{0.5}Cd_{0.5}S$ | 3.17 | 0.07 |
| $AgGaSe_2^b$ | 1.83 | 2.03 | 2.29 | 0.25 | 0.31 | $Zn_{0.5}Cd_{0.5}Se$ | 2.33 | 0.43 |
| $AgGaTe_2$ | 1.316 | 1.425 | 2.26[b] | 0.165 | 0.84[b] | $Zn_{0.5}Cd_{0.5}Te$ | 2.18 | 0.93 |
| $AgInS_2$ | 1.87 | 2.02 | 2.02 | 0.15 | 0 | CdS | 2.53 | 0.07 |
| $AgInSe_2$ | 1.24 | 1.33 | 1.60 | 0.12 | 0.30 | CdSe | 1.85 | 0.43 |
| $AgInTe_2$ | 1.04[b] | — | — | — | — | CdTe | 1.80 | 0.92 |

[a] Units: electron volts.
[b] At 77 K.

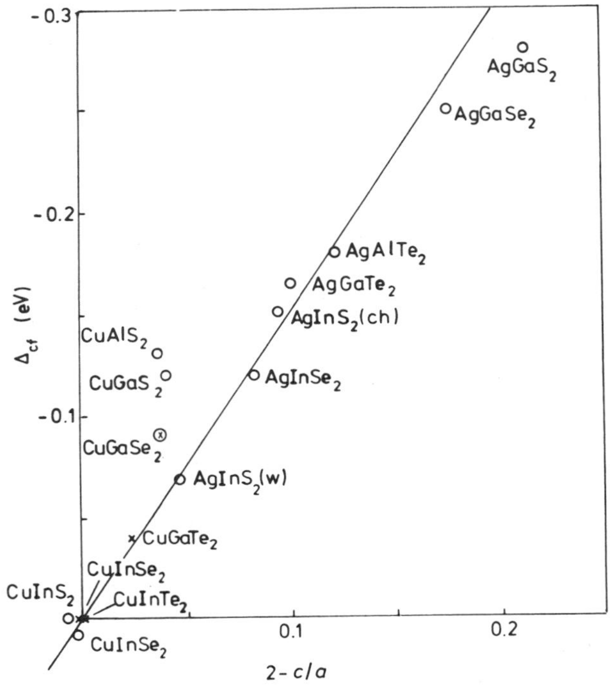


FIG. 19. The crystal field splitting parameter plotted against the tetragonal compression for several I–III–$VI_2$ compounds (Ref. 2 and 125).

functions for these states will accompany the change in position of the bands, and an estimate of the amount of hybridization can be obtained either from the downshift of the energy gap relative to the cubic binary analog or from the observed spin–orbit splittings of p-like binary analog and atomic d-levels.[128] A plot of these two quantities gives the straight line shown in Fig. 21 for I–III–$VI_2$ compounds.

Because of the difficulty in dealing with the d-level mixing in I–III–$VI_2$ compounds, no satisfactory band structure calculation exists for these materials. The only calculations for $CuGaS_2$, $CuGaSe_2$, $CuGaTe_2$,[129] $CuInS_2$, $CuInSe_2$, and $CuInTe_2$[130] ignored the d-levels and were unable to predict the observed valence band splittings.

[128] J. L. Shay, B. Tell, H. M. Kasper, and L. M. Schiavane, *Phys. Rev. B: Solid State* [3] **5**, 5003 (1972).
[129] A. S. Poplavnoi and Yu. I. Polygalov, *Inorg. Mater.* **7**, 1527 (1971).
[130] A. S. Poplavnoi and Yu. I. Polygalov, *Inorg. Mater.* **7**, 1531 (1971).

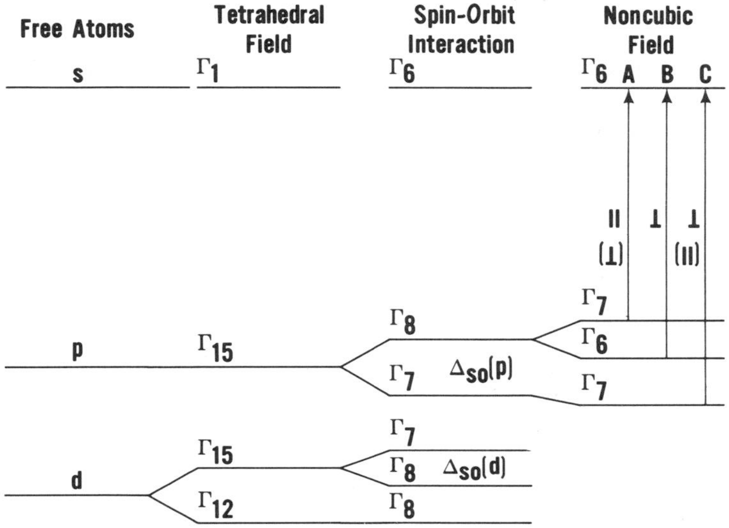


FIG. 20. Zone-center band splitting of I–III–$VI_2$ compounds including the d-levels introducted by the Group I atom (Ref. 108).

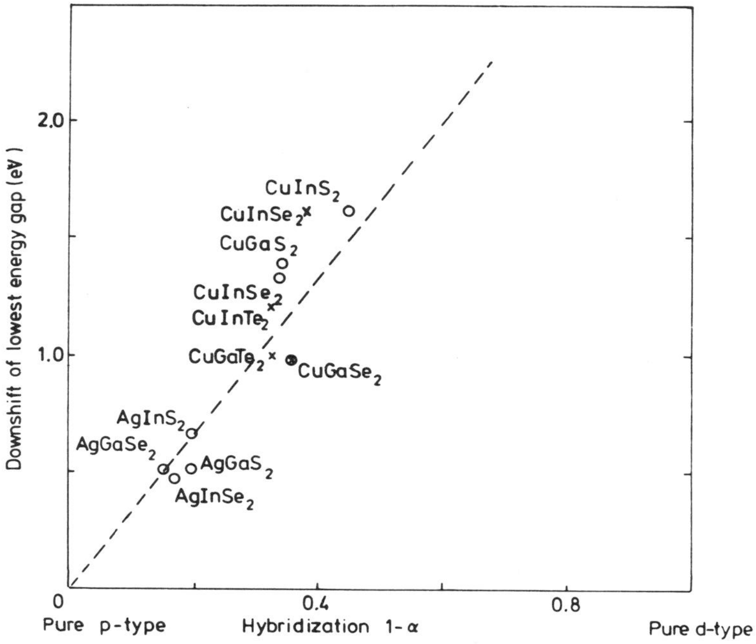


FIG. 21. Hybridization against downshift of energy of the lowest energy gap of I–III–$VI_2$ compounds (Ref. 2 and 125).

Electroreflectance measurements have shown evidence of transitions between the d-states of the copper atoms and the conduction band in $CuInS_2$,[131] $CuInSe_2$,[126] $CuGaS_2$,[131] and $CuGaSe_2$.[132] x-Ray and UV photoemission studies have given the energies of the noble metal atomic d-levels below the valence band edge in $AgGaS_2$,[133,134] $CuAlS_2$,[134,135] $CuGaS_2$,[134] $CuInS_2$,[134,136] and $AgInTe_2$.[136]

## 10. Ordered Vacancy Compounds

Only a few of the ordered vacancy semiconductors have been studied so far for lowest band gap transitions, despite the wealth of interesting properties that must exist for these materials. In Section 5, we argued that the crystal field splitting of $I\bar{4}2m$, $I\bar{4}$, and $P\bar{4}2m$ vacancy structures should be opposite to that normally found for chalcopyrite. This has indeed been found to be the case for $HgIn_2\square Te_4$ ($I\bar{4}2m$) and $HgGa_2\square Se_4$ ($I\bar{4}$). Optical measurements using oriented crystals and polarized light have yet to be performed on most of the family of $I\text{–}III_2\text{–}\square\text{–}VI_4$ semiconductors however, so the nature of the energy band structures in the vicinity of the band gap have not been established at present. In this section, we review those materials in which some of the electronic interband transition properties are known.

### a. $HgIn_2\square Te_4$

Unpolarized optical absorption measurements determined a band gap of 0.9 eV for $HgIn_2\square Te_4$ at room temperature.[137–139] Subsequent polarized absorption and electroreflectance measurements have determined a direct band gap at 0.96 eV with polarization $E \perp c$ and a weaker structure in both polarizations at 1.09 eV.[38] This is the opposite polarization dependence to chalcopyrite structured semiconductors and was interpreted as being due to a positive crystal field splitting of the valence bands. This is consistent

[131] B. Tell, J. L. Shay, and H. M. Kasper, *Phys. Rev. B: Solid State* [3] **4,** 2463 (1971).
[132] M. J. Thwaites, R. D. Tomlinson, and M. J. Hampshire, *Solid State Commun.* **27,** 727 (1978).
[133] M. J. Luciano and C. J. Vesely, *Appl. Phys. Lett.* **23,** 60 (1973).
[134] S. Kano and M. Okusawa, *J. Phys. Soc. Jpn.* **37,** 1301 (1974).
[135] M. J. Luciano, and C. J. Vesely, *Appl. Phys. Lett.* **23,** 453 (1973).
[136] W. Braun, A. Goldmann, and M. Cardona, *Phys. Rev. B: Solid State* [3] **10,** 5069 (1974).
[137] P. M. Spencer, B. R. Pamplin, and D. A. Wright, *Proc. Int. Conf. Phys. Semicond., (Exeter) Inst. Phys., 1962* p. 244.
[138] P. M. Spencer, *Br. J. Appl. Phys.* **15,** 625 (1964).
[139] S. L. Dahake, *Br. J. Appl. Phys.* **18,** 1340 (1967).

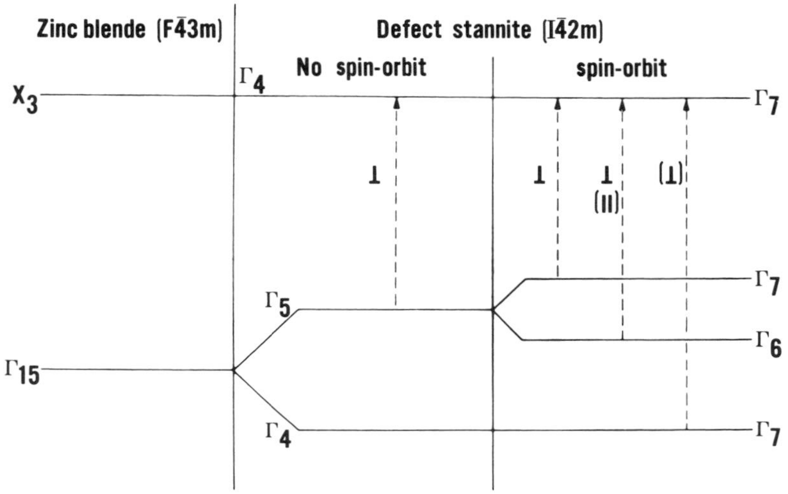


FIG. 22. Zone-center band structure of $HgIn_2\square Te_4$ (Ref. 38).

with the zone-center band structure shown in Fig. 22 where, like chalcopyrite pseudodirect compounds, the lowest conduction band derives from $X_3$ ($X_1$) and becomes $\Gamma_4$ in $I\bar{4}2m$ symmetry. However in this case the transitions are not pseudodirect (Section 5). We also note that, in contrast to chalcopyrite compounds, the absorption edge of $HgIn_2\square Te_4$ is anisotropic, even though zero tetragonal compression is reported for this material. This is a consequence of the first-order mixing of $\Gamma$ and $X$-derived symmetries in this structure.

*b. $HgGa_2\square Se_4$*

Positive crystal field splitting has been indicated by photoconductivity measurements of $HgGa_2\square Se_4$ crystals by Lebedev *et al.*[140] The maximum response was obtained at 1.97 and 2.02 eV for $E \perp c$ and $E \parallel c$, respectively, and a similar shift was observed in weaker structure in the region of 1.6 eV. Beun *et al.*[141] had previously reported unpolarized absorption measurements giving a band gap of either 1.79 or 1.95 eV depending on whether the data were interpreted as being due to indirect or direct transitions.

[140] A. A. Lebedev, P. N. Metlinskii, Yu. V. Rud, and V. G. Tyrziu, *Sov. Phys.—Semicond.* (*Engl. Transl.*) **11,** 615 (1977).

[141] J. A. Beun, R. Nitsche, and M. Lichtensteiger, *Physica* (*Amsterdam*) **27,** 448 (1961).

*c. $CdGa_2\square S_4$ and $CdGa_2\square Se_4$*

Several optical investigations of band gap transitions in $CdGa_2\square S_4$[141–153] and $CdGa_2\square Se_4$[141–148,154–159] have been reported because of promising photoconductor applications. Both crystals have space group $I\bar{4}$.

Early measurements on $CdGa_2\square S_4$ gave band gap energies of 3.40 (indirect) and 3.44 eV (direct) from absorption studies,[141] 3.58 eV from reflection spectra,[142] and 3.54 eV from the photoconductivity response.[153]

For $CdGa_2\square Se_4$, absorption measurements gave 1.9,[154] 2.25,[141] and 2.27 eV[155] for indirect transitions, and 2.1,[154] 2.41,[155] and 2.43 eV[141] for direct transitions. Reflection studies gave structure at 2.55 eV[142] and photoconductivity indicated a band gap of 2.48,[156,157] 2.20 (indirect),[158] and 2.50 eV (direct).[158]

Considerable variations are thus apparent in these reported values for the lowest optical transitions, and comparative studies have been carried out more recently on these two materials in an attempt to resolve these

[142] G. B. Abdullaev, D. A. Guseinova, T. G. Kirimova, and R. Kh. Nani, *Sov. Phys.—Semicond.* (*Engl. Transl.*) **7,** 575 (1973).

[143] S. T. Kshirsagar and A. P. B. Sinha, *J. Mater. Sci.* **12,** 1614 (1977).

[144] S. T. Kshirsagar, *Thin Solid Films* **45,** L5 (1977).

[145] S. I. Radautsan, N. N. Syrbu, I. I. Nebola, V. G. Tyrziu, and D. M. Bercha, *Sov. Phys.—Semicond.* (*Engl. Transl.*) **11,** 38 (1977).

[146] D. A. Guseinova, T. G. Kerimova, and R. Kh. Nani, *Sov. Phys.—Semicond.* (*Engl. Transl.*) **11,** 670 (1977).

[147] P. Kivits, *J. Phys. C* **9,** 605 (1976).

[148] P. Kivits, J. Reulen, J. Hendricx, F. Van Empel, and J. Van Kleef, *J. Lumin.* **16,** 145 (1978).

[149] P. Kivits, M. Wignakker, J. Claasen, and J. Geerts, *J. Phys. C* **11,** 2361 (1978).

[150] M. Springford, *Proc. Phys. Soc., London* **82,** 1029 (1963).

[151] L. Krausbauer, R. Nitsche, and P. Wild, *Proc. Int. Conf. Lumin., 1966* p. 1107 (1968).

[152] A. G. Karipides, A. V. Cafiero, *Inorg. Synth.* **11,** 5 (1968).

[153] S. I. Radautsan, V. F. Zhitar', and V. S. Donu, *Sov. Phys.—Semicond.* (*Engl. Transl.*) **9,** 670 (1975)

[154] L. N. Strel'tsov, Y. Ya. Chernykh, and V. M. Petrov. *Sov. Phys.—Semicond.* (*Engl. Transl.*) **1,** 656 (1967).

[155] G. B. Abdullaev, V. G. Agaev, V. B. Antonov, R. Kh. Nani, and E. Yu Salaev, *Sov. Phys.—Semicond.* (*Engl. Transl.*) **5,** 1854 (1971).

[156] G. B. Abdullaev, V. G. Agaev, V. B. Antonov, R. Kh. Nani, and E. Yu. Salaev, *Sov. Phys.—Semicond.* (*Engl. Transl.*) **6,** 1492 (1972).

[157] G. B. Abdullaev, V. G. Agaev, V. B. Antonov, A. A. Mamedov, R. Kh. Nani, and E. Yu. Salaev, *Sov. Phys.—Semicond.* (*Engl. Transl.*) **7,** 717 (1973).

[158] S. I. Radautsan, *Sov. Phys.—Semicond.* (*Engl. Transl.*) **5,** 1959 (1971).

[159] S. I. Radautsan, N. N. Syrbu, V. E. Tezlevan, N. P. Baran, V. D. Vlas, and V. A. Titov, "Properties of Some Semiconductor Materials and Devices" (in Russian). Shtiintsa, Kishinev, USSR. 1974.

difficulties. Kshirsagar and Sinha[143] have studied absorption and photoconductivity in a range of alloys with compositions between $CdGa_2\square S_4$ and $CdGa_2\square Se_4$. Results for both parent compounds indicated direct band gaps with no evidence of indirect transitions from absorption measurements: 3.25 eV for the sulfide and 2.25 eV for the selenide. Photoconductivity gave a peak at 3.35 eV apparently owing to direct transitions in $CdGa_2\square S_4$ and at 2.25 eV in $CdGa_2\square Se_4$. The latter material also showed evidence of electronic transitions from lower valence bands to the conduction band by additional photoconductive maxima at 2.92 and 3.3 eV. Absorption in thin films of these compounds by Kshirsagar[144] indicated direct band gaps at 3.5 and 2.51 eV in the sulfide and selenide, respectively.

The conclusions of Refs. 143 and 144 that the lowest transitions are direct for $CdGa_2\square S_4$ are in disagreement with polarized absorption measurements of Radautsan *et al.*,[145] in which fine structure observed just below the fundamental absorption edge is most likely due to phonon assisted indirect transitions. This interpretation gives the forbidden gap as 3.05 ($E \parallel c$) and 3.07 eV ($E \perp c$). The minimum direct gap was observed to be 3.25 ($E \parallel c$) and 3.27 eV ($E \perp c$). The 20 meV valence band splitting deduced from these results is small compared to that observed in most chalcopyrites and $HgIn_2\square Te_4$, and of opposite sign to that predicted in Section 5. On the other hand, the polarized reflection spectra measured by Radautsan *et al.*[145] shows structure at 3.2 eV in $E \perp c$, but no structure below 3.6 eV for $E \parallel c$, which is consistent with positive crystal field splitting of the valence bands.

A comparison of unpolarized absorption in the two compounds by Guseinova *et al.*[146] shows that the slope of the absorption edge of $CdGa_2\square S_4$ is much shallower than that of $CdGa_2\square Se_4$. This suggests that the lowest transitions for the selenide are direct at 2.32 eV. These authors made a group theoretical comparison of the energy bands in the sulfide with its approximate isoelectronic analogs $ZnGeP_2$ and GaP and concluded that the lowest conduction band of the sulfide would derive from the $X_3$ ($X_1$) zone boundary gap of GaP (Fig. 14). However, the ordered vacancy of the defect chalcopyrite structure should cause considerable differences in the band structures of these materials, and the pseudodirect transitions of chalcopyrite no longer exist for $I\bar{4}$ symmetry. It is thus most likely that $CdGa_2\square S_4$ is an indirect semiconductor with a forbidden gap in the range 3.05–3.4 eV.

In spite of the large number of publications cited here, it is still not clear what the lowest interband transitions and valence band splittings are for $CdGa_2\square S_4$ and $CdGa_2\square Se_4$. Inconsistency in the measured values of the band gap energy may be due to nonstoichiometric composition of crystals prepared by the chemical transport reaction method as pointed out by Strel'tsov *et al.*[154] Polarized absorption measurements combined with modulation spectroscopy are required for a complete interpretation of the energy band structure in the vicinity of the band gap.

*d. $CdIn_2\square Se_4$ and $CdIn_2\square Te_4$*

$CdIn_2\square Se_4$[141,160–170] is a fascinating example of a $II-III_2-\square-VI_4$ semiconductor because of its unique pseudocubic $P\bar{4}2m$ structure. Koval *et al.*[170] reported that this compound can be grown in both $\alpha(P\bar{4}2m)$ and $\beta(I\bar{4})$ modifications, and studied the absorption edge for each structure type. (The existence of the $\beta$ form has not been confirmed by other authors.) The results indicated indirect gaps at 1.51 ($\alpha$) and 1.30 eV ($\beta$) and direct transitions at 1.73 ($\alpha$) and 1.49 eV ($\beta$). The reflectance spectra above the direct energy gaps were the same for both structures.

Photoconductivity measurements in $\alpha$-$CdIn_2\square Se_4$ have given three features that were believed to be caused by valence band to donor impurities at 0.94 eV,[163] indirect transitions at 1.55 eV,[163,164] and direct transitions at 1.73 eV.[163,164] Absorption measurements gave the indirect gap at 1.48 eV and a direct gap at 1.64 eV, whereas thermoreflectance gave the direct gap as 1.65 eV.[163]

Pseudopotential calculations by Baldereschi *et al.*[165–167] predict an indirect band gap for $CdIn_2\square Se_4$ of unusual nature. The calculated conduction band minimum is at the zone center, but the valence band maximum is at the $R$ point of the Brillouin zone (Fig. 7). It is pointed out that this may be somewhat unreliable because of the flatness of the topmost valence band, however, it is consistent with experimental results. A prediction of the calculation is a 0.5 eV gap in the density of states just below the top of the valence band not present in zinc blende or chalcopyrite semiconductors as a consequence of the vacant tetrahedral

[160] B. T. Kolomiets and A. A. Malkora, *Fiz. Tverd. Tela, Sb.* **2,** 32 (1959).

[161] H. Suzuki and S. Mori, *J. Phys. Soc. Jpn.* **19,** 1082 (1964).

[162] O. P. Derid, L. S. Koval, and M. M. Marcus, "Sloznye Poluprovodiniki i ich fizičeskie Svoistva," p. 140. Shtiitsa, Kishinev, USSR, 1971.

[163] R. Trykozko, *Conf. Ser.—Inst. Phys.* **35,** 249 (1977).

[164] E. Fortin and F. Raga, *Solid State Commun.* **14,** 847 (1974).

[165] A. Baldereschi, F. Meloni, F. Aymerich, and G. Mula, *Solid State Commun.* **21,** 113 (1977).

[166] F. Meloni, F. Aymerich, G. Mula, and Baldereschii, *Helv. Phys. Acta* **49,** 687 (1976).

[167] F. Aymerich, A. Baldereschi, F. Meloni, and G. Mula, *Phys. Semicond., Proc. Int. Conf., 13th,* 1976 p. 154 (1976).

[168] P. Picco, I. Abbati, L. Braicovich, F. Cerrina, F. Levy, and G. Margaritondo, *Phys. Lett. A* **65,** 447 (1978).

[169] C. Quaresima, F. Cerrina, *Conf. Ser.—Inst. Phys.* **43,** 1199 (1978).

[170] L. S. Koval, E. K. Arushanov, and S. I. Radautsan, *Phys. Status Solidi* **A 9,** K.73 (1972).

site. This has not been confirmed by UV photoemission results,[168,169] perhaps because the anion displacement was not accounted for in the calculation.[168] Agreement between calculations and photoemission experiments does exist on the larger width of the valence band density of states than for equivalent binaries, believed to be a consequence of the presence of the group III cations whose s-electrons are rather deep in energy.[169]

Thermoelectric power measurements of the alloy range $CdIn_2\square Se_4$–$CdIn_2\square Te_4$ give a large electron effective mass of approximately 0.15 almost independent of composition. The conduction band states are thus likely to be similar in these two compounds.[170] Pseudopotential calculations also result in similar band structures for both materials.[166]

*e. $ZnIn_2\square Se_4$ and $ZnIn_2\square Te_4$*

There has been some doubt as to the energy band gap in $ZnIn_2\square Se_4$, $I\bar{4}$.[141,151,164,166,169,171–174] Busch *et al.*[171] measured an activation energy of 2.6 eV, whereas Beun *et al.*[141] gave the absorption edge at 1.8 eV. Photoconductivity and reflectivity structure at 2.0 eV[172,173] showed that this compound probably has a direct gap. Analysis of intensity-dependent photoconductive measurements in polycrystalline material indicates that a quasi-continuous distribution of traps gives rise to the main observed response. In single-crystals, intrinsic photoconductivity was found to be dominant, peaking near 2.1 eV.[164]

Luminescence was reported in $ZnIn_2\square Se_4$ at 1.29 eV at 77 K by Krausbauer *et al.*,[151] and a more detailed study was carried out by Grilli *et al.*[174] by luminescence, excitation, infrared stimulation, and luminescence decay. This led to a proposed energy scheme involving two types of recombination center and trap levels. These authors suggest that the band gap is probably at 2.25–2.35 eV and that this type of luminescence is characteristic of $II–III_2–\square–VI_4$ compounds.

UV photoemission measurements on $ZnIn_2\square Se_4$[169] give results very similar to $CdIn_2\square Se_4$. Comparison with results for $ZnIn_2\square S_4$ showed the effect of the octohedral coordinated cation sites in the sulfide.

The photoconductive response of $ZnIn_2\square Te_4$, $I\bar{4}$, has given evidence of the lowest interband transitions in this material.[172,173,175] Boltovets *et al.*[175] found the maximum photoconductive response to be at 1.3 eV. Manca *et al.*[172] deduced that $ZnIn_2\square Te_4$ has an indirect band gap at 1.4 eV from photoconductivity, absorption, and reflectivity results. Later work by the

[171] G. A. Busch, E. Mooser, and W. B. Person, *Helv. Phys. Acta* **29**, 192 (1956).
[172] P. Manca. F. Raga, and A. Spiga, *Phys. Status Solidi* **A 16**, K105 (1973).
[173] P. Manca, F. Raga, and A. Spiga, *Nuovo Cimento Soc. Ital. Fis. B* **19B**, 15 (1974).
[174] E. Grilli, M. Guzzi, and R. Molteni, *Phys. Status Solidi* **A 37**, 399 (1976).
[175] N. S. Boltovets, V. P. Drobyazko, and V. K. Mityurev, *Sov. Phys.—Semicond. (Engl. Transl.)* **2**, 867 (1969).

same authors[173] give the lowest direct transition at 1.87 eV from reflectivity measurements. A careful analysis of the intensity-dependent photoconductivity[173] led to similar conclusions as for $ZnIn_2\square Se_4$ regarding the role of transitions from traps as the principal contribution to the response.

Pseudopotential calculations of the energy band structures of $ZnIn_2\square Se_4$ and $ZnIn_2\square Te_4$ by Meloni *et al.*[166] have given results very similar to $CdIn_2\square Se_4$, but not details were reported.

### *f. $HgGa_2\square S_4$, $ZnGa_2\square S_4$, and $ZnGa_2\square Se_4$*

Beun *et al.*[141] made unpolarized absorption edge measurements of the fundamental absorption edge of $HgGa_2\square S_4$. Band gap energies were either 2.79 or 2.84 eV depending on whether the data were interpreted as being due to indirect or direct transitions. The maximum photoconductive response was found at 2.53 eV.

These authors also investigated $ZnGa_2\square S_4$ and $ZnGa_2\square Se_4$ by photoconductive techniques. The maximum responses were found to be at 3.18 eV for the sulfide and 2.17 eV for the selenide.

## VI. Appendix: Character Tables

In the character tables (Tables A.1–A.3) the following symbols have been used:

| | |
|---|---|
| $x, y, z$ | Displacement, and so on, in $x$, $y$, $z$ direction |
| $R_\alpha$ | Rotation about the axis $\alpha$ |
| $E$ | Identity |
| $C_{2\alpha}$ | Rotation of 180° about the axis $\alpha$ |
| $C_3$ | Rotation of 120° |
| $S_{4\alpha}$ | Improper rotation of 90° about the axis $\alpha$ |
| $\sigma_{\alpha\beta}$ | Reflection in the plane $\alpha$–$\beta$ |

TABLE A. 1. CHARACTER TABLE FOR $\bar{4}$; $S_4$ (e.g., DEFECT CHALCOPYRITE)[a]

| | | $E$ | $S_{4z}^{-1}$ | $C_{2z}$ | $S_{4z}$ | |
|---|---|---|---|---|---|---|
| $W_1$ | $A$ | 1 | 1 | 1 | 1 | $x^2 + y^2; x^2; R_z$ |
| $W_2$ | $B$ | 1 | $-1$ | 1 | $-1$ | $z; x^2 - y^2; xy$ |
| $W_3$ | $E$ | 1 | $-i$ | $-1$ | $i$ | $x + iy; xz - iyz; R_x - iR_y$ |
| $W_4$ | $E$ | 1 | $i$ | $-1$ | $-1$ | $x - iy; xz + iyz; R_x + iR_y$ |

[a] The IRs $W_3$ and $W_4$ ($E$) are degenerate in the absence of time reversal symmetry.

TABLE A.2. CHARACTER TABLE FOR $\bar{4}2m$; $D_{4h}$ (e.g., STANNITE)

| | | $E$ | $2D_{4z}$ | $C_{2z}$ | $2C_{2x}$ | $2\sigma_{xy}$ | |
|---|---|---|---|---|---|---|---|
| $X_1$ | $A_1$ | 1 | 1 | 1 | 1 | 1 | $x^2 + y^2$; $z^2$ |
| $X_4$ | $A_2$ | 1 | 1 | 1 | $-1$ | $-1$ | $R_z$ |
| $X_2$ | $B_1$ | 1 | $-1$ | 1 | 1 | $-1$ | $x^2 - y^2$ |
| $X_3$ | $B_2$ | 1 | $-1$ | 1 | $-1$ | 1 | $z$; $xy$ |
| $X_5$ | $E$ | 2 | 0 | $-2$ | 0 | 0 | $(x, y)$; $(xz, yz)$; $(R_x, R_y)$ |

TABLE A.3. CHARACTER TABLE FOR $\bar{4}3m$; $T_d$ (e.g., ZINC BLENDE)

| | | $E$ | $8C_3$ | $3C_{2x}$ | $6S_{4x}$ | $6\sigma_{xy}$ | |
|---|---|---|---|---|---|---|---|
| $\Gamma_1$ | $A_1$ | 1 | 1 | 1 | 1 | 1 | $x^2 + y^2 + z^2$ |
| $\Gamma_2$ | $A_2$ | 1 | 1 | 1 | $-1$ | $-1$ | $R_xR_yR_z$ |
| $\Gamma_{12}$ | $E$ | 2 | $-1$ | 2 | 0 | 0 | $(x^2 - y^2, x^2 + y^2 - 2z^2)$ |
| $\Gamma_{25}$ | $T_1$ | 3 | 0 | $-1$ | 1 | $-1$ | $(R_x, R_y, R_z)$ |
| $\Gamma_{15}$ | $T_2$ | 3 | 0 | $-1$ | $-1$ | 1 | $(x, y, z)$; $(xy, yz, xz)$ |

SOLID STATE PHYSICS, VOLUME 36

# Correlations in Electron Liquids*

K. S. SINGWI

*Department of Physics and Astronomy, Northwestern University, Evanston, Illinois*

AND

M. P. TOSI

*International Center for Theoretical Physics, Trieste, Italy*



## I. Introduction

A highly degenerate fluid of electrons exists in metals. The behavior of these electrons is very complicated, since they not only interact among themselves, but also with the ionic lattice and the lattice vibrations. As a first approximation, one considers the positive ions to be at rest and smeared out so that the electrons move against a rigid uniform positive background. Such a model of a metal is referred to as the jellium model

* Based on work supported in part under the NSF-MRL program through the Materials Research Center of Northwestern University, Evanston, Illinois, and in part under the NSF Grant No. DMR77-09937.




ISBN 0-12-607736-3

and has long been of interest for the evaluation of correlations between electrons in delocalized states.

At zero temperature, the properties of this system of electrons depend only on its density $n$, which is conveniently expressed in terms of a dimensionless parameter $r_s$ through the relation

$$n = \left(\frac{4\pi}{3} r_s^3 a_B^3\right)^{-1},$$

where $a_B$ is the Bohr radius. For metallic plasmas $n \sim 10^{23}\ cm^{-3}$ and $r_s$ lies between 2 (for Al) and 6 (for Cs). In terms of $r_s$, the ground-state energy per electron is[1–4]

$$\varepsilon = \left(\frac{2.21}{r_s^2} - \frac{0.916}{r_s} + \varepsilon_c\right) \text{ Ry},$$

where the first term is the kinetic energy, and the second is the exchange energy arising as a result of the Pauli principle. The last term is what is called the correlation energy; actually it is the difference between the true energy $\varepsilon$ and the first two terms of the above equation. The correlation energy can be calculated exactly only in the two extreme limits: one[5] of very high density, i.e., $r_s \to 0$ and the other[6] of very low density, i.e., $r_s \to \infty$. In the former limit it is the kinetic energy that dominates, and the system behaves like a gas of charged fermions; whereas in the latter limit the Coulomb forces dominate the motion, and the ground state is expected to be a zero-point lattice.[6] In the intermediate region of $r_s$ which is really of interest in metal physics, an exact calculation of $\varepsilon_c(r_s)$ has not been possible, and one must resort to approximate schemes, which we shall discuss later. Wigner has given the following simple formula for $\varepsilon_c(r_s)$ which turns out to be reasonably good for all $r_s$ values:

$$\varepsilon_c = -0.88/(r_s + 7.8) \text{ Ry}.$$

[1] D. Pines, "Elementary Excitations in Solids." Benjamin, New York, 1963.

[2] D. Pines and P. Nozières, "The Theory of Quantum Fluids." Benjamin, New York, 1966.

[3] A. L. Fetter and J. D. Walecka, "Quantum Theory of Many-Particle Systems." McGraw-Hill, New York, 1971; W. Jones and N. H. March, "Theoretical Solid State Physics," Vol. I. Wiley, New York, 1973. G. D. Mahan, "Many-Particle Physics." Plenum, New York and London, 1981. This book contains an excellent discussion of the electron liquid problem (Chapter 5).

[4] L. Hedin and S. Lundqvist, *Solid State Phys.* **23,** 1 (1969).

[5] M. Gell-Mann and K. A. Brueckner, *Phys. Rev.* [2] **106,** 364 (1957); see also Refs. 1 and 2.

[6] E. P. Wigner, *Phys. Rev.* [2] **46,** 1002 (1934); *Trans. Faraday Soc.* **34,** 678 (1938). Crystallization has been unambiguously observed in computer simulations of the classical plasma in the appropriate *high* density limit [S. G. Brush, H. L. Sahlin, and E. Teller, *J. Chem. Phys.* **45,** 2102 (1966); J. P. Hansen, *Phys. Rev. A* [3] **8,** 3096 (1973)]. For a recent review on Wigner lattice, see C. M. Care and N. H. March, *Adv. Phys.* **24,** 101 (1975).

The main theme of this article is restricted to approximate theories for the fluid in the intermediate, metallic density range.

In developing the formalism, we have intentionally adopted the equation-of-motion approach for the density fluctuation $n(\mathbf{r}, t)$ in the presence of a weak external perturbation. Although the formalism is fully quantum mechanical, it has the "flavor" of a classical fluid theory and thus has the advantage of familiarity and appeals very easily to one's intuition. Besides, our approach may hopefully bridge the gap that exists between more familiar methods and the Feynman diagram "technology." We have laid special emphasis on the frequency- and wave number-dependent dielectric function, which offers a simple, compact way to discuss a number of properties of the electron gas in metals, e.g., the density-fluctuation excitation spectrum $S(q, \omega)$ and the ground-state energy of the system. To illustrate the use and the power of the method, we have also discussed some of its applications.

## II. Homogeneous Electron Liquid

### 1. Density Response and Dielectric Function

It is both useful and instructive to open the discussion by reviewing the characteristic collective behavior of the charged fluid in long wavelength phenomena ($kr_s a_B \ll 1$), which arises from the long-range nature of the Coulomb interaction and manifests itself in the twin properties of perfect screening in static phenomena and plasmon resonance in dynamic phenomena. Let us consider first the static perturbation produced by a distribution $n_e(\mathbf{r})$ of external charge, which generates a redistribution $\delta n(\mathbf{r})$ of electronic density and an internal electric field $\mathbf{E}(\mathbf{r})$ given by Poisson's equation,

$$\nabla \cdot \mathbf{E}(\mathbf{r}) = 4\pi e[n_e(\mathbf{r}) + \delta n(\mathbf{r})]. \tag{1.1}$$

The inhomogeneity of the electron distribution causes a gradient of chemical potential $\nabla\mu(\mathbf{r})$, and current does not flow when the internal forces balance each other, that is,

$$e\mathbf{E}(\mathbf{r}) - \nabla\mu(\mathbf{r}) = 0. \tag{1.2}$$

Using for slow variations in space the thermodynamic relation $\delta\mu(\mathbf{r}) = (\partial\mu/\partial n)\, \delta n(\mathbf{r})$, we find the displaced electronic density in Fourier transform

$$n(\mathbf{k}) = -n_e(\mathbf{k})(1 + k^2/k_s^2)^{-1} \qquad (kr_s a_B \ll 1) \tag{1.3}$$

and the internal (Hartree) potential

$$V_H(\mathbf{k}) = \frac{4\pi e^2}{k^2}[n_e(\mathbf{k}) + n(\mathbf{k})] = \frac{4\pi e^2}{k_s^2 + k^2} n_e(k) \qquad (k r_s a_B \ll 1), \quad (1.4)$$

where

$$k_s^2 = 4\pi e^2/(\partial\mu/\partial n). \quad (1.5)$$

Thus, the total displaced charge exactly balances the total external charge, and $V_H$ is a screened potential tending to a constant value for $k \to 0$. Furthermore, on the assumption that the background has no stiffness, the expression (1.5) for the inverse screening length $k_s$ can be related to the compressibility $K$ of the system,

$$k_s^2 = 4\pi n^2 e^2 K. \quad (1.6)$$

The Thomas–Fermi theory of the electron gas assumes (1.4) to be valid for all values of $k$ and evaluates $k_s$ from (1.6) using the compressibility of the ideal Fermi gas. It is worth recalling that $V_H(\mathbf{r})$ is then incorrect even asymptotically for the quantal plasma, since it omits the long-range Friedel oscillations[7] which arise from the discontinuity in the momentum distribution at the Fermi surface.

The parallel phenomenon of dynamic screening of a time-dependent external charge is simply illustrated in the long-wavelength limit by considering the equation of motion for the longitudinal particle current density $\mathbf{j}(\mathbf{r}, t)$,

$$\frac{m}{n}\frac{\partial \mathbf{j}(\mathbf{r}, t)}{\partial t} = e\mathbf{E}(\mathbf{r}, t). \quad (1.7)$$

The current is undamped for $k \to 0$ since total momentum is conserved in electron–electron collisions. Use of the continuity equation and Poisson's equation yields the displaced electron density in Fourier transform,

$$n(\omega) = n_e(\omega)\omega_p^2(\omega^2 - \omega_p^2)^{-1}, \quad (1.8)$$

where $\omega_p = (4\pi n e^2/m)^{1/2}$ is the plasma frequency. Evidently, dynamic screening exhibits a resonance at this frequency.

The notion of screening can be extended to arbitrary values of $k$ and $\omega$ by introducing the dielectric function $\varepsilon(k, \omega)$ through

$$1/\varepsilon(\hat{k}, \omega) = \hat{k} \cdot \mathbf{E}(\mathbf{k}, \omega)/\hat{k} \cdot \mathbf{D}(\mathbf{k}, \omega) = 1 + n(\mathbf{k}, \omega)/n_e(\mathbf{k}, \omega). \quad (1.9)$$

The evaluation of the dielectric function of the electron fluid thus involves the study of the density response of the system to an external Coulomb

[7] J. Friedel, *Nuovo Cimento, Suppl.* **7**, 287 (1958).

potential. The static response can be determined by a suitable equilibrium condition, and the generalization of (1.2) forms the basis of the Hohenberg–Kohn–Sham (HKS) theory.[8] The generalization of (1.7) provides, instead, a tool for the study of the dynamic behavior of the fluid and of interparticle correlations arising from its dynamics.

*a. Density Response*

We, therefore, review the main results of linear response theory in the specific example of the density response of a fluid. A fuller discussion can be found, for instance, in the books by Pines and Nozières[2] and by Martin,[9] and in the article by Kubo.[10]

Consider the response of the fluid to an external potential $V_e(\mathbf{r}, t)$ due to a probe of density $n_e(\mathbf{r}, t)$ which is coupled to the density fluctuations in the fluid. In a linear response regime, the induced density change is related to $V_e$ by

$$\delta n(\mathbf{r}, t) = \int\int d\mathbf{r}'\, dt' K(|\mathbf{r} - \mathbf{r}'|, t - t') V_e(\mathbf{r}', t'), \qquad (1.10)$$

where the response function $K$ is determined by the state of the fluid in the absence of $V_e$ and thus reflects its space–time invariance properties. We can then write

$$n(\mathbf{k}, \omega) = \chi(k, \omega) V_e(\mathbf{k}, \omega), \qquad (1.11)$$

where $\chi(k, \omega)$ is the Fourier transform of $K(|\mathbf{r} - \mathbf{r}'|, t - t')$. The physical requirements that the response be causal and finite are satisfied if $K$ vanishes for $t < t'$ and is finite for $t > t'$. These properties ensure that $\chi(k, \omega)$ has no singularities in the upper half of the complex $\omega$ plane and obeys Kramers–Kronig relations.

Explicit calculation by time-dependent perturbation theory yields the expression

$$K(r, t) = (-i/\hbar)\theta(t) \langle [\rho(\mathbf{0}, 0), \rho(\mathbf{r}, t)] \rangle \qquad (1.12)$$

where $\rho(\mathbf{r}, t)$ is the particle density operator, $\theta(t)$ is the Heaviside step function accounting for causality, and the brackets denote the average over the equilibrium ensemble. The response function is thus seen to be intimately related to the time-dependent pair correlation function

$$G(r, t) = N^{-1} \int d\mathbf{r}' \, \langle \rho(\mathbf{r}', 0) \rho(\mathbf{r}' + \mathbf{r}, t) \rangle \qquad (1.13)$$

[8] P. Hohenberg and W. Kohn, *Phys. Rev.* [2] **136,** B864 (1964); W. Kohn and L. J. Sham, *ibid.* **140,** A1133 (1965).

[9] P. C. Martin, *in* "Many-Body Physics" (C. deWitt and R. Balian, eds.), p. 37. Gordon & Breach, New York, 1968.

[10] R. Kubo, *Rep. Prog. Phys.* **29,** 255 (1966); see also D. Forster, "Hydrodynamic Fluctuations, Broken Symmetry and Correlation Functions." Benjamin, New York, 1975.

introduced by van Hove[11] in connection with the theory of neutron inelastic scattering from liquids. Precisely,

$$\chi(k, \omega) = (n/2\pi\hbar) \int_{-\infty}^{\infty} d\omega' [S(\mathbf{k}, \omega') - S(-\mathbf{k}, -\omega')](\omega - \omega' + i\eta)^{-1} \tag{1.14}$$

where $\eta$ is a positive infinitesimal accounting for causality and $S(\mathbf{k}, \omega)$ is the so-called dynamic structure factor,

$$S(\mathbf{k}, \omega) = \int\int d\mathbf{r}\, dt\, G(r, t) \exp[-i(\mathbf{k} \cdot \mathbf{r} - \omega t)]. \tag{1.15}$$

As shown by van Hove, $S(\mathbf{k}, \omega)$ has a precise experimental meaning in Born approximation scattering theory, where the probability per unit volume that the fluid absorbs a momentum $\hbar\mathbf{k}$ and an energy $h\omega$ by the scattering of an incidend particle is given by

$$P(\mathbf{k}, \omega) = n|v(\mathbf{k})|^2 S(\mathbf{k}, \omega) \tag{1.16}$$

if $v(\mathbf{k})$ is the interaction potential between the incident particle and a particle of the fluid. Equation (1.14) yields

$$\operatorname{Im} \chi(k, \omega) = -(n/2\hbar)[S(\mathbf{k}, \omega) - S(-\mathbf{k}, -\omega)].$$

As $T = 0$, since the system can only absorb energy from the probe, the above equation becomes

$$\operatorname{Im} \chi(k, \omega) = -(n/2\hbar) S(k, \omega) \qquad (\omega > 0). \tag{1.17}$$

Thus, the imaginary part of the response function is directly related to the inelastic, weak-scattering properties of the system and, through (1.15), to the correlations between its density fluctuations.

We apply these results to the electron fluid by taking the external potential as the potential of the field $\mathbf{D}(\mathbf{r}, t)$ *in vacuo*,

$$V_{\mathrm{e}}(\mathbf{k}, \omega) = \frac{4\pi e^2}{k^2} n_{\mathrm{e}}(\mathbf{k}, \omega), \tag{1.18}$$

and we have from Eqs. (1.9) and (1.11)

$$\frac{1}{\varepsilon(k, \omega)} = 1 + \frac{4\pi e^2}{k^2} \chi(k, \omega). \tag{1.19}$$

From Eqs. (1.17) and (1.19) we have

$$S(k, \omega) = -\frac{\hbar k^2}{2\pi n e^2} \operatorname{Im} \frac{1}{\varepsilon(, \omega)}, \tag{1.20}$$

[11] L. van Hove, *Phys. Rev.* [2] **95**, 249 (1954).

showing that the dielectric function, originally introduced in connection with screening, also describes the time-dependent pair correlations and the weak scattering properties.

The static pair correlations between density fluctuations, in particular, follow from the knowledge of $\varepsilon(k, \omega)$. These are described by the static structure factor $S(k)$ of X-ray diffraction theory,

$$S(k) = \int_{-\infty}^{\infty} \frac{d\omega}{2\pi} S(k, \omega), \tag{1.21}$$

or by the radial distribution function $g(r)$,

$$g(r) = 1 + \sum_{\mathbf{k}} [S(k) - 1] \exp(i\mathbf{k} \cdot \mathbf{r}). \tag{1.22}$$

The latter function, being by definition the probability that two electrons be found at a relative distance $r$ at any given time, averaged over the relative spin orientations, allows an evaluation of the mean potential energy per particle, which can be written

$$\begin{aligned} E_{\mathrm{pot}} &= \tfrac{1}{2}n \int d\mathbf{r} \frac{e^2}{r} [g(r) - 1] \\ &= \sum_{\mathbf{k}} \frac{2\pi n e^2}{k^2} [S(k) - 1]. \end{aligned} \tag{1.23}$$

One can then pass from the interaction energy to the total energy of the ground state by accounting for correlation effects in the kinetic energy through a well-known integration over the strength of the interaction.[12] The final expression for the ground-state energy of the electron fluid is[2,13]

$$E_{\mathrm{gs}} = E_0 - \sum_{\mathbf{k}} \left[\frac{2\pi n e^2}{k^2} + \frac{\hbar}{2\pi} \int_0^{e^2} \frac{d\lambda}{\lambda} \int_0^{\infty} d\omega \, \mathrm{Im} \frac{1}{\varepsilon_\lambda(k, \omega)}\right] \tag{1.24}$$

where $E_0$ is the ground-state energy of the ideal Fermi gas, and $\varepsilon_\lambda(k, \omega)$ is the dielectric function corresponding to the strength $\lambda$ of the interaction. The equation of state and the compressibility follow, of course, by differentiation with respect to the density. The strength parameter $\lambda$ is in fact directly related to $r_s$.

### *b. Local Fields*

The development in the preceding subsection should be supplemented by a discussion of local fields—an aspect of the physics of charged fluids

[12] See, e.g., K. Sawada, *Phys. Rev.* [2] **106**, 372 (1957).

[13] P. Nozières and D. Pines, *Nuovo Cimento* **9**, 470 (1958).

crucial in evaluating their response. In addition to $\chi(k, \omega)$, which relates the polarization to the field *in vacuo*, it is first useful to introduce a second response function, $\tilde{\chi}(k, \omega)$, which relates it to the Hartree potential associated with the internal field. That is, we write

$$n(\mathbf{k}, \omega) = \tilde{\chi}(k, \omega) V_{\mathrm{H}}(\mathbf{k}, \omega), \tag{1.25}$$

whence, using Eqs. (1.9) and (1.11), we find

$$\chi(k, \omega) = \tilde{\chi}(k, \omega)/\varepsilon(k, \omega) \tag{1.26}$$

and

$$\varepsilon(k, \omega) = 1 - (4\pi e^2/k^2)\, \tilde{\chi}(k, \omega). \tag{1.27}$$

In the language of many-body perturbation theory,[14,15] $\tilde{\chi}(k,\omega)$ is given by the sum of the "proper" polarization diagrams, which cannot be broken into separate diagrams by cutting a single Coulomb interaction line. The long-range effects of the Coulomb interaction are in this way accounted for through the Hartree potential, and in particular the plasmon resonance is brought out explicitly in the denominator of (1.26).

The introduction of the response to the Hartree field is a major step forward and is the basis of the random phase approximation[16] (RPA), which then proceeds to replace $\tilde{\chi}$ by the density response function of an ideal Fermi gas. However, local field effects are soon met when one attempts a more refined evaluation of $\tilde{\chi}$. The interaction between an electron and the displaced charge is not merely given by the Hartree term but includes exchange and correlation terms[14] that in fact depend on the electron's momentum and energy. The detailed motion of each electron does not, of course, enter explicitly the equation of motion for the charge density, but this equation still contains local field terms, since a local deficiency of charge is induced by exchange and correlation around the charge density at a given point. This so-called Pauli and Coulomb hole is described instantaneously by the radial distribution function $g(r)$ and dynamically by the van Hove function $G(r, t)$. Clearly, the distortion of the hole by the applied field will contribute to the internal forces determining the density response.

To illustrate the usefulness of these ideas in the static case, let us write, following Ballentine,[17] the internal potential felt by the particle density as

[14] J. Hubbard, *Proc. R. Soc. London, Ser. A* **240,** 539 (1957); **243,** 336 (1957).

[15] D. F. DuBois, *Ann. Phys.* (*N. Y.*) **7,** 174 (1959); **8,** 24 (1959).

[16] D. Bohm and D. Pines, *Phys. Rev.* [2] **92,** 609 (1953); H. Ehrenreich and M. Cohen, *ibid.* **115,** 786 (1959).

[17] L. E. Ballentine, *Phys. Rev.* [2] **158,** 670 (1967); see also C. A. Kukkonen and A. W. Overhauser, *Phys. Rev.* **B20,** 550 (1979).

$$V_p(\mathbf{k}) = V_H(\mathbf{k}) + V_{xc}(\mathbf{k}) \equiv V_e(\mathbf{k})/\varepsilon_p(k), \tag{1.28}$$

where $V_{xc}$ represents the local field correction due to exchange and correlation, and $\varepsilon_p(k)$ is a new dielectric function that describes the screening of the external field as measured by the electron density [whereas $\varepsilon(k)$ describes the screening as measured by an infinitesimal external test charge]. The HKS theory[8] shows that $V_{xc}$ can be related to the polarization by

$$V_{xc}(\mathbf{k}) = f_{xc}(k)n(\mathbf{k}), \tag{1.29}$$

where the exchange and correlation factor $f_{xc}(k)$ can in principle be constructed to include also the correlation effects in the kinetics of the electrons. We can therefore relate the polarization to $V_p(\mathbf{k})$ through the density response function $\chi_0(k)$ of the ideal Fermi gas,

$$n(\mathbf{k}) = \chi_0(k)V_p(\mathbf{k}), \tag{1.30}$$

and we find

$$\varepsilon(k) = -\frac{4\pi e^2}{k^2}\chi_0(k)/[1 - f_{xc}(k)\chi_0(k)] \tag{1.31}$$

and

$$\varepsilon_p(k) = 1 - \left[\frac{4\pi e^2}{k^2} + f_{xc}(k)\right]\chi_0(k). \tag{1.32}$$

The burden of the theory of static screening thus lies in the determination of $f_{xc}(k)$. We shall return to its evaluation in Section 4.

Similar arguments have recently been given by Kukkonen and Overhauser[17] to derive an expression for the effective interaction between two electrons in the electron gas. This interaction, which depends on the relative spin alignment of the two electrons can be expressed in terms of $f_{xc}(k)$ and the difference between its exchange and correlation parts. This difference determines the static magnetic response, as we shall see in Section 9.

### *c. Generalized Conductivity*

The extension of results such as (1.31) to the dynamic case presents obvious difficulties because of the dissipative effects associated with current flow that are not present in the ideal Fermi gas (multipair excitations). Dissipative effects are brought out explicitly in the structure of the dynamic response function $\chi(k, \omega)$ by introducing the generalized longitudinal conductivity $\sigma(k, \omega)$ through

$$e\mathbf{j}(\mathbf{k}, \omega) = \sigma(k, \omega)\mathbf{E}_j(\mathbf{k}, \omega), \tag{1.33}$$

where $E_j(\mathbf{k}, \omega)$ is the current-driving local field. The condition that this vanish in the static limit is satisfied in the approach developed by Mori[18] (see also Kubo[10]) by the choice

$$e\mathbf{E}_j(\mathbf{k}, \omega) = -i\mathbf{k}[V_e(\mathbf{k}, \omega) - n(\mathbf{k}, \omega)/\chi(k, 0)] \tag{1.34}$$

whence, using Eq. (1.11) and the continuity equation, we find

$$\chi(k, \omega) = \chi(k, 0)\Big/\left[1 + \frac{i\omega e^2}{k^2}\frac{\chi(k, 0)}{\sigma(k, \omega)}\right]. \tag{1.35}$$

In particular, by evaluating $\chi(k, 0)$ for $k \to 0$ from Eq. (1.3) [see Eq. (2.1)], it is easily seen that at low frequencies

$$\lim_{\mathbf{k}\to 0} e\mathbf{E}_j(\mathbf{k}, \omega) = -i\mathbf{k}\left[V_{\mathrm{H}}(\mathbf{k}, \omega) + \left(\frac{\partial\mu}{\partial n}\right) n(\mathbf{k}, \omega)\right] \tag{1.36}$$

and that Eq. (1.35) reduces for $k = 0$ to the well-known relation between complex dielectric function and complex conductivity,

$$\varepsilon(\omega) = 1 + \frac{4\pi i}{\omega}\sigma(\omega).$$

The physical content of Eq. (1.35) can be illustrated by writing $\sigma(k, \omega)$ in a generalized Drude form,

$$\sigma(k, \omega) = (ne^2/m)[-i\omega + \gamma(k, \omega)]^{-1}, \tag{1.37}$$

which we shall derive in Section 3. We then have

$$\chi(k, \omega) = (nk^2/m)[\omega^2 - \omega_0^2(k) + i\omega\gamma(k, \omega)]^{-1} \tag{1.38}$$

with

$$\omega_0^2(k) = -nk^2/[m\chi(k, 0)]. \tag{1.39}$$

Equation (1.38) clearly displays a collective mode at frequency $\omega_0(k)$, with a damping and a frequency shift determined by the real and the imaginary part of the generalized resistivity $\gamma(k, \omega)$. This property involves the processes of pair excitations in the electron fluid.

## 2. Limiting Behaviors and Sum Rules

It is convenient to summarize at this point the exact limiting forms for the response functions at long wavelength that follow from the results

[18] H. Mori, *Prog. Theor. Phys.* **33**, 423 (1965); **34**, 399 (1965).

derived at the beginning of the preceding section. From the thermo dynamic treatment leading to Eq. (1.3) we have

$$\lim_{k\to 0} \chi(k, 0) = -(k^2/4\pi e^2)(1 + k^2/k_s^2)^{-1} \tag{2.1}$$

and

$$\lim_{k\to 0} \tilde{\chi}(k, 0) = -n^2 K. \tag{2.2}$$

The latter relation is also obeyed by the density response function of a neutral fluid. From the analogous relation for $\chi_0(k)$, involving the ideal-gas compressibility $K_0 = 3/(2n\varepsilon_F)$, we also have

$$\lim_{k\to 0} f_{xc}(k) = -(2\varepsilon_F/3n)(1 - K_0/K) \tag{2.3}$$

and

$$\lim_{k\to 0} \varepsilon_p(k) = 1 + 4\pi n^2 e^2 K_0/k^2. \tag{2.4}$$

It follows that the potential $V_{ei}(k)/\varepsilon_p(k)$ exerted by a screened ion on an electron in a metal tends to the value $-\frac{2}{3}Z\varepsilon_{F/n}$, irrespective of exchange and correlation effects.

In the dynamic case we obtain from Eq. (1.8) the long-wavelength expression

$$1/\varepsilon(0, \omega) = 1 + \omega_p^2(\omega^2 - \omega_p^2 + i\eta)^{-1}, \tag{2.5}$$

where we have introduced an infinitesimal imaginary part in the denominator in accord with the causality requirement. This also serves to represent the resonant absorption of energy at the plasma frequency, since Eq. (1.20) then yields

$$\lim_{k\to 0} S(k, \omega) = (\pi\hbar k^2/m\omega_p)\delta(\omega - \omega_p). \tag{2.6}$$

We consequently find

$$\lim_{k\to 0} S(k) = \tfrac{1}{2}\hbar\omega_p k^2/4\pi n e^2. \tag{2.7}$$

This expression illustrates the deep-lying relation between dynamics and structure: in the classical limit, the zero-point energy of the plasmon is simply replaced by the thermal energy $k_B T$. Equation (2.5) also yields

$$\lim_{k\to 0} \tilde{\chi}(k, \omega) = nk^2/m\omega^2, \tag{2.8}$$

which is the free-particle value. The validity of this result at *all* frequencies is a consequence of the vanishing electrical resistivity of the model system: In the presence of internal scatterers yielding a finite dc conductivity $\sigma$, $\tilde{\chi}(k, \omega)$ at long wavelength and *low* frequency would instead obey the Kubo relation[19]

$$\lim_{\omega\to 0} \omega \left[ \lim_{k\to 0} k^{-2} \operatorname{Im} \tilde{\chi}(k, \omega) \right] = -\sigma/e^2 \tag{2.9}$$

as follows from Eq. (1.35).

The asymptotic behaviors noted above can be recast in the form of sum rules on the imaginary part of the response functions, through the use of Kramers–Kronig relations: An exhaustive discussion has been given by Martin.[19] We discuss here instead another set of sum rules which relate to the short-time behavior of the response, as described by successive frequency moments of the dissipation spectrum. To this end we note that the density response function introduced in Section 1,a can be written

$$\chi(k, \omega) = \int_0^\infty dt\, \Phi(k, t) \exp(-i\omega t) \tag{2.10}$$

with

$$\Phi(k, t) = (-i/\hbar) \int d\mathbf{r}\, \langle [\rho(\mathbf{0}, 0), \rho(\mathbf{r}, t)] \rangle \exp(i\mathbf{k}\cdot\mathbf{r}). \tag{2.11}$$

It is easily seen that $\Phi(k, t)$ is an odd function of time, so that (2.10) can be inverted to

$$\Phi(k, t) = (i/\pi) \int_{-\infty}^{\infty} d\omega \operatorname{Im} \chi(k, \omega) \exp(i\omega t). \tag{2.12}$$

The timed derivatives of $\Phi(k, t)$ at $t = 0$, if they exist, are then given by

$$\left.\frac{d^{(2n+1)}\Phi(k, t)}{dt^{(2n+1)}}\right|_{t=0} = -(-1)^n \int_{-\infty}^{\infty} \frac{d\omega}{\pi} \omega^{2n+1} \operatorname{Im} \chi(k, \omega) \equiv (-1)^n M_{2n+1}(k). \tag{2.13}$$

The odd moments of the dissipation spectrum can therefore be expressed through static correlation properties of the fluid at equilibrium if Eq. (2.11) is used to evaluate the initial time derivatives of $\Phi(k, t)$.

Let us consider the first moment, which correspnds to the *f*-sum rule expressing particle conservation, as the simplest example of such a calculation. We have

$$M_1(k) = \dot{\Phi}(k, 0) = \hbar^{-2} \int d\mathbf{r}\, \langle \{\rho(\mathbf{0}), [H, \rho(\mathbf{r})]\} \rangle \exp(i\mathbf{k}\cdot\mathbf{r}), \tag{2.14}$$

[19] R. Kubo, *J. Phys. Soc. Jpn.* **12,** 570 (1957); P. C. Martin, *Phys. Rev.* [2] **161,** 143 (1967).

and it is evident that only the kinetic part of the Hamiltonian enters the evaluation of the commutator, leading to a free-particle value for $M_1(k)$ irrespective of the interactions. Precisely,

$$[H, \rho(\mathbf{r})] = -(\hbar^2/2m) \sum_i [\nabla_i^2 \delta(\mathbf{r} - \mathbf{r}_i) + 2\nabla_i \delta(\mathbf{r} - \mathbf{r}_i) \cdot \nabla_i] \quad (2.15)$$

whence

$$M_1(k) = m^{-1} \left\langle \int d\mathbf{r} \exp(i\mathbf{k} \cdot \mathbf{r}) \sum_i \nabla_i \delta(\mathbf{r} - \mathbf{r}_i) \cdot \nabla_i \delta(\mathbf{r}_i) \right\rangle = nk^2/m. \quad (2.16)$$

Evidently, the calculation of the moments involves the repeated use of the commutation relations between position and momentum operators, and the results are therefore formally independent of the statistics. The third moment was first derived by Puff,[20] with the result

$$M_3(k) = \frac{nk^2}{m} \left\{ \omega_p^2 + \frac{2k^2}{m} E_{\text{kin}} + \left(\frac{\hbar k^2}{2m}\right)^2 \right.$$
$$\left. + \, \omega_p^2 \sum_{\mathbf{q}(\neq \mathbf{k})} (\hat{q} \cdot \hat{k})^2 [S(\mathbf{k} - \mathbf{q}) - S(q)] \right\}, \quad (2.17)$$

where $E_{\text{kin}}$ is the exact kinetic energy per electron. The expression for the fifth moment $M_5(k)$ in the classical fluid has been given by Forster *et al.*[21]

To illustrate the usefulness of the moment sum rules, we use them to evaluate the plasmon dispersion on the assumption that the plasmon remains a sharp excitation to order $k^2$. That is, we write

$$S(k, \omega) \simeq \frac{\pi \hbar k^2}{m\omega_p(k)} \delta[\omega - \omega_p(k)], \quad (2.18)$$

where the spectral strength has already been chosen so as to satisfy the sum rule (2.16). From (2.17) we find

$$\lim_{k \to 0} M_3(k) = \frac{nk^2}{m} \left[ \omega_p^2 + \frac{k^2}{m} \left( 2E_{\text{kin}} + \frac{4}{15} E_{\text{pot}} \right) \right] \quad (2.19)$$

where $E_{\text{pot}}$ is the exact potential energy per electron, given by (1.23). By requesting that the approximate spectrum (2.18) satisfy this sum rule, we find

$$\omega_p^2(k) = \omega_p^2 + \frac{k^2}{m} \left[ 2E_{\text{kin}} + \frac{4}{15} E_{\text{pot}} \right] + \cdots. \quad (2.20)$$

[20] R. D. Puff, *Phys. Rev.* [2] **137,** A406 (1965); see also B. Goodman and A. Sjölander, *Phys. Rev. B Solid State* [3] **8,** 200 (1973).

[21] D. Forster, P. C. Martin, and S. Yip, *Phys. Rev.* [2] **170,** 155 (1968).

The first term in the bracket is dominant for small $r_s$ and leads to a positive plasmon dispersion given by

$$\lim_{r_s \to 0} \omega_p^2(k) = \omega_p^2 + \tfrac{6}{5}\, \varepsilon_F k^2/m + \cdots, \tag{2.21}$$

which is the RPA result[16] for all $r_s$. At large $r_s$, on the other hand, the potential energy term in (2.20) becomes dominant, and the dispersion of the plasmon may thus be expected to become negative, as observed[22] by molecular dynamics for the classical plasma in the appropriate limit of strong interactions.

From a rigorous standpoint, the spectral form (2.18) leading to the result (2.20) omits pair excitations which add a continuum to the collective excitation spectrum and cause plasmon damping as well as a shift in the plasmon dispersion. Plasmon decay into single-pair excitations is allowed by the conservation laws only above a cutoff wave number $k_c \sim \omega_p/v_F$, owing to the sharp drop of the ground-state momentum distribution at the Fermi surface.[16,23] Multipair processes[15] are instead operative at all nonvanishing wave numbers and lead to a plasmon lifetime that behaves as $k^{-2}$ for $k \to 0$. Experimentally, the long wavelength plasmon is a rather narrow excitation in many metals, even though single-pair decay may become allowed through Umklapp processes.[24] Pair excitations may be empirically allowed for in the theory by the use of Eq. (1.38), which can be adapted to satisfy the third moment sum rule by rewriting it, for instance in the form

$$\chi(k, \omega) = \frac{nk^2}{m}\left[\omega^2 - \omega_0^2(k) + i\omega\,\frac{\omega_\infty^2(k) - \omega_0^2(k)}{-i\omega + \tau^{-1}(k, \omega)}\right]^{-1} \tag{2.22}$$

where $\omega_\infty^2(k) = mM_3(k)/nk^2$, and $\tau(k, \omega)$ has the meaning of a generalized relaxation time. Equation (2.22) shows, in particular, that the plasmon dispersion would shift from the result (2.20) in the "collisionless" regime to the value $\omega_0^2(k) \to \omega_p^2 + k^2/nmK$ in a strong collision regime.

As a final comment on the moment sum rules (2.13), we recall that they are commonly used in liquid-state theory to express the high-frequency dynamics either through a straightforward inverse frequency expansion of the response or through the more powerful continued-fraction expansion

[22] J. P. Hansen, E. Pollock, and I. R. McDonald, *Phys. Rev. Lett.* **32,** 277 (1974); M. C. Abramo and M. P. Tosi, *Nuovo Cimento* **21,** 363 (1974).

[23] R. A. Ferrell, *Phys. Rev.* [2] **107,** 450 (1957); K. Sawada, K. A. Brueckner, N. Fukuda, and R. Brout, *ibid.* **108,** 507 (1957).

[24] The experimental evidence suggests that this could be an important decay mechanism in the alkali metals [see P. C. Gibbsons and S. E. Schnatterly, *Phys. Rev. B: Solid State* [3] **15,** 2420 (1977).

of the spectrum proposed by Mori,[18] of which Eq. (2.22) is in fact the first step. Recent work[25] suggests, however, that the fifth moment may diverge for the degenerate electron fluid over the range of $r_s$ for which $g(r = 0) \neq 0$. The fifth moment in fact determines the $\omega^{-6}$ term in an inverse frequency expansion, so that its divergence would herald the presence of a term of the form $k^4\omega^{-11/2}$ in $S(k, \omega)$ at long wavelength, as reported[25] for the high-frequency behavior of the spectrum. Great caution is thus indicated in the use of moment sum rules beyond the first two reported in Eqs. (2.16) and (2.17).

Finally, it should be pointed out that Eq. (2.18), applied at *all* values of $k$, forms the basis for the so-called "plasmon pole approximation" for the dielectric response function,[4,25a]

$$1/\varepsilon(k, \omega) = 1 + \omega_p^2/[\omega^2 - \omega_p^2(k)]. \tag{2.23}$$

The above form because of its simplicity has found use in various calculations of many-body effects. Hedin and Lundqvist[4] choose the parameter $\omega_p(k)$ in the form

$$\omega_p^2(k) = \omega_p^2 + \frac{4}{3}\frac{\hbar^2 k_F^2}{m^2}k^2 + \frac{\hbar^2 k^4}{4m^2}, \tag{2.24}$$

whereas Overhauser[25a] adjusts it to the static dielectric function through the relation

$$\omega_p^2(k) = \omega_p^2 \varepsilon(k)/[\varepsilon(k) - 1]. \tag{2.25}$$

Both these choices are correct in the two limiting cases $k \rightarrow 0$ and $k \rightarrow \infty$.

## 3. Equations of Motion for Density Matrices

As we move away from the long-wavelength limit dealt with at the beginning of Sections 1 and 2, two techniques offer themselves for the evaluation of the microscopic density response, namely the Feynman diagrams approach and the equations-of-motion approach. Both techniques have been applied widely to the problem of the electron fluid, and either presents special advantages. Thus, the diagrammatic approach has a precise theoretical meaning in the context of a perturbative treatment of the interactions, but one has to sum an infinite set of diagrams even in the simplest approximation, and both the choice and the evaluation of the relevant sets of diagrams present difficulties as one goes beyond the RPA.

[25] F. Family, *Phys. Rev. Lett.* **36,** 1374 (1975).

[25a] A. W. Overhauser, *Phys. Rev. B: Solid State* [3] **3,** 1888 (1971).

On the other hand, an approximate solution of the hierarchy of equations of motion can be given by decoupling schemes that compactly account for some effects of the interaction to infinite order, but one may find a justification on intuitive grounds and in the verification of sum rules and limiting behaviors rather than in a precise theoretical development. It should also be stressed at this point that no complete theory of either kind exists as yet, and that one is at times forced to introduce different approximate schemes for dealing with different aspects of the physics of the system. We shall concentrate the discussion on the equations-of-motion approach, which is the basis of most recent progress in the problem of the electron fluid as well as for other fluids both quantal and classical, and shall make contact with the diagrammatic approach wherever possible.

We collect in this section the basic equations of motion that we shall need. The starting point of the development is the equation of motion for the one-particle density matrix $\rho_\sigma(\mathbf{x}, \mathbf{x}'; t)$ in an external potential $V_e(\mathbf{x}, t)$, defined by

$$\rho_\sigma(\mathbf{x}, \mathbf{x}'; t) = \langle \psi_\sigma^+(\mathbf{x}, t)\psi_\sigma(\mathbf{x}', t)\rangle, \tag{3.1}$$

where $\psi_\sigma^+(\mathbf{x}, t)$ and $\psi_\sigma(\mathbf{x}, t)$ are the creation and annihilation operators for an electron of spin $\sigma$ at the space–time point $(\mathbf{x}, t)$ and obey the usual anticommutation relations for fermions. We shall also need the mixed density matrix or Wigner distribution function $f_{\mathbf{p}\sigma}(\mathbf{R}, t)$ defined by

$$f_{\mathbf{p}\sigma}(\mathbf{R}, t) = \int d\mathbf{r}\, \exp(-i\mathbf{p}\cdot\mathbf{r}/\hbar)\, \langle \psi_\sigma^+(\mathbf{R} - \tfrac{1}{2}\mathbf{r}, t)\psi_\sigma(\mathbf{R} + \tfrac{1}{2}\mathbf{r}, t)\rangle \tag{3.2}$$

as well as its Fourier transform

$$f_{\mathbf{p}\sigma}(\mathbf{k}, t) = \int d\mathbf{R}\, \exp(-i\mathbf{k}\cdot\mathbf{R}) f_{\mathbf{p}\sigma}(\mathbf{R}, t) = \langle a^+_{\mathbf{p}-\mathbf{k}/2,\sigma}(t)\, a_{\mathbf{p}+\mathbf{k}/2,\sigma}(t)\rangle \tag{3.3}$$

where $a_{\mathbf{p}\sigma}(t)$ and $a_{\mathbf{p}\sigma}(t)$ are the creation and annihilation operators for an electron with momentum $\mathbf{p}$ and spin $\sigma$ at time $t$. From the Hamiltonian of the system in the external potential,

$$\begin{aligned} H(t) &= \sum_\sigma \int d\mathbf{x}\, \psi_\sigma^+(\mathbf{x}, t)\left[-\frac{\hbar^2}{2m}\nabla_{\mathbf{x}}^2 + V_e(\mathbf{x}, t)\right]\psi_\sigma(\mathbf{x}, t) \\ &\quad + \frac{1}{2}\sum_{\sigma\sigma'}\int\int d\mathbf{x}\, d\mathbf{x}'\, \psi_\sigma^+(\mathbf{x}, t)\psi_{\sigma'}^+(\mathbf{x}', t)v(\mathbf{x} - \mathbf{x}')\psi_{\sigma'}(\mathbf{x}', t)\psi_\sigma(\mathbf{x}, t) \\ &= \frac{\hbar^2}{2m}\sum_{\mathbf{k}\sigma} k^2 a_{\mathbf{k}\sigma}^+(t)a_{\mathbf{k}\sigma}(t) + \sum_{\mathbf{q}} V_e(\mathbf{q}, t)\sum_{\mathbf{k}\sigma} a^+_{\mathbf{k}+\mathbf{q}/2,\sigma}(t)a_{\mathbf{k}-\mathbf{q}/2,\sigma}(t) \\ &\quad + \frac{1}{2}\sum_{\mathbf{q}\neq 0} v(q)\sum_{\mathbf{k}\sigma}\sum_{\mathbf{k}'\sigma'} a^+_{\mathbf{k}-\mathbf{q}/2,\sigma}(t)a^+_{\mathbf{k}'-\mathbf{q}/2,\sigma'}(t)a_{\mathbf{k}'+\mathbf{q}/2,\sigma'}(t)a_{\mathbf{k}+\mathbf{q}/2,\sigma'}(t), \end{aligned} \tag{3.4}$$

we have the equation of motion

$$\left[i\hbar\frac{\partial}{\partial t} - \frac{\hbar^2}{2m}(\nabla_{\mathbf{x}}^2 - \nabla_{\mathbf{x}'}^2) + V_e(\mathbf{x}, t) - V_e(\mathbf{x}', t)\right]\langle\psi_\sigma^+(\mathbf{x}, t)\psi_\sigma(\mathbf{x}', t)\rangle$$

$$= -\int d\mathbf{x}''[v(\mathbf{x} - \mathbf{x}'') - v\mathbf{x}' - \mathbf{x}'')]\,\langle\psi_\sigma^+(\vec{\mathbf{x}}, t)\rho(\vec{\mathbf{x}}'', t)\psi_\sigma(\vec{\mathbf{x}}', t)\rangle \quad (3.5)$$

where

$$\rho(\mathbf{x}, t) = \sum_\sigma \psi_\sigma^+(\mathbf{x}, t)\psi_\sigma(\mathbf{x}, t) \quad (3.6)$$

is the particle density operator. Hartree field terms are brought out explicitly by introducing the cumulant part of the two-body function in Eq. (3.5),

$$\langle\psi_\sigma^+(\mathbf{x}, t)\rho(\mathbf{x}'', t)\psi_\sigma(\mathbf{x}', t)\rangle_c$$

$$= \langle\psi_\sigma^+(\mathbf{x}, t)\rho(\mathbf{x}'', t)\psi_\sigma(\mathbf{x}', t)\rangle - n(\mathbf{x}'', t)\langle\psi_\sigma^+(\mathbf{x}, t)\psi_\sigma(\mathbf{x}', t)\rangle, \quad (3.7)$$

so that, defining the Hartree field by

$$V_H(\mathbf{x}, t) = V_e(\mathbf{x}, t) + \int d\mathbf{x}''\, v(\mathbf{x} - \mathbf{x}'')n(\mathbf{x}'', t), \quad (3.8)$$

Eq. (3.5) can be written

$$\left[i\hbar\frac{\partial}{\partial t} - \frac{\hbar^2}{2m}(\nabla_{\mathbf{x}}^2 - \nabla_{\mathbf{x}'}^2) + V_H(\mathbf{x}, t) - V_H(\mathbf{x}', t)\right]\langle\psi_\sigma^+(\mathbf{x}, t)\psi_\sigma(\mathbf{x}', t)\rangle$$

$$= -\int d\mathbf{x}''[v(\mathbf{x} - \mathbf{x}'') - v(\mathbf{x}' - \mathbf{x}'')]\langle\psi_\sigma^+(\mathbf{x}, t)\rho(\mathbf{x}'', t)\psi_\sigma(\mathbf{x}', t)\rangle_c. \quad (3.9)$$

The RPA solution is obtained by neglecting the right-hand side of this equation, and obviously corresponds to a system of free fermions in an "external" potential given by $V_H(\mathbf{x}, t)$.

Equation (3.9) is equivalent to an infinite set of equations of motion for observable physical quantities with classical equivalents. This follows from the fact that the Wigner distribution function (3.2) has properties[26] analogous to the classical phase space distribution function $f(\mathbf{R}, \mathbf{p}; t)$ and in particular can be used to construct physical observables such as the particle density

$$n(\mathbf{R}, t) = \sum_{\mathbf{p}\sigma} f_{\mathbf{p}\sigma}(\mathbf{R}, t), \quad (3.10)$$

the particle current density

$$\mathbf{j}(\mathbf{R}, t) = m^{-1}\sum_{\mathbf{p}\sigma} \mathbf{p}f_{\mathbf{p}\sigma}(\mathbf{R}, t) \quad (3.11)$$

[26] See, e.g., R. Balescu, "Equilibrium and Nonequilibrium Statistical Mechanics." Wiley, New York, 1975.

and the kinetic stress tensor

$$\boldsymbol{\pi}(\mathbf{R}, t) = m^{-1} \sum_{\mathbf{p}\sigma} \mathbf{pp} f_{\mathbf{p}\sigma}(\mathbf{R}, t). \tag{3.12}$$

By expanding Eq. (3.9) about its diagonal, i.e., in powers of $\mathbf{r} = \mathbf{x}' - \mathbf{x}$, we have

$$\left[i\hbar \frac{\partial}{\partial t} + \frac{\hbar^2}{m} \boldsymbol{\nabla}_{\mathbf{R}} \cdot \boldsymbol{\nabla}_{\mathbf{r}} - \mathbf{r} \cdot \boldsymbol{\nabla}_{\mathbf{R}} V_{\mathrm{H}}(\mathbf{R}, t) + O(r^3)\right] \sum_{\mathbf{p}\sigma} \exp(i\mathbf{p} \cdot \mathbf{r}/\hbar) f_{\mathbf{p}\sigma}(\mathbf{R}, t)$$
$$= \int d\mathbf{x}''[\mathbf{r} \cdot \boldsymbol{\nabla}_{\mathbf{R}} v(\mathbf{R} - \mathbf{x}'') + O(r^3)][\langle \rho(\mathbf{R}, t)\rho(\mathbf{x}'', t)\rangle_{\mathrm{c}}$$
$$+ (im/\hbar)\mathbf{r} \cdot \langle \mathbf{j}(\mathbf{R}, t)\rho(\mathbf{x}'', t)\rangle_{\mathrm{c}} + O(r^2)]. \tag{3.13}$$

We now equate the coefficients of equal powers of $\mathbf{r}$ on the two sides of this equation to obtain the first three equations of the one-particle hierarchy, which are first the usual continuity equation,

$$\frac{\partial n(\mathbf{R}, t)}{\partial t} + \boldsymbol{\nabla} \cdot \mathbf{j}(\mathbf{R}, t) = 0; \tag{3.14}$$

next, the equation of motion for the current density,

$$m \frac{\partial \mathbf{j}(\mathbf{R}, t)}{\partial t} = -\boldsymbol{\nabla} \cdot \boldsymbol{\pi}(\mathbf{R}, t) - n(\mathbf{R}, t)\boldsymbol{\nabla} V_{\mathrm{H}}(R, t)$$
$$- \int d\mathbf{x} \boldsymbol{\nabla} v(\mathbf{R} - \mathbf{x})\langle \rho(\mathbf{R}, t)\rho(\mathbf{x}, t)\rangle_{\mathrm{c}}; \tag{3.15}$$

third, the equation of motion for the kinetic stress tensor,

$$\frac{1}{2} \frac{\partial \boldsymbol{\pi}(\mathbf{R}, t)}{\partial t} = -(2m^2)^{-1} \sum_{\mathbf{p}\sigma} \mathbf{pp}(\mathbf{p} \cdot \boldsymbol{\nabla}) f_{\mathbf{p}\sigma}(\mathbf{R}, t) - \mathbf{j}(\mathbf{R}, t)\boldsymbol{\nabla} V_{\mathrm{H}}(\mathbf{R}, t)$$
$$- \int d\mathbf{x}\, \boldsymbol{\nabla} v(\mathbf{R} - \mathbf{x})\langle \mathbf{j}(\mathbf{R}, t)\rho(\mathbf{x}, t)\rangle_{\mathrm{c}}. \tag{3.16}$$

Equation (3.15) is the generalized Navier–Stokes equation for the fluid, which after linearization tends to Eq. (1.7) in the long-wavelength limit where the Hartree term is dominant. The trace of Eq. (3.16) can be written[27] in the form of an energy transport equation, when one notes that the trace of $\frac{1}{2}\boldsymbol{\pi}(\mathbf{R}, t)$ is the density of kinetic energy and that the trace of the first term on the right-hand side of (3.16) is simply related to the flux of kinetic energy.

[27] N. H. March and M. P. Tosi, *Ann. Phys.* (*N. Y.*) **81**, 414 (1973).

*a. Static Limit*

In the static limit the vanishing of the right-hand side of Eq. (3.15),

$$\nabla_\beta \pi_{\alpha\beta}(\mathbf{R}) + n(\mathbf{R})\nabla_\alpha V_H(\mathbf{R}) + \int d\mathbf{x}\ \nabla_\alpha v(\mathbf{R}-\mathbf{x})\langle\rho(\mathbf{R})\rho(\mathbf{X})\rangle_c = 0 \quad (3.17)$$

where $\alpha$ and $\beta$ are Cartesian indices, and summation over repeated indices is implied, may be viewed as an equilibrium condition which requests that the total generalized force at each point **R** vanish. This is thus the microscopic extension of the equilibrium condition (1.2) already used in the static, long-wavelength limit, as we shall see later. We can now obtain a formal solution of Eq. (3.17), yielding the static response $\chi(k, 0)$, by making use of the Hohenberg–Kohn theorem[8] according to which the state of the system in the static external potential $V_e(\mathbf{R})$ is uniquely determined by its density distribution $n(\mathbf{R})$. This allows us to write, for a weak external potential, the change in the kinetic stress tensor and in the two-body correlation function as

$$\delta\pi_{\alpha\beta}(\mathbf{R}) = \int d\mathbf{y}\ \frac{\delta\pi_{\alpha\beta}(\mathbf{R})}{\delta n(\mathbf{y})}\bigg|_{V_e=0}\delta n(\mathbf{y}) \equiv \int d\mathbf{y}\ \lambda_{\alpha\beta}(\mathbf{R}-\mathbf{y})\delta n(\mathbf{y}) \quad (3.18)$$

and

$$\delta\langle\rho(\mathbf{R})\rho(\mathbf{x})\rangle_c = \int d\mathbf{y}\ \frac{\delta\langle\rho(\mathbf{R})\rho(\mathbf{x})\rangle_c}{\delta n(\mathbf{y})}\bigg|_{V_e=0}\delta n(\mathbf{y})$$

$$\equiv \int d\mathbf{y}\ F(\mathbf{R}-\mathbf{x}, \mathbf{x}-\mathbf{y})\delta n(\mathbf{y}), \quad (3.19)$$

where we have used the standard symbol for a functional derivative. The solution of the linearized equilibrium condition is then easily found in Fourier transform, with the result

$$\chi(k, 0) = -\frac{nk^2}{m}\left[\omega_p^2 + \frac{1}{m}k_\alpha k_\beta \lambda_{\alpha\beta}(\mathbf{k})\right.$$

$$\left. + \frac{1}{m}\sum_{\mathbf{k}'}\mathbf{k}\cdot\mathbf{k}'v(k')F(\mathbf{k}-\mathbf{k}', \mathbf{k})\right]^{-1}. \quad (3.20)$$

This yields the exchange and correlation factor $f_{xc}(k)$ introduced in Eq. (1.31),

$$f_{xc}(k) = (nk^2)^{-1}\{k_\alpha k_\beta[\lambda_{\alpha\beta}(\mathbf{k}) - \lambda^0_{\alpha\beta}(\mathbf{k})]$$

$$+ \sum_{\mathbf{k}'}\mathbf{k}\cdot\mathbf{k}'v(k')F(\mathbf{k}-\mathbf{k}', k)\}, \quad (3.21)$$

where the superscript 0 denotes the value for the ideal Fermi gas.

Let us now evaluate more explicitly the functional derivatives introduced in Eqs. (3.18) and (3.19). Using the fact that the tensor $\lambda_{\alpha\beta}(\mathbf{R}-\mathbf{y})$ must reflect the symmetry properties of the fluid in the absence of the external field, we have

$$\lambda_{\alpha\beta}(\mathbf{R}-\mathbf{y}) = \delta_{\alpha\beta}\,\tfrac{1}{3}\,\mathrm{tr}[\lambda(|\mathbf{R}-\mathbf{y}|)] = \tfrac{2}{3}\delta_{\alpha\beta}\,\frac{\delta t(\mathbf{R})}{\delta n(\mathbf{y})}\bigg|_{V_e=0}, \tag{3.22}$$

where $t(\mathbf{R})$ is the density of kinetic energy at point $\mathbf{R}$. On the other hand, the function $F$ in (3.19) can be related to the nonlinear density response, by writing

$$\begin{aligned}\frac{\delta\langle\rho(\mathbf{R})\rho(\mathbf{x})\rangle_c}{\delta n(\mathbf{y})} &= \iiint d\mathbf{y}'\,d\tau\,d\tau'\,\frac{\delta\langle\rho(\mathbf{R},t)\rho(\mathbf{x},t)\rangle_c}{\delta V_e(\mathbf{y}',\tau')}\,\frac{\delta V_e(\mathbf{y}',\tau')}{\delta n(\mathbf{y},\tau)}\\ &= -(i/\hbar)\iiint d\mathbf{y}'\,d\tau\,d\tau'\,\vartheta(t-\tau')\\ &\quad\times\langle(\rho(\mathbf{R},t)\rho(\mathbf{x},t),\rho(\mathbf{y}',\tau')]\rangle_c\chi^{-1}(\mathbf{y}'-\mathbf{y},\tau'-\tau)\end{aligned} \tag{3.23}$$

[cf. with Eq. (1.12)], whence

$$F(\mathbf{k}-\mathbf{k}',\mathbf{k}) = \chi^{(2)}(\mathbf{k}-\mathbf{k}',\mathbf{k},0)/\chi(k,0), \tag{3.24}$$

where $\chi^{(2)}(\mathbf{k}-\mathbf{k}',\mathbf{k},0)$ is the Fourier transform of the dynamic nonlinear response introduced in Eq. (3.23), taken at zero frequency. The appearance of the inverse linear response function in the above equations indicates that the functional derivative in Eq. (3.19) is in essence a triplet function with a screened vertex at point $\mathbf{y}$.

In the long-wavelength limit, the functional derivatives in Eqs. (3.18) and (3.19) become local in space. From Eq. (3.22) we have at once

$$\lim_{\mathbf{k}\to 0} k_\alpha k_\beta \lambda_{\alpha\beta}(\mathbf{k}) = \frac{2}{3}\,k^2\,\frac{dt(n)}{dn} \tag{3.25}$$

where $t(n) = nE_{\mathrm{kin}}$ is the density of kinetic energy as a function of the average electron density $n$. We can also write in the same limit[28,29]

$$\delta\langle\rho(\mathbf{R})\rho(\mathbf{x})\rangle_c = \tfrac{1}{2}[n(\mathbf{R}) + n(\mathbf{x})]\,\frac{\partial}{\partial n}\,\{n^2[g(\mathbf{R}-\mathbf{x};n) - 1]\}, \tag{3.26}$$

where $g(\mathbf{R}-\mathbf{x};n)$ is the pair correlation function of the unperturbed fluid with average density $n$. Hence,

[28] T. Schneider, *Physica (Amsterdam)* **52,** 481 (1971).

[29] G. Niklasson, A. Sjölander, and K. S. Singwi, *Phys. Rev. B: Solid State* [3] **11,** 113 (1975).

$$\lim_{\mathbf{k}\to 0} \sum_{\mathbf{k}'} \mathbf{k}\cdot\mathbf{k}' v(k')F(\mathbf{k}-\mathbf{k}', \mathbf{k})$$
$$= -\tfrac{1}{2}\int d\mathbf{r}\,\frac{\partial}{\partial n}\{n^2[g(r;\, n) - 1]\}(\mathbf{k}\cdot\mathbf{r})\left(\mathbf{k}\cdot\frac{dv(\mathbf{r})}{d\mathbf{r}}\right)$$
$$= \tfrac{1}{6}\int d\mathbf{r}\,\frac{e^2}{r}\frac{\partial}{\partial}\, n\{n^2[g(r;\, n) - 1]\}$$
$$= \tfrac{1}{3}k^2\frac{du(n)}{dn} \tag{3.27}$$

where $u(n) = nE_{\text{pot}}$ is the density of potential energy. Use of the virial theorem,

$$\tfrac{2}{3}t + \tfrac{1}{3}u = P, \tag{3.28}$$

thus yields

$$\lim_{\mathbf{k}\to 0}\chi(k, 0) = -(nk^2/m)[\omega_p^2 + k^2/mnK]^{-1}, \tag{3.29}$$

in agreement with Eq. (2.1).

A brief comment on the relation between the above treatment of the static response and the HKS theory[8] seems called for at this point, although we shall return to this topic in more detail in Section 6. In the HKS approach one imposes the equilibrium condition that the total energy as a functional of the density $n(\mathbf{R})$ be a minimum against variations of $n(\mathbf{R})$ which conserve the total number of particles:

$$\frac{\delta E_0}{\delta n(\mathbf{R})} + \frac{\delta E_{\text{xc}}}{\delta n(\mathbf{R})} + V_{\text{H}}(\mathbf{R}) = \mu, \tag{3.30}$$

where $E_0$ is the energy of the ideal Fermi gas of density $n(\mathbf{R})$,

$$E_{\text{xc}} = E_{\text{kin}} - E_0 + \tfrac{1}{2}\iint d\mathbf{R}\, d\mathbf{x}\, v(\mathbf{R}-\mathbf{x})\langle\rho(\mathbf{R})\rho(\mathbf{x})\rangle_c, \tag{3.31}$$

and $\delta E_{\text{xc}}/\delta n(\mathbf{R})$ is the exchange and correlation potential $V_{\text{xc}}(\mathbf{R})$ introduced in Eq. (1.28). By linearizing Eq. (3.30) in the density change produced by the external potential, it is easy to show that the exchange and correlation factor in Eq. (1.31) is then given by

$$f_{\text{xc}}(k) = \int d(\mathbf{R}-\mathbf{x})\exp[ik\cdot(\mathbf{R}-\mathbf{x})]\left.\frac{\delta^2 E_{\text{xc}}}{\delta n(\mathbf{R})\,\delta n(\mathbf{x})}\right|_{V_e=0} \tag{3.32}$$

and that the form (3.29) of the static response is recovered in the long-wavelength limit, with the compressibility being expressed through the second density derivative of the energy. The two approaches thus yield the same result in the long-wavelength limit[29] by virtue of the virial theorem, of which Eq. (3.17) provides the microscopic extension.

*b. Dynamic Response*

In the dynamic case it is again of primary interest to examine the equation of motion (3.15), or the equation of motion for the particle density $n(\mathbf{R}, t)$ which is obtained by combining Eqs. (3.14) and (3.15),

$$m \frac{\partial^2 n(\mathbf{R}, t)}{\partial t^2} = \nabla_\alpha \nabla_\beta \pi_{\alpha\beta}(\mathbf{R}, t) + \nabla_\alpha[n(\mathbf{R}, t)\nabla_\alpha V_{\mathrm{H}}(\mathbf{R}, t)]$$

$$+ \int d\mathbf{x}\, \nabla_\alpha[\nabla_\alpha v(\mathbf{R} - \mathbf{x})\langle \rho(\mathbf{R}, t)\rho(\mathbf{x}, t)\rangle_{\mathrm{c}}]. \tag{3.33}$$

After linearization of the Hartree term, this becomes in Fourier transform

$$\omega^2 n(\mathbf{k}, \omega) = \frac{nk^2}{m} V_{\mathrm{H}}(\mathbf{k}, \omega) + \frac{1}{m} k_\alpha k_\beta \pi_{\alpha\beta}(\mathbf{k}, \omega)$$

$$+ \frac{1}{m} \sum_{\mathbf{k}'} \mathbf{k} \cdot \mathbf{k}' v(k') n^{(2)}(\mathbf{k} - \mathbf{k}', \mathbf{k}', \omega). \tag{3.34}$$

Denoting for brevity the sum of the second and third term on the right-hand side of Eq. (3.33) by $C(\mathbf{R}, t)$, we can linearize these terms by writing

$$\delta C(\mathbf{R}, t) = \int \int dy\, d\tau \left.\frac{\delta C(\mathbf{R}, t)}{\delta V_{\mathrm{e}}(\mathbf{y}, \tau)}\right|_{V_{\mathrm{e}}=0} V_{\mathrm{e}}(\mathbf{y}, \tau) \tag{3.35}$$

and this yields for Eq. (3.34)

$$\omega^2 n(\mathbf{k}, \omega) = \frac{nk^2}{m} V_{\mathrm{H}}(\mathbf{k}, \omega) + \frac{1}{m} \Gamma(k, \omega) V_{\mathrm{e}}(\mathbf{k}, \omega)$$

$$= \frac{nk^2}{m} V_{\mathrm{H}}(\mathbf{k}, \omega) + \frac{1}{m}\tilde{\Gamma}(k, \omega) n(\mathbf{k}, \omega), \tag{3.36}$$

where $\Gamma(k, \omega)$ is the Fourier transform of the functional derivative entering Eq. (3.35), and $\tilde{\Gamma}(k, \omega) = \Gamma(k, \omega)/\chi(k, \omega)$. The formal solution for the dynamic response function is

$$\chi(k, \omega) = \frac{nk^2}{m} \left[\omega^2 - \omega_{\mathrm{p}}^2 - \frac{1}{m} \tilde{\Gamma}(k, \omega)\right]^{-1} \tag{3.37}$$

which has the Mori structure already given in Eq. (1.38), with

$$\gamma(k, \omega) = i[\tilde{\Gamma}(k, \omega) - \tilde{\Gamma}(k, 0)/m\omega. \tag{3.38}$$

The highly complex nature of this function, which must include couplings between the various excitations and processes such as backflow and energy dissipation involved in microscopic currents, need not be stressed. The achievement of the formal calculation is thus to have put the dynamic response in a form appropriate for a system that can sustain a collective excitation, and we shall then (fortunately!) be forced to inject some physics into the problem to develop suitable approximations. We have already indicated in Eq. (2.22) a possible approximation for $\gamma(k, \omega)$ in connection with the long-wavelength plasmon, and we shall discuss recent work based on the Mori approach in Section 7. However, before we turn to approximate schemes, it will be useful to complete this section by giving the equation of motion for the two-particle density matrix. This will provide some more insight into the problem and will allow a full determination[30] of the density response in the limit of large $k$ or large $\omega$.

### c. *Two-Particle Density Matrix*

We return first to Eq. (3.9), which we rewrite as an equation of motion for the Fourier transform $f_{\mathbf{p}\sigma}(\mathbf{k}\ t)$ of the Wigner distribution function, defined in Eq. (3.3). This is linearized by setting

$$f_{\mathbf{p}\sigma}(\mathbf{k}, t) = \delta_{\mathbf{k},0} n_{\mathbf{p}\sigma} + \bar{f}_{\mathbf{p}\sigma}(\mathbf{k}, t) \tag{3.39}$$

where $n_{\mathbf{p}\sigma}$ is the momentum distribution function in the unperturbed ground state and $\bar{f}_{\mathbf{p}\sigma}(\mathbf{k}, t)$ is the change caused by the external potential. We also introduce the corresponding two-particle function $f^{(2)}_{\mathbf{p}\sigma,\mathbf{p}'\sigma'}(\mathbf{R}, \mathbf{R}', t)$ by

$$\begin{aligned} f^{(2)}_{\mathbf{p}\sigma,\mathbf{p}'\sigma'}(\mathbf{R}, \mathbf{R}', t) &= \iint d\mathbf{r}\, d\mathbf{r}' \exp[-i(\mathbf{p}\cdot\mathbf{r} + \mathbf{p}'\cdot\mathbf{r}')\hbar]\langle \psi^{+}_{\sigma}(\mathbf{R} - \tfrac{1}{2}\mathbf{r}, t)\psi^{+}_{\sigma'}(\mathbf{R}' - \tfrac{1}{2}\mathbf{r}'\ t) \\ &\quad \times \psi_{\sigma'}(\mathbf{R}' + \tfrac{1}{2}\mathbf{r}', t)\psi_{\sigma}(\mathbf{R} + \tfrac{1}{2}\mathbf{r}, t)\rangle - f_{\mathbf{p}\sigma}(\mathbf{R}, t) f_{\mathbf{p}'\sigma'}(\mathbf{R}', t), \end{aligned} \tag{3.40}$$

which is related to the pair correlation function that we have met in the preceding development by

$$\sum_{\mathbf{p}\sigma}\sum_{\mathbf{p}'\sigma'} f^{(2)}_{\mathbf{p}\sigma,\mathbf{p}'\sigma'}(\mathbf{R}, \mathbf{R}', t) = \langle \rho(\mathbf{R}, t)\rho(\mathbf{R}', t)\rangle_c. \tag{3.41}$$

The Fourier transform of Eq. (3.40) is

$$\begin{aligned} f^{(2)}_{\mathbf{p}\sigma,\mathbf{p}'\sigma'}(\mathbf{k}, \mathbf{k}', t) = \langle a^{+}_{\mathbf{p}-\mathbf{k}/2,\sigma}(t) a^{+}_{\mathbf{p}'-\mathbf{k}'/2,\sigma'}(t) a_{\mathbf{p}'+\mathbf{k}'/2,\sigma'}(t) a_{\mathbf{p}+\mathbf{k}/2,\sigma'}(t)\rangle \\ - f_{\mathbf{p}\sigma}(\mathbf{k}, t) \mathrm{f}_{\mathbf{p}'\sigma'}(\mathbf{k}', t), \end{aligned} \tag{3.42}$$

[30] G. Niklasson, *Phys. Rev. B: Solid State* [3] **10**, 3052 (1974).

that we again separate into an equilibrium part and a perturbation part,

$$f^{(2)}_{\mathbf{p}\sigma,\mathbf{p}'\sigma'}(\mathbf{k},\mathbf{k}',t) = \delta_{\mathbf{k}+\mathbf{k}',0} f^{(2)}_{\mathbf{p}\sigma,\mathbf{p}'\sigma'}(\mathbf{k}) + \bar{f}^{(2)}_{\mathbf{p}\sigma,\mathbf{p}'\sigma'}(\mathbf{k},\mathbf{k}',t). \tag{3.43}$$

With these definitions we find that Eq. (3.9) is equivalent to

$$\begin{aligned}(\hbar\omega - \frac{\hbar}{m}\mathbf{p}\cdot\mathbf{k})\bar{f}_{\mathbf{p}\sigma}(\mathbf{k},\omega) &= (n_{\mathbf{p}-\mathbf{k}/2,\sigma} - n_{\mathbf{p}+\mathbf{k}/2,\sigma})V_{\mathrm{H}}(\mathbf{k},\omega)\\ &+ \sum_{\mathbf{q}} v(q) \sum_{\mathbf{p}'\sigma'}[\bar{f}^{(2)}_{\mathbf{p}-\mathbf{q}\sigma/2,\mathbf{p}'\sigma'}(\mathbf{k}-\mathbf{q},\mathbf{q},\omega)\\ &- \bar{f}^{(2)}_{\mathbf{p}+\mathbf{q}\sigma/2,\mathbf{p}'\sigma'}(\mathbf{k}-\mathbf{q},\mathbf{q},\omega)].\end{aligned} \tag{3.44}$$

Summing both sides of this equation over $\mathbf{p}$ and $\sigma$ we obtain the continuity equation,

$$\omega n(\mathbf{k},\omega) - \sum_{\mathbf{p}\sigma} \mathbf{k}\cdot\mathbf{p}\bar{f}_{\mathbf{p}\sigma}(\mathbf{k},\omega) = 0, \tag{3.45}$$

while multiplying both sides by $\mathbf{k}\cdot\mathbf{p}/m$ and summing over $\mathbf{p}$ and $\sigma$ we recover the equation of motion (3.34).

The analogous equation of motion for $\bar{f}^{(2)}$ has been derived by Niklasson[30] and can be compactly written as

$$\begin{aligned}&\left(\hbar\omega - \frac{\hbar}{m}\mathbf{p}\cdot\mathbf{k} - \frac{\hbar}{m}\mathbf{p}'\cdot\mathbf{k}'\right)\bar{f}^{(2)}_{\mathbf{p}\sigma,\mathbf{p}'\sigma'}(\mathbf{k},\mathbf{k}',\omega)\\ &\quad = [f^{(2)}_{\mathbf{p}-(\mathbf{k}+\mathbf{k}')\sigma/2,\mathbf{p}'\sigma'}(-\mathbf{k}') + f^{(2)}_{\mathbf{p}'-(\mathbf{k}+\mathbf{k}')\sigma/2,\mathbf{p}\sigma}(-\mathbf{k})\\ &\quad -\bar{f}^{(2)}_{\mathbf{p}'+(\mathbf{k}+\mathbf{k}')\sigma/2,\mathbf{p}\sigma}(-\mathbf{k}) - f^{(2)}_{\mathbf{p}+(\mathbf{k}+\mathbf{k}')\sigma/2,\mathbf{p}'\sigma'}(-\mathbf{k}')]V_{\mathrm{H}}(\mathbf{k}+\mathbf{k}',\omega)\\ &\quad + \bar{F}^{(2)}_{\mathbf{p}\sigma,\mathbf{p}'\sigma'}(\mathbf{k},\mathbf{k}',\omega) + \bar{F}^{ms}_{ps,\mathbf{p}'\sigma'}(\mathbf{k},\mathbf{k}',\omega) + \bar{F}^{ms}_{p'\sigma',\mathbf{p}\sigma}(\mathbf{k},\mathbf{k}',\omega)\end{aligned} \tag{3.46}$$

where on the right-hand side we have reported explicitly only the Hartree term, while the other terms are just shorthand notations for more complicated expressions which can be found in the original paper. In a quasi-classical interpretation of $f^{(2)}$ as the equal-time correlation between two particles with moments $\mathbf{p}$ and $\mathbf{p}'$, $\bar{F}^{(2)}$ arises from the mutual interactions between these two particles, whereas the last two terms in Eq. (3.46) contain the many-body aspects of the interaction except for those already included in the Hartree term. As one expects, these terms involve also the change in the three-particle density matrix produced by the external field.

Only the free-particle-like terms in Eqs. (3.44) and (3.46) survive in the limit of large $k$ or large $\omega$. These are the first terms on the right-hand side of the above equations, with $V_{\mathrm{H}}$ replaced by $V_{\mathrm{e}}$. From Eq. (3.44) one finds

$$\chi(k,\omega) \to \sum_{\mathbf{p}\sigma} \frac{n_{\mathbf{p}-\mathbf{k}/2,\sigma} - n_{\mathbf{p}+\mathbf{k}/2,\sigma}}{\hbar\omega - \hbar\mathbf{p}\cdot\mathbf{k}/m}$$
$$\to \frac{nk^2}{m}\left[\omega^2 - \left(\frac{\hbar k^2}{2m}\right)^2\right]^{-1} \qquad \left(\left|\omega \pm \frac{\hbar k^2}{2m}\right| \gg \varepsilon_F/\hbar\right) \tag{3.47}$$

where $n_{\mathbf{p}\sigma}$ is still the true momentum distribution in the ground state. The second term on the right-hand side of Eq. (3.44) can be evaluated in the same limits from Eq. (3.46). Niklasson recovers for large $\omega$ the first two terms of the inverse frequency expansion discussed in Section 2,

$$\lim_{\omega\to\infty} \chi(k,\omega) = M_1(k)\omega^{-2} + M_3(k)\omega^{-4} + \cdots, \tag{3.48}$$

whereas in the limit of large $k$ he finds[8]

$$\lim_{k\to\infty} \sum_{\mathbf{q}} v(q) \sum_{\mathbf{p}\sigma} \sum_{\mathbf{p}'\sigma'} [\bar{f}^{(2)}_{\mathbf{p}-\mathbf{q}\sigma/2,\mathbf{p}'\sigma'}(\mathbf{k}-\mathbf{q},\mathbf{q},\omega)$$
$$- \bar{f}^{(2)}_{\mathbf{p}+\mathbf{q}\sigma/2,\mathbf{p}'\sigma'}(\mathbf{k}-\mathbf{q},\mathbf{q},\omega)]/[\hbar\omega - \hbar\mathbf{p}\cdot\mathbf{k}/m]$$
$$= -\tfrac{2}{3}[1 - g(r=0)]v(k)\chi(k,\omega)n(\mathbf{k},\omega). \tag{3.49}$$

This equation provides the large $k$ value of the local field term that we shall discuss in the following sections. These results are also used by Niklasson[30] to prove that the logarithmic derivative of $g(r)$ at the origin is simply given by

$$\left[\frac{dg(r)}{dr}(g(r))^{-1}\right]_{r=0} = a_B^{-1}, \tag{3.50}$$

a relation previously obtained by Kimball[31] by an analysis of the two-particle collision problem at short distances.

### *d. Hartree–Fock Approximation*

The infinite hierarchy of equations for the Wigner distribution functions can be closed if we adopt the Hartree–Fock decoupling scheme for $f^{(2)}$ This is

$$\langle a^+_{\mathbf{p}-\mathbf{k}/2,\sigma} a^+_{\mathbf{p}'-\mathbf{q}/2,\sigma'} a_{\mathbf{p}'+\mathbf{q}/2,\sigma'} a_{\mathbf{p}+\mathbf{k}/2-\mathbf{q},\sigma}\rangle$$
$$\simeq f^{(1)}_{\mathbf{p}-\mathbf{q}/2,\sigma}(\mathbf{k}-\mathbf{q};t) f^{(1)}_{\mathbf{p}'\sigma'}(\mathbf{q},t)$$
$$- \delta_{\sigma\sigma'} f^{(1)}_{\mathbf{p}/2+\mathbf{p}'/2-\mathbf{k}/4+\mathbf{q}/4,\sigma}(-\mathbf{p}+\mathbf{p}'+\tfrac{1}{2}\mathbf{k}+\tfrac{1}{2}\mathbf{q},t)$$
$$\times f^{(1)}_{\mathbf{p}/2+\mathbf{p}'/2+\mathbf{k}/4-3\mathbf{q}/4,\sigma}(\mathbf{p}-\mathbf{p}'+\tfrac{1}{2}\mathbf{k}-\tfrac{1}{2}\mathbf{q};t). \tag{3.51}$$

[31] J. C. Kimball, *Phys. Rev. A* [3] 7, 1648 (1973).

Equation (3.44) then reduces to

$$\left(\hbar\omega - \frac{\hbar}{m}\,\mathbf{p}\cdot\mathbf{k}\right)\bar{f}^{(1)}_{\mathbf{p}\sigma}(\mathbf{k},\omega)$$
$$= (n_{\mathbf{p}-\mathbf{k}/2,\sigma} - n_{\mathbf{p}+\mathbf{k}/2,\sigma})V_{\mathrm{H}}(\mathbf{k},\omega)$$
$$+ \sum_{\mathbf{q}} v(q)(n_{\mathbf{p}-\mathbf{q}-\mathbf{k}/2,\sigma} - n_{\mathbf{p}-\mathbf{q}+\mathbf{k}/2,\sigma})\bar{f}^{(1)}_{\mathbf{p}\sigma}(\mathbf{k},\omega)$$
$$- \sum_{\mathbf{q}} v(q)(n_{\mathbf{p}-\mathbf{k}/2,\sigma} - n_{\mathbf{p}+\mathbf{k}/2,\sigma})\bar{f}^{(1)}_{\mathbf{p}-\mathbf{q},\sigma}(\mathbf{k},\omega). \tag{3.52}$$

On inspection it is seen that the second term on the right-hand side of Eq. (3.52) plays the role of modifying the single-particle energies, while the third introduces a local field correction arising from exchange; that is, equation (3.52) can be rewritten as

$$(\hbar\omega - \tilde{\varepsilon}_{\mathbf{p}+\mathbf{k}/2} + \tilde{\varepsilon}_{\mathbf{p}-\mathbf{k}/2})\bar{f}^{(1)}_{\mathbf{p}\sigma}(\mathbf{k},\omega) = (n_{\mathbf{p}-\mathbf{k}/2,\sigma} - n_{\mathbf{p}+\mathbf{k}/2,\sigma})V^{\mathrm{eff}}_{p}(\mathbf{k},\omega), \tag{3.53}$$

where

$$\tilde{\varepsilon}_{\mathbf{p}} = \frac{\hbar^2 p^2}{2m} - \sum_{\mathbf{k}} v(k)n_{\mathbf{p}-\mathbf{k}}, \tag{3.54}$$

$$V^{\mathrm{eff}}_{p}(\mathbf{k},\omega) = V_{\mathrm{e}}(\mathbf{k},\omega) + v(k)\bar{n}(\mathbf{k},\omega) - \sum_{\mathbf{q}} v(q)\bar{f}^{(1)}_{\mathbf{p}-\mathbf{q},\sigma}(\mathbf{k},\omega). \tag{3.55}$$

It is thus seen that in the Hartree–Fock approximation, there is a shift in the single-particle energies but no effects of lifetime of single-particle states. Also the local field correction depends on the momentum of the particle.

If we write the proper polarizability as

$$\tilde{\chi}_{\mathrm{HF}}(k,\omega) = \sum_{\mathbf{q}\sigma} \frac{n_{\mathbf{q}\sigma} - n_{\mathbf{q}+\mathbf{k},\sigma}}{\hbar\omega - \tilde{\varepsilon}_{\mathbf{k}+\mathbf{q}} + \tilde{\varepsilon}_{\mathbf{q}}}\,\Lambda^{\mathrm{HF}}_{\mathbf{q}\sigma}(\mathbf{k},\omega), \tag{3.56}$$

then the so-called vertex function $\Lambda^{\mathrm{HF}}$ is found from Eq. (3.53) to satisfy the following integral equation

$$\Lambda^{\mathrm{HF}}_{\mathbf{q}\sigma}(\mathbf{k},\omega) = 1 + \sum_{\mathbf{q}'\sigma'} v(\mathbf{q}-\mathbf{q}')\,\frac{n_{\mathbf{q}'\sigma'} - n_{\mathbf{q}'+\mathbf{k},\sigma'}}{\hbar\omega - \tilde{\varepsilon}_{\mathbf{k}+\mathbf{q}'} + \tilde{\varepsilon}_{\mathbf{q}'}}\,\Lambda^{\mathrm{HF}}_{\mathbf{q}'\sigma'}(\mathbf{k},\omega) \tag{3.57}$$

Equation (3.57) can be solved iteratively. To the first order in the potential, one replaces $\Lambda^{\mathrm{HF}}$ by unity on the right-hand side, and one finds

$$\tilde{\chi}(\mathbf{k},\omega) = \chi_0(\mathbf{k},\omega) + \chi_1(\mathbf{k},\omega) \tag{3.58}$$

where $\chi_0$ is the response function of an ideal Fermi gas [see Eq. (5.3)], and $\chi_1$ in the diagrammatic language corresponds to the first-order bubble diagrams representing the self-energy correction and the exchange local field. A detailed discussion of these results will be given in Section 7.

## 4. Mean-Field Theory

The quantum-mechanical equation of motion (3.34) for the density of the homogeneous electron fluid in the presence of an external field is exact but impossible to solve, since it involves solving an infinite hierarchy of equations for higher-order correlation functions. Under such circumstances, the only course open to us is to resort to approximations guided by physical considerations. There are, in particular, a class of approximations that go under the name of mean-field theory and that have been used with some success both for electron gas and liquid $^3$He problems. These can in essence be viewed as microscopic extrapolations of the known behavior of the liquid in the collisionless regime where $\omega\tau \gg 1$, $\tau$ being the quasi-particle lifetime.

If we neglect the last term in Eq. (3.34), we get what is called the random phase approximation. In this approximation, it is then obvious that the effective field in which the density fluctuation moves is the Hartree field, which is the sum of the external field and the field produced by the induced density fluctuation, i.e., the polarization field. Therefore, in the mean-field approximation, the restoring force responsible for the density oscillation (plasmon, or zero sound) is the averaged self-consistent field of all the particles moving in concert. This is just the reverse of what happens in the "hydrodynamic" regime where the frequent collisions between the particles provide the restoring force for the density oscillations (first sound). An important question that arises is: Is it possible to approximate the last term in Eq. (3.34) so that it acts as a part of an effective field for the density oscillations? In the technical parlance it would mean going beyond the RPA, thus taking exchange and correlation corrections to the Hartree field into account. How is this to be achieved will be for the moment relegated to the background, and we shall assume that it can be done and proceed to develop a theory purely phenomenologically. Such an approach is originally due to Pines.[32]

Consider an interacting Fermi liquid perturbed by a weak external scalar potential $V_e(\mathbf{k}, \omega)$ which induces a particle density polarization $n(\mathbf{k}, \omega)$. Let us formally write the polarization potential as

$$V_{\text{pol}}(\mathbf{k}, \omega) = f_s(k)n(\mathbf{k}, \omega) \tag{4.1}$$

[32] D. Pines, *in* "Quantum Fluids" (D. F. Brewer, ed.), p. 338. North-Holland Publ., Amsterdam, 1966; see also C. H. Aldrich, III, Ph.D. Thesis, University of Illinois, Urbana, 1974 (unpublished).

and introduce a "screened" response function through the following equation,

$$n(\mathbf{k}, \omega) = \chi_{sc}(k, \omega)[V_e(\mathbf{k}, \omega) + f_s(k)n(\mathbf{k}, \omega)]. \tag{4.2}$$

This is clearly an intuitive extension of the development presented in Section 1,b, where $f_s(k)$ was the Coulomb potential $4\pi e^2/k^2$ and $\chi_{sc}(\mathbf{k}, \omega)$ was the proper polarizability $\tilde{\chi}(k, \omega)$. From Eq. (4.2) and the definition of $\chi(k, \omega)$, it follows that

$$\chi(k, \omega) = \chi_{sc}(k, \omega)/[1 - f_s(k)\chi_{sc}(k, \omega)]. \tag{4.3}$$

$\chi_{sc}(k, \omega)$ has two contributions, one coming from single particle–hole excitations and the other coming from simultaneous excitations of more than one particle–hole pair.

In addition to the scalar polarization potential (4.1), we shall also have in the dynamic case a vector polarization potential $\mathbf{A}_{pol}(\mathbf{k}, \omega)$ generated by the current density $\mathbf{j}(\mathbf{k}, \omega)$ induced by $V_e$. Now consider an external vector potential $\mathbf{A}_e(\mathbf{k}, \omega)$ which couples to the current density fluctuation via

$$H_e = \mathbf{A}_e(\mathbf{k}, \omega) \cdot \mathbf{J}(\mathbf{k}) \exp(-i\omega t) \tag{4.4}$$

where the current density operator $\mathbf{J}(\mathbf{k})$ is

$$\mathbf{J}(\mathbf{k}) = \frac{1}{2m} \sum_i (\mathbf{p}_i \exp(-i\mathbf{k} \cdot \mathbf{r}_i) + \exp(-i\mathbf{k} \cdot \mathbf{r}_i)\mathbf{p}_i). \tag{4.5}$$

For a weak longitudinal perturbation, the induced density change is

$$n(\mathbf{k}, \omega) = \boldsymbol{\chi}^{dc}(k, \omega) \cdot \mathbf{A}_e(\mathbf{k}, \omega), \tag{4.6}$$

where the density-current response $\boldsymbol{\chi}^{dc}$ in standard second-order perturbation theory is given by

$$\boldsymbol{\chi}^{dc}(\mathbf{r} - \mathbf{r}', \mathbf{t} - \mathbf{t}') = (-i/\hbar)\vartheta(t - t')\langle[\mathbf{J}(\mathbf{r}, t), \rho(\mathbf{r}', t')]\rangle \tag{4.7}$$

[cf. with Eq. (1.12) ]. Use of the continuity equation yields

$$\mathbf{k} \cdot \boldsymbol{\chi}^{dc}(k, \omega) = \omega\chi(k, \omega). \tag{4.8}$$

Let us now formally write the vector polarization potential generated under application of an external scalar potential $V_e(\mathbf{k}, \omega)$ as

$$\mathbf{A}_{pol}(\mathbf{k}, \omega) = f_v(k)\mathbf{j}(\mathbf{k}, \omega). \tag{4.9}$$

In accord with the preceding calculation, we can write the induced polarization as

$$n(\mathbf{k}, \omega) = \chi_{sc}(k, \omega) [V_e(\mathbf{k}, \omega) + f_s(k)n(\mathbf{k}, \omega)] + \boldsymbol{\chi}_{sc}^{dc}(k, \omega) \cdot \mathbf{A}_{pol}(\mathbf{k}, \omega). \tag{4.10}$$

Equation (4.8) also applies for the screened response because of the continuity equation, i.e.,

$$\mathbf{k} \cdot \boldsymbol{\chi}_{\mathrm{sc}}^{\mathrm{dc}}(k, \omega) = \omega \chi_{\mathrm{sc}}(k, \omega). \tag{4.11}$$

From the above equations, it is straightforward to show that

$$n(\mathbf{k}, \omega) = \chi_{\mathrm{sc}}(k, \omega)[V_{\mathrm{e}}(\mathbf{k}, \omega) + f_{\mathrm{s}}(k) n(\mathbf{k}, \omega) + (\omega^2/k^2) f_{\mathrm{v}}(k) n(\mathbf{k}, \omega)], \tag{4.12}$$

which then gives

$$\chi(k, \omega) = \chi_{\mathrm{sc}}(k, \omega)/\{1 - [f_{\mathrm{s}}(k) + (\omega^2/k^2) f_{\mathrm{v}}(k)]\chi_{\mathrm{sc}}(k, \omega)\}. \tag{4.13}$$

Let us again emphasize the fact that the functions $f_{\mathrm{s}}(k)$ and $f_{\mathrm{v}}(k)$ have been introduced purely phenomenologically, and no attempt has so far been made to derive them from a microscopic theory. The physical meaning of these functions will become clearer as we proceed further. Nor do we have as yet a microscopic description of $\chi_{\mathrm{sc}}(k, \omega)$. Equation (4.13) for the density response function is quite general and is applicable to a condensed fluid.[33] However, the validity of the $\omega^2$ dependence of the vector potential term is limited to low frequencies.

*Connection with the Landau–Fermi Liquid Theory*

In the presence of a weak external potential $V_{\mathrm{e}}(\mathbf{r}, t)$ which varies slowly in space and time, the quasi-particle distribution function is

$$n_{\mathbf{p}}(\mathbf{r}, t) = n_{\mathbf{p}}^0 + \delta n_{\mathbf{p}}(\mathbf{r}, t), \tag{4.14}$$

where $n_{\mathbf{p}}^0$ is the equilibrium quasi-particle distribution, and the subscript **p** subsumes a spin index. In the absence of collisions the flow of quasi-particles in phase space is given by[2]

$$\frac{\partial n_{\mathbf{p}}(\mathbf{r}, t)}{\partial t} + \nabla_{\mathbf{r}} n_{\mathbf{p}}(\mathbf{r}, t) \cdot \nabla_{\mathbf{p}} \tilde{\varepsilon}_{\mathbf{p}}(\mathbf{r}, t) - \nabla_{\mathbf{p}} n_{\mathbf{p}}(\mathbf{r}, t) \cdot \nabla_{\mathbf{r}} \tilde{\varepsilon}_{\mathbf{p}}(\mathbf{r}, t) = 0, \tag{4.15}$$

where $\tilde{\varepsilon}_{\mathbf{p}}$ is the local quasi-particle energy,

$$\tilde{\epsilon}_{\mathbf{p}} = \epsilon_{\mathbf{p}}^0 + V_{\mathrm{H}}(\mathbf{r}, t) + \delta\epsilon_{\mathbf{p}}(\mathbf{r}, t) \tag{4.16}$$

and

$$\delta\epsilon_{\mathbf{p}} = \sum_{p'} f_{\mathbf{p}\mathbf{p}'} \delta n_{\mathbf{p}'}(\mathbf{r}, t). \tag{4.17}$$

[33] Similar expressions have been used in interpretations of the dynamical behavior of liquid $^3$He [C. H. Aldrich, III, C. J. Pethick, and D. Pines, *Phys. Rev. Lett.* **37**, 845 (1976)] and of classical liquids [K. S. Singwi, K. Sköld, and M. P. Tosi, *Phys. Rev. Lett.* **21**, 881 (1968); *Phys. Rev. A* [3] **1**, 454 (1970)].

Equation (4.15) is the form taken by Eq. (3.44) in the Landau theory, and Eq. (4.17) embodies the essential point of the theory according to which the interaction energy of a single excited quasi-particle of momentum **p** depends on the distribution of all other quasi-particles. $f_{\mathbf{pp}'}$ denotes the interaction between quasi-particles of momenta **p** and **p**′, and

$$\epsilon_{\mathbf{p}}^{0} = p^{2}/2m^{*} \tag{4.18}$$

is the energy of a free quasi-particle of mass $m^{*}$.

Linearizing Eq. (4.15) we have

$$\frac{\partial}{\partial t}\delta n_{\mathbf{p}}(\mathbf{r}, t) + \nabla_{\mathbf{r}}\delta n_{\mathbf{p}}(\mathbf{r}, t)\cdot\nabla_{\mathbf{p}}\varepsilon_{\mathbf{p}}^{0} - \nabla_{\mathbf{p}} n_{\mathbf{p}}^{0}\cdot\nabla_{\mathbf{r}}[V_{\mathrm{H}}(\mathbf{r}, t) + \delta\varepsilon_{\mathbf{p}}(\mathbf{r}, t)] = 0. \tag{4.19}$$

Taking the Fourier transform of Eq. (4.19) and using Eq. (4.17) we have

$$\delta n_{\mathbf{p}}(\mathbf{k}, \omega) = \frac{\mathbf{k}\cdot\nabla_{\mathbf{p}} n_{\mathbf{p}}^{0}}{\mathbf{k}\cdot\nabla_{\mathbf{p}}\varepsilon_{\mathbf{p}}^{0} - \omega}\left[V_{\mathrm{H}}(\mathbf{k}, \omega) + \sum_{\mathbf{p}'} f_{\mathbf{pp}'}\delta n_{\mathbf{p}'}(\mathbf{k}, \omega)\right]. \tag{4.20}$$

Now

$$\nabla_{\mathbf{p}} n_{\mathbf{p}}^{0} = \nabla_{\mathbf{p}}\varepsilon_{\mathbf{p}}^{0}\partial n_{\mathbf{p}}^{0}/\partial\varepsilon_{\mathbf{p}}^{0} \tag{4.21}$$

and since $n_{\mathbf{p}}^{0}$ is the usual Fermi distribution, $\partial n_{\mathbf{p}}^{0}/\partial\varepsilon_{\mathbf{p}}^{0}$ is a delta function centered around the Fermi momentum and both **p** and **p**′ are equal to $p_{\mathrm{F}}$. This enables us to expand $f_{\mathbf{pp}'}$ in sperical harmonics,

$$f_{\mathbf{pp}'} = f_0 + f_1\hat{p}\cdot\hat{p}' + \cdots. \tag{4.22}$$

Substituting Eq. (4.22) in Eq. (4.20) and summing over **p**, we have

$$n(\mathbf{k}, \omega) = \chi_{\mathrm{s}}(k, \omega)[V_{\mathrm{H}}(\mathbf{k}, \omega) + f_0 n(\mathbf{k}, \omega)] + \sum_{\mathbf{p}}\frac{\mathbf{k}\cdot\nabla_{\mathbf{p}}\varepsilon_{\mathbf{p}}^{0}}{\mathbf{k}\cdot\nabla_{\mathbf{p}}\varepsilon_{\mathbf{p}}^{0} - \omega}\frac{\partial n_{\mathbf{p}}^{0}}{\partial\varepsilon_{\mathbf{p}}^{0}} f_1\hat{p}\cdot\sum_{\mathbf{p}'}\hat{p}'\delta n_{\mathbf{p}'}(\mathbf{k}, \omega), \tag{4.23}$$

where

$$\chi_{\mathrm{s}}(k, \omega) = \sum_{\mathbf{p}}\frac{\mathbf{k}\cdot\nabla_{\mathbf{p}}\varepsilon_{\mathbf{p}}^{0}}{\mathbf{k}\cdot\nabla_{\mathbf{p}}\varepsilon_{\mathbf{p}}^{0} - \omega}\frac{\partial n_{\mathbf{p}}^{0}}{\partial\varepsilon_{\mathbf{p}}^{0}} \to \lim_{\substack{\omega\to 0\\ k\to 0}} - \nu(0). \tag{4.24}$$

This function is the appropriate form of the density response function for a system of free particles with mass $m^{*}$, with a limiting value determined by $\nu(0) = 3m^{*}/p_{\mathrm{F}}^{2}$, the density of states per particle at the Fermi surface. Using the fact that the distortion from the equilibrium distribution is restricted to values of **p**′ lying on the Fermi surface, we can write in the last term of Eq. (4.23)

$$\sum_{\mathbf{p}'}\hat{p}'\delta n_{\mathbf{p}'}(\mathbf{k}, \omega) = m\mathbf{j}(\mathbf{k}, \omega)/p_{\mathrm{F}},$$

whence, using the continuity equation and performing the integration, we have

$$n(\mathbf{k}, \omega) = \chi_s(k, \omega)\left[V_H(\mathbf{k}, \omega) + f_0 n(\mathbf{k}, \omega) + f_1 \frac{mm^*}{p_F^2} \frac{\omega^2}{k^2} n(\mathbf{k}, \omega)\right]. \quad (4.25)$$

This equation clearly leads to a form of $\chi(k, \omega)$ which agrees with the result (4.13) of the mean-field theory, and by comparison one notices that the polarization potential theory of Pines corresponds to the Landau theory if

$$\lim_{k\to 0} f_s(k) = \frac{4\pi e^2}{k^2} + f_0, \quad (4.26a)$$

$$\lim_{k\to 0} f_v(k) = mm^* f_1/p_F^2, \quad (4.26b)$$

and

$$\lim_{k\to 0}\lim_{\omega\to 0} \chi_{sc}(k, \omega) = -\nu(0). \quad (4.26c)$$

The polarization potential theory can, therefore, be considered as an extension of the Landau theory for finite values of $k$ and $\omega$ subject to the proviso that the mean field is a good description in that region of wave number and frequency. In the absence of a microscopic theory, one can use sum rule arguments to determine $f_s(k)$ and $f_v(k)$. Such an approach for an electron fluid has been examined by Gupta *et al.*[34]

The foregoing discussion of the Landau theory has ignored the effect of collisions between quasi-particles. However, it is possible to take collisions into account within a constant-relaxation-time approximation, as done by Mermin[35] in the RPA.

## III. Approximate Schemes

### 5. Self-Consistent Theory and Local Field Correction

If in Eqs. (3.33) and (3.34) we neglect the last term entirely and approximately relate $\delta\pi_{\alpha\beta}(\mathbf{R}, t)$ to the Hartree potential via a free particle calculation,

$$\pi_{\alpha\beta}(\mathbf{k}, \omega) = \sum_{\mathbf{p}\sigma} \frac{P_\alpha P_\beta}{m} \frac{n^0_{\mathbf{p}-\mathbf{k}/2,\sigma} - n^0_{\mathbf{p}+\mathbf{k}/2,\sigma}}{\omega - \mathbf{p}\cdot\mathbf{k}/m + i\eta} V_H(\mathbf{k}, \omega), \quad (5.1)$$

[34] A. K. Gupta, P. K. Aravind, and K. S. Singwi, *Solid State Commun.* **26,** 49 (1978).
[35] D. N. Mermin, *Phys. Rev. B: Solid State* [3] **1,** 3262 (1970).

where $n^0_{\mathbf{p}\sigma}$ is the usual Fermi distribution function, then the density response function is the usual RPA result,

$$\chi_{\text{RPA}}(k, \omega) = \chi_0(k, \omega)\left[1 - \frac{4\pi e^2}{k^2}\chi_0(k, \omega)\right]^{-1}. \tag{5.2}$$

Here $\chi_0(k, \omega)$ is the Lindhard polarizability[36] of an ideal Fermi gas,

$$\chi_0(k, \omega) = \sum_{\mathbf{p}\sigma} \frac{n^0_{\mathbf{p}-\mathbf{k}/2,\sigma} - n^0_{\mathbf{p}+\mathbf{k}/2,\sigma}}{\omega - \mathbf{p}\cdot\mathbf{k}/m + i\eta}. \tag{5.3}$$

Explicit expressions for the real and imaginary parts of this function can be found, for instance, in Ref. 2.

It is apparent from the above derivation that RPA corresponds to a mean-field theory in which $f_s(k) = 4\pi e^2/k^2$ and $f_v(k) = 0$, and $\chi_{sc}(k, \omega)$ is rplaced by $\chi_0(k, \omega)$. It is a very simple approximation that accounts for both the collective and the singl electron–hole pair excitations, and has been sidely used. However, a number of quantitative deficiencies of the RPA have been noted by many authors over the years. For example, it leads through Eq. (2.1) to the free particle value of the compressibility[37] and hence of the screening length—a major defect from the point of view of determining the effective interionic potential in metals. Another glaring deficiency is that it leads through Eqs. (1.20)–(1.22) to a pair correlation function $g(r)$ that becomes negative[38] for small values of $r$ over the entire metallic density range. This fact has the consequence of overestimating the correlation energy. In this section we shall be concerned mainly with these properties and shall defer the discussion of the static dielectric function to Section 6 and of the dynamical aspects to Section 7.

From the diagrammatic point of view, RPA corresponds to taking for the proper polarizability $\tilde{\chi}(k, \omega)$, introduced in Eq. (1.25), only the empty electron–hole bubble diagram,[39] whose value is given by Eq. (5.3). Later on Hubbard[40] discussed the leading exchange diagrams and was in a

[36] J. Lindhard, *Mat.-Fys. Medd.—K. Dan. Vidensk. Selsk.* **28** (8) (1954).

[37] Of course, the value of the compressibility that one obtains from the RPA ground state energy differs from the free particle value. The requirement that the two values of the compressibility (one obtained from the long-wavelength limit of the static response and the other from differentiating the ground-state energy) should agree, provides a self-consistency condition for the theory. This is usually referred to as the compressibility sum rule.

[38] A. J. Glick and R. A. Ferrell, *Ann. Phys.* (*N.Y.*) **11,** 359 (1960); L. Hedin, *Phys. Rev.* [2] **139,** A796 (1965); F. Brouers, *Phys. Status Solidi* **19,** 867 (1967).

[39] See, e.g., P. Nozières, "Theory of Interacting Fermi Systems." Benjamin, New York, 1964.

[40] J. Hubbard, *Proc. R. Soc. London, Ser. A* **240,** 539 (1957); **243,** 336 (1957).

certain approximation able to sum the infinite set of ladder diagrams. His expression for the density response is

$$\chi_H(k, \omega) = \chi_0(k, \omega)/\{1 - 4\pi e^2/k^2)[1 - G_H(k)]\chi_0(k, \omega)\}, \tag{5.4}$$

with

$$G_H(k) = \tfrac{1}{2}k^2/(k^2 + k_F^2). \tag{5.5}$$

Hubbard's result has again the structure of a static mean-field theory in which the bare Coulomb potential is reduced by the factor $[1 - G_H(k)]$. The factor $G_H(k)$plays the role of a local field correction, which vanishes for $k \to 0$ since the electrons are completely delocalized and tends according to Eq. (5.5) to a value $\frac{1}{2}$ for $k \gg k_F$, implying that the exchange terms are taken to cancel one-half of the direct Coulomb term in this limit. Physically the Hubbard local field correction thus accounts for the presence of the Pauli hole around each electron, but clearly neglects the effect of Coulomb correlations in the local field correction.

Historically, a number of authors[41,42] have attempted to account empirically for Coulomb correlations by modifying Eq. (5.5) to

$$G_H(k) = \tfrac{1}{2}k^2/(k^2 + \xi k_F^2), \tag{5.6}$$

where $\xi$ is a parameter. In particular Geldart and Vosko[42] stressed the importance of satisfying the compressibility sum rule, which would require $\xi \simeq 2$ in Eq. (5.6). They analyzed the diagrams relevant in the long-wavelength, static limit, showing that a proper theory must include correlation diagrams in addition to the exchange diagrams considered by Hubbard.

### *a. Self-Consistency on the Pair Correlation Function*

The last term in Eq. (3.33) describes through the two-body correlation function $\langle \rho(\mathbf{R}, t)\rho(\mathbf{x}, t)\rangle_c$ all the complicated effects of the Pauli and Coulomb hole surrounding each electron, in the presence of the external field. A rigorous solution of the problem, as mentioned earlier, is impossible, and our first aim is to extract a local field correction. Noting that

$$\langle \rho(\mathbf{R}, t)\rho(\mathbf{x}, t)\rangle_c = n(\mathbf{R}, t)n(\mathbf{x}, t)[g(\mathbf{R}, \mathbf{x}; t) - 1], \tag{5.7}$$

where $\mathbf{g}(\mathbf{R}, \mathbf{x}; t)$ is the nonequilibrium pair correlation function, the simplest approximation[43] is to replace this function by its equilibrium

[41] L. J. Sham, *Proc. R. Soc. London, Ser. A* **283,** 33 (1965); W. M. Shyu and G. D. Gaspari, *Phys. Rev.* [2] **163,** 667 (1967),; **170,** 687 (1968).

[42] D. J. W. Geldart and S. H. Vosko, *Can. J. Phys.* **44,** 2137 (1966).

[43] K. S. Singwi, M. P. Tosi, R. H. Land, and A. Sjölander, *Phys. Rev.* [2] **176,** 589 (1968).

value $\mathbf{g}(\mathbf{R} - \mathbf{x})$, thus accounting for the local depletion of charge density but neglecting the dynamics of the hole. This approximation replaces the Hartree potential by a local effective potential given by

$$V_{\text{eff}}(\mathbf{k}, \omega) = V_{\text{H}}(\mathbf{k}, \omega) - (4\pi e^2/k^2)G(k)n(\mathbf{k}, \omega) \tag{5.8}$$

with

$$G(k) = -\frac{1}{n}\int \frac{d\mathbf{q}}{(2\pi)^3}\frac{\mathbf{k}\cdot\mathbf{q}}{q^2}[S(\mathbf{k}-\mathbf{q}) - 1]. \tag{5.9}$$

Relating $\pi_{\alpha\beta}(\mathbf{k}, \omega)$ to this effective potential by an equation analogous to (5.1), one has

$$\chi(k, \omega) = \chi_0(k, \omega)/\{1 - (4\pi e^2/k^2)[1 - G(k)]\chi_0(k, \omega)\}. \tag{5.10}$$

Although $G(k)$ contains the unknown structure factor $S(k)$ of the electron fluid, this can be determined by requiring consistency with the structure factor obtained through the use of the fluctuation-dissipation theorem, Eqs. (1.19)–(1.21). In this scheme, therefore, both exchange and correlation corrections to the Hartree field are automatically included. For example, it is straightforward to recover[43] the Hubbard result (5.5) by substituting for the structure factor in Eq. (5.9) its Hartree–Fock value.

The reasonableness of the above approximation coupled with the self-consistency requirement yields a pair distribution function that shows a remarkable improvement over those calculated from previous approximate dielectric functions (see Figs. 1 and 2). Using the above self-consistent scheme, Hubbard[44] first calculated the pair correlation function and demonstrated its reasonableness compared with that obtained from earlier theories. Recently, two entirely different approaches have been used to calculate $g(r)$, one based on the method of correlated basis wave function[45] and the other based on diagrammatic perturbation theory in which an infinite sum of electron–electron ladders are taken into account.[46,47] Furthermore, Chihara[48] has suggested on the basis of Mori's theory[18] that $S(\mathbf{k} - \mathbf{q})$ in Eq. (5.9) should be replaced by $\chi(\mathbf{k} - \mathbf{q})/\chi_0(\mathbf{k} - \mathbf{q})$, these two quantities being equivalent in the classical case but not in the quantum case. The pair correlation function calculated by Chihara remains positive even for $r_s = 6$. It is gratifying to note that all these

[44] J. Hubbard, *Phys. Lett. A* **25,** 709 (1967).

[45] S. Chakravarty and C. W. Woo, *Phys. Rev. B: Solid State* [3] **13,** 4815 (1976). References to previous work by this technique can be found in this paper.

[46] H. Yasuhara, *Solid State Commun.* **11,** 1481 (1972); *J. Phys. Soc. Jpn.* **36,** 361 (1974); *Physica (Amsterdam)* **78,** 420 (1974).

[47] D. N. Lowy and G. E. Brown, *Phys. Rev. B: Solid State* [3] **12,** 2138 (1975).

[48] J. Chihara, *Prog. Theor. Phys.* **50,** 1156 (1973); **53,** 400 (1975).

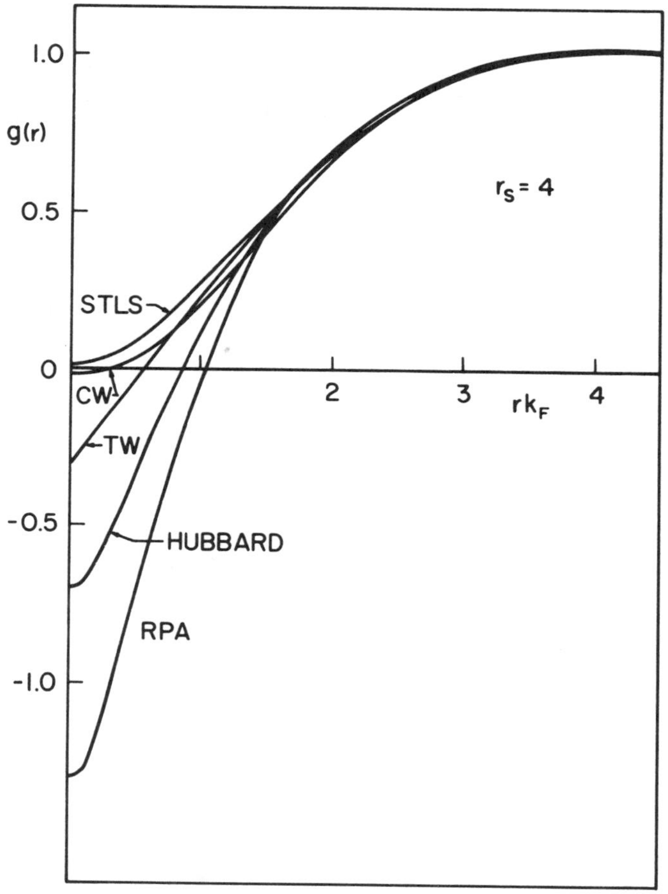


FIG. 1. The pair correlation function $g(r)$ versus $rk_F$ for $r_s = 4$. The abbreviations refer to the following articles: STLS: K. S. Singwi, M. P. Tosi, R. H. Land, and A. Sjölander, *Phys. Rev.* [2] **176**, 589 (1968); TW: F. Toigo and T. O. Woodruff, *Phys. Rev. B: Solid State* [3] **4**, 371 (1971); CW: S. Chakravarty and C. W. Woo, *Phys. Rev. B: Solid State* [3] **13**, 4814 (1976).

methods give values of $g(r)$ that are very close to those calculated by the self-consistent dielectric scheme.

Using Eqs. (1.23) and (1.24) and remembering the fact that the integral over the interaction strength parameter $\lambda$ is equivalent to an integral over the parameter $r_s$, the ground-state energy can be written as[43]

$$E_{gs} = \left\{\frac{2.2099}{r_s^2} + \frac{2}{\pi r_s^2}\left(\frac{9\pi}{4}\right)^{1/3} \int_0^{r_s} dr_s \int_0^{\infty} [S(q;\, r_s) - 1]d(q/q_F)\right\} \text{ Ry.} \quad (5.11)$$

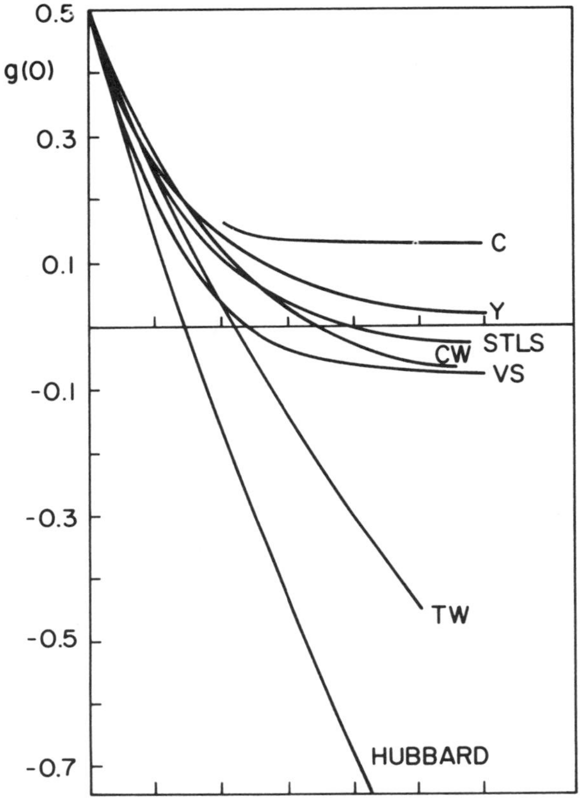


FIG. 2. The pair correlation function $g(r = 0)$ versus $r_s$. The abbreviations refer to the following articles: STLS: K. S. Singwi, M. P. Tosi, R. H. Land, and A. Sjölander, *Phys. Rev.* [2] **176,** 589 (1968); TW: F. Toigo and T. O. Woodruff, *Phys. Rev. B: Solid State* [3] **4,** 371 (1971); VS: P. Vashishta and K. S. Singwi, *Phys. Rev. B: Solid State* [3] **6,** 875 (1972); Y: H. Yasuhara, *J. Phys. Soc. Jpn.* **36,** 361 (1974); CW: S. Chakravarty and C. W. Woo, *Phys. Rev. B: Solid State* [3] **13,** 4815 (1976); C: J. Chihara, *Prog. Theor. Phys.* **53,** 400 (1975). The curves Y and C have been drawn from rough readings of graphs in the original articles.

In Table I are given the values of the correlation energy at various values of $r_s$ for different approximate schemes. It is clear from the table that they all agree within about 20% over the whole range of metallic densities, with the exception of the RPA.[48a] These values yield rather good agreement with the cohesive energies of simple metals.[2,49] The density dependence of the

[48a] Note the remarkable agreement between what is presently considered to be the most reliable values of the correlation energy (Ref. *i* of Table I) and those given by the *self*-consistent scheme (STLS, Ref. *e* of Table I) for values of $r_s$ as large as 20.

[49] N. W. Ashcroft and D. C. Langreth, *Phys. Rev.* [2] **155,** 682 (1967).

TABLE I. CORRELATION ENERGY[a]

| $r_s$ | 1 | 2 | 3 | 4 | 5 | 6 | 10 | 20 |
|---|---|---|---|---|---|---|---|---|
| Wigner[b] | −0.100 | −0.090 | −0.082 | −0.075 | −0.069 | −0.064 | −0.049 | −0.032 |
| RPA[c] | −0.157 | −0.124 | −0.105 | −0.094 | −0.085 | −0.078 | — | — |
| Hubbard[d] | −0.131 | −0.102 | −0.086 | −0.076 | −0.069 | −0.064 | — | — |
| Nozières and Pines[c] | −0.115 | −0.094 | −0.081 | −0.072 | −0.065 | −0.060 | — | — |
| STLS[e] | −0.124 | −0.092 | −0.075 | −0.064 | −0.056 | −0.050 | −0.036 | −0.022 |
| Vashishta and Singwi[f] | −0.130 | −0.098 | −0.081 | −0.070 | −0.062 | −0.056 | — | — |
| Toigo and Woodruff[g] | −0.120 | −0.092 | −0.077 | −0.068 | −0.061 | −0.056 | — | — |
| Chakravarty and Woo[h] | — | −0.0913 | — | −0.0679 | — | — | — | — |
| Ceperley[i] | −0.120 | −0.090 | −0.074 | −0.064 | −0.056 | −0.051 | −0.037 | −0.023 |

[a] In rydbergs per electron.

[b] E. P. Wigner, *Phys. Rev.* [2] **46,** 1002 (1934); *Trans. Faraday Soc.* **34,** 678 (1938).

[c] D. Pines and P. Nozières, "The Theory of Quantum Fluids." Benjamin, New York, 1966.

[d] J. Hubbard, *Proc. London, Ser. Soc. A* **240,** 539 (1957); **243,** 336 (1957).

[e] K. S. Singwi, M. P. Tosi, R. H. Land, and A. Sjölander, *Phys. Rev.* [2] **176,** 589 (1968).

[f] P. Vashishta and K. S. Singwi, *Phys. Rev. B: Solid State* [3] **6,** 873, 4883 (1972).

[g] F. Toigo and T. O. Woodruff, *Phys. Rev. B: Solid State* [3] **2,** 3958 (1970); **4,** 371 (1971).

[h] S. Chakravarty and C. W. Woo, *Phys. Rev. B: Solid State* [3] **13,** 4815 (1976).

[i] D. Ceperley, *Phys. Rev. B: Condens. Matter* [3] **18,** 3126 (1978). These results are obtained by Monte Carlo variational calculations based on a trial Jastrow function. See also D. Ceperley and B. J. Alder, *Phys. Rev. Lett.* **45,** 566 (1980); S. H. Vosko, L. Wilk, and M. Nusair, *Can. J. Phys.* **58,** 1200 (1980).

correlation energy in the various schemes over this range of $r_s$ is such that the compressibility of the electron fluid derived from differentiating the ground-state energy does not differ by more than a few percent from the Hartree–Fock value,

$$(K_0/K)_{\mathrm{HF}} = 1 - \tfrac{1}{4}(k_{\mathrm{TF}}/k_{\mathrm{F}})^2 \tag{5.12}$$

where $k_{\mathrm{TF}} = (6\pi ne^2/\varepsilon_{\mathrm{F}})^{1/2}$ is the Thomas–Fermi inverse screening length.

In the self-consistent scheme discussed above, the compressibility sum rule is grossly violated. In subsequent work[50] the effect of screening the interaction in the integral (5.9) was examined with the result that the compressibility sum rule was nearly satisfied without much detriment to the pair correlation function. Such screening arises from the dynamic deformation of the hole, neglected in arriving at Eq. (5.9), and leads in principle to a frequency-dependent local field correction.

These schemes as well as analogous schemes[51] have been successfully used to calculate the pair correlation function and the thermodynamic properties of a classical one-component plasma, for which a direct comparison against computer simulation data is possible. An extension of the STLS (Singwi *et al.*, 1968; see Fig. 1) scheme to the calculation of the structure factor of liquid metals has recently been given by Ailawadi *et al.*[52] For details we refer the reader to the original articles.

### *b. Extended Self-Consistent Scheme*

Our primary concern is now to improve the above scheme in such a manner that it will also satisfy the compressibility sum rule self-consistently. If we compare the expression (5.10) for $\chi(k, \omega)$ in the appropriate limit with the exact expression (3.29) of the static response at long wavelength, we find

$$\lim_{k\to 0} G(k) = \frac{k^2}{4\pi e^2}\left\{\frac{1}{3}\int d\mathbf{r}\, r\,\frac{dv(r)}{dr}\left(1 + \frac{1}{2}n\frac{\partial}{\partial n}\right)[g(r;n) - 1] - \frac{2}{3n}\frac{d}{dn}[t(n) - t_0(n)]\right\}. \tag{5.13}$$

[50] K. S. Singwi, A. Sjölander, M. P. Tosi, and R. H. Land, *Phys. Rev. B: Solid State* [3] **1,** 1044 (1970).

[51] K. F. Berggren, *Phys. Rev. A* [3] **1,** 1783 (1970); S. Ichimaru, *ibid.* **2,** 494 (1970); M. C. Abramo and M. P. Tosi, *Nuovo Cimento Sor. Ital. Fis-B* [11] **10B,** 21 (1972); **21B,** 363 (1974); H. Totsuji and S. Ichimaru, *Prog. Theor. Phys.* **50,** 753 (1973); **52,** 42 (1974); K. I. Golden, G. Kalman, and M. B. Silevitch, *Phys. Rev. Lett.* **33,** 1544 (1974); G. Kalman, T. Datta, and K. I. Golden, *Phys. Rev. A* **12,** 1125 (1975). See also G. Katman (ed.), "Strongly Coupled Plasmas." Plenum, New York, 1978.

[52] N. K. Ailawadi, D. E. Miller, and J. Naghizadeh, *Phys. Rev. Lett.* **36,** 1494 (1976).

From the virial theorem, we can derive the following relation[29]

$$\frac{5}{3}\,t(n) - n\,\frac{dt(n)}{dn} = \frac{1}{2}\,n^2 \int d\mathbf{r}\; v(r)\left(1 + n\,\frac{\partial}{\partial n}\right)[g(r;\,n) - 1] + \frac{1}{6}\,n^2 \int d\mathbf{r}\; r\,\frac{dv(r)}{dr}\,[g(r;\,n) - 1], \tag{5.14}$$

which can be integrated to give

$$t(n) - t_0(n) = -\frac{1}{2}\,n^2 \int d\mathbf{r}\; v(r)[g(r;\,n) - 1] - \frac{1}{6}\,n^{5/3} \int d\mathbf{r}\; v(r) \int_\infty^n dn\; n^{-2/3}[g(r;\,n) - 1]. \tag{5.15}$$

Using the above expression, Eq. (5.13) becomes

$$\lim_{k\to 0} G(k) = \frac{k^2}{4\pi e^2}\left\{\frac{1}{3}\int d\mathbf{r}\; r\,\frac{dv(r)}{dr}\left(1 + \frac{1}{2}\,n\,\frac{\partial}{\partial n}\right)[g(r;\,n) - 1] + \frac{1}{3}\int d\mathbf{r}\; v(r)\left(\frac{7}{3} + n\,\frac{\partial}{\partial n}\right)[g(r;\,n) - 1] + \frac{5}{27}\,n^{-1/3}\int d\mathbf{r}\; v(r)\int_\infty^n dn\; n^{-2/3}[g(r;\,n) - 1]\right\}. \tag{5.16}$$

It is clear that the long-wavelength limit of the local field factor in Eq. (5.10) is completely determined by the density dependence of $g(r)$.

In extending the self-consistent scheme, Vashishta and Singwi[53] took the following form of $G(k)$,

$$G(k) = \left(1 + a\,\frac{\partial}{\partial n}\right)\left\{-\frac{1}{n}\int \frac{d\mathbf{q}}{(2\pi)^3}\,\frac{\mathbf{k}\cdot\mathbf{q}}{q^2}\,[S(\mathbf{k} - \mathbf{q};\,n) - 1]\right\} \tag{5.17}$$

with the requirements that the structure factor be determined self-sonsistently through the fluctuation-dissipation theorem and that the parameter $a$ be determined self-consistently from the compressibility sum rule. This expression reduces in the long-wavelength limit to the first term of Eq. (5.13) with the coefficient $\frac{1}{2}$ replaced by the parameter $a$, and the calculation yielded $a \simeq \frac{2}{3}$ for all values of $r_s$ in the metallic density range. It was later verified[54] that this value agrees closely with what one obtains from Eq. (5.16) over the same range. In fact, $a$ is a slowly increasing function of $r_s$ for $1 \le r_s \le 6$, and tends to the value $\frac{1}{2}$ for $r_s \to 0$.[55]

[53] P. Vashishta and K. S. Singwi, *Phys. Rev. B: Solid State* [3] **6,** 875, 4883 (1972).
[54] J. S. Vaishya and A. K. Gupta, *Phys. Rev. B: Solid State* [3] **7,** 4300 (1973).
[55] A. K. Gupta and K. S. Singwi, *Phys. Rev. B: Solid State* [3] **15,** 1801 (1977).

This scheme which is tailored to satisfy the compressibility sum rule also yields values of $g(r)$ in the metallic density range that are of comparable quality to those obtained in the earlier scheme. The values of the correlation energy are given in Table I and can actually be represented very closely by the expression

$$\varepsilon_c(r_s) = \left(-0.112 + 0.0335 \ln r_s - \frac{0.02}{0.1 + r_s}\right) \text{ Ry} \tag{5.18}$$

for $1 \leq r_s \leq 6$. The correlation contributions to the kinetic and potential energy are reported separately in Table II together with the free particle kinetic energy and the Hartree–Fock potential energy. Values of the compressibility can be obtained easily using the analytic expression (5.18).

A common result of all the theories is that the compressibility of the quantal electron fluid on an inert background goes negative for $r_s \simeq 5.2$. A negative electronic compressibility implies, according to the discussion given in Section 1, that the screening length has become imaginary. The same behavior has been observed in computer simulation studies[56] of the classical one-component plasma, where the appearance of negative values of $k_s^2$ is soon followed by the emergence of oscillations in $g(r)$. One is therefore tempted to speculate that a similar transition may occur in the quantum electron fluid, well before the value of $r_s$ at which Wigner crystallization is believed to take place.

### c. Connection with Diagrammatic Approach

The proper polarizability $\tilde{\chi}(\mathbf{k}, \omega)$ is given[39] in terms of an irreducible vertex function $\tilde{\Lambda}_k(p)$ by

$$\tilde{\chi}(\mathbf{k}, \omega) = 2 \int \frac{d^4p}{(2\pi)^4 i} G(p)G(p + k)\tilde{\Lambda}_k(p) \tag{5.19}$$

where $k = (\mathbf{k}, \omega)$, $p = (\mathbf{p}, p_0)$, and $G(p)$ is the exact propagator for an electron. The vertex function is to be determined through the integral equation

$$\tilde{\Lambda}_k(p) = 1 + \int \frac{d^4p'}{(2\pi)^4 i} \tilde{I}_k(p, p')G(p')G(p' + k)\tilde{\Lambda}_k(p'), \tag{5.20}$$

where $\tilde{I}_k(p, p')$ is the irreducible particle–hole interaction, for which a precise prescription is provided in terms of diagrams. This program

[56] P. Vieillefosse and J. P. Hansen, *Phys. Rev. A* [3] **12**, 1106 (1975).

TABLE II. CONTRIBUTIONS TO THE GROUND-STATE ENERGY[a]

| $r_s$ | $E_0$ | $-0.9163/r_s$ | $E_{\text{kin}} - E_0$ | $\varepsilon_c^{\text{pot}}$ |
|---|---|---|---|---|
| 1 | 2.2099 | −0.9163 | 0.083 | −0.213 |
| 2 | 0.5525 | −0.4582 | 0.055 | −0.153 |
| 3 | 0.2455 | −0.3054 | 0.041 | −0.122 |
| 4 | 0.1381 | −0.2291 | 0.032 | −0.102 |
| 5 | 0.0884 | −0.1833 | 0.027 | −0.089 |
| 6 | 0.0614 | −0.1527 | 0.023 | −0.079 |

[a] In rydbergs per electron.

obviously presents great difficulties, but fortunately the average over $p$ involved in Eq. (5.19) suggests that the dielectric function may not be overly sensitive to the detailed dependence of $\tilde{\Lambda}$ and $\tilde{I}$ on the four-momenta of the particles.

From Eqs. (5.19) and (5.20), it follows that

$$\int \frac{d^4p}{(2\pi)^4 i} G(p)G(p+k)\tilde{\Lambda}_k(p) = \int \frac{d^4p}{(2\pi)^4 i} G(p)G(p+k) + \iint \frac{d^4p\, d^4p'}{(2\pi)^8 i^2} \tilde{I}_k(p, p')G(p)G(p+k)G(p')G(p'+k)\tilde{\Lambda}_k(p'), \tag{5.21}$$

where, in fact, the first term on the right-hand side is one-half of the free particle polarizability with self-energy corrections included, $\tilde{\chi}_0(\mathbf{k}, \omega)$. If one now replaces $\tilde{I}_k(p, p')$ by some average over the four-momenta of the two particles, Eq. (5.21) trivially yields

$$\tilde{\chi}(\mathbf{k}, \omega) = \tilde{\chi}_0(\mathbf{k}, \omega)/[1 - \langle \tilde{I}_k(p, p')\rangle \tilde{\chi}_0(\mathbf{k}, \omega)], \tag{5.22}$$

which can be cast in a mean-field form with a local field factor given by

$$G(\mathbf{k}, \omega) = \frac{k^2}{4\pi e^2}\left[-\frac{1}{2}\langle \tilde{I}_k(p, p')\rangle + \frac{1}{\tilde{\chi}_0(\mathbf{k}, \omega)} - \frac{1}{\chi_0(\mathbf{k}, \omega)}\right]. \tag{5.23}$$

In fact, Hubbard[40] used the Hartree–Fock value of $\tilde{I}_k(p, p')$.

$$\tilde{I}_k(p, p') = -4\pi e^2/(\mathbf{p} - \mathbf{p}')^2 \tag{5.24}$$

and noticed that important contributions to the integral in Eq. (5.21) come from $(\mathbf{p} - \mathbf{p}')^2 \simeq \mathbf{k}^2 + k_F^2$. This suggests $\langle \tilde{I}_k(p, p')\rangle = -4\pi e^2/(\mathbf{k}^2 + k_F^2)$ whence, neglecting self-energy corrections, the value (5.5) for the local field factor is obtained.

The approach based on Eqs. (5.19) and (5.20) has been followed by several authors.[57–61] Kleinman,[57] working with a screened Hartree–Fock form for $\tilde{I}$, has extended Hubbard's work by accounting for two values of $(\mathbf{p} - \mathbf{p}')^2$ which yield important contributions to the last integral in Eq. (5.21), and at the same time including self-energy corrections. Langreth,[58] on the other hand, has sought a variational solution of Eq. (5.20) such that $\tilde{\Lambda}_k(p)$ is independent of the four-momentum $p$, under the assumption that $\tilde{I}_k(p, p')$ is independent of the energies of the incoming particles. His result can trivially be obtained from Eq. (5.21) by assuming $\tilde{\Lambda}_k(p)$ to be independent of $p$. Assumptions analogous to those of Kleinman on the momentum dependence of $\tilde{I}_k(p, p')$ are then made to arrive at an estimate of the dielectric function.

In this context an important point to note is that, although Kleinman's approximations lead to a value of $\langle \tilde{I}_k(p, p') \rangle$ which is a constant at large values of $k$, this divergent contribution to $G(\mathbf{k}, \omega)$ in Eq. (5.23) is canceled by the self-energy corrections contained in the remaining two terms.[62] That such a cancellation must occur exactly at large $k$ is clear from the work of Niklasson[30] discussed in Section 3,c [Eq. (3.49)], which yields the exact result

$$\lim_{k\to\infty} G(k, \omega) = \tfrac{2}{3}[1 - g(0)]. \qquad (5.25)$$

The need to preserve this cancellation between effective interactions and self-energy effects imposes a restriction on the mean-field theory discussed in Section 4.

Because of the complexity of the problem, the soundness of the assumptions which go into the calculation of a dielectric function should be judged by the results it yields. The pair correlation function obtained from the Kleinman–Langreth approximation is of the same quality[62] as is obtained in the simple Hubbard approximation. Thus, a screened Hartree–Fock approximation seems unable to yield a reasonable $g(r)$.

[57] L. Kleinman, *Phys. Rev.* [2] **160,** 585 (1967); **172,** 383 (1968).

[58] D. C. Langreth, *Phys. Rev.* [2] **181,** 753 (1969).

[59] J. W. F. Woo and S. S. Jha, *Phys. Rev. B: Solid State* [3] **3,** 87 (1971); S. S. Jha, K. K. Gupta, and J. W. F. Woo, *ibid.* **4,** 1005 (1971).

[60] A. K. Rajagopal and K. P. Jain, *Phys. Rev. A* [3] **5,** 1475 (1972); A. K. Rajagopal, *ibid.* **6,** 1239 (1972); *Pramana* **4,** 140 (1975).

[61] F. Brosens, L. F. Lemmens, and J. T. Devreese, *Phys. Status Solidi B* **74,** 45 (1975).

[62] P. R. Antoniewicz and L. Kleinman, *Phys. Rev. B: Solid State* [3] **2,** 2808 (1970). The statement by these authors, to the effect that the RPA value of $g(0)$ at $r_s = 2$ in Ref. 43 differs from theirs because of a cutoff in Fourier transform integration at smaller values of $q/q_F$, is incorrect. In fact, $S(q)$ was calculated in Ref. 43 up to $q = 100q_F$ and an analytic expansion at large values of $q$ was used to carry out the integration. An independent calculation by Hubbard[44] gives results very close to those given in Ref. 43.

A systematic study of the static dielectric response function using the standard diagrammatic perturbation theory was undertaken by Geldart and Taylor,[63] following previous work by Geldart and Vosko[42] and Geldart.[64] They first considered the lowest-order Hartree–Fock terms which consist of a bubble with a single electron–hole interaction and of two bubbles each with the simplest self-energy insertion. They demonstrated the occurrence of an exact cancellation at large $k$ between the two types of diagrams, as mentioned earlier, and calculated numerically the static response to this order. They next considered higher order diagrams, showing that the above cancellation at large $k$ is still present in the Hartree–Fock terms to all orders. They proposed the following approximate form for the static polarizability,

$$\tilde{\chi}(k, 0) = \chi_0(k, 0) + \chi_1(k, 0)/[1 - \tfrac{1}{2}\bar{\gamma}\chi_0(k, 0)], \qquad (5.26)$$

where $\chi_1(k, 0)$ is estimated from the lowest-order Hartree–Fock diagrams with static screening and from relevant second-order correlation diagrams, and $\bar{\gamma}$ is a constant that is adjusted to the compressibility sum rule. This result has the merit of reducing to the first-order Hartree–Fock result discussed in Section 3,d in the weak interaction limit. Since the above theory is purely static, it cannot be used to obtain the pair correlation function. We shall return to discuss its results in Section 6.

### *d. Moment-Conserving Schemes*

Another line of approach, based on the equations of motion for Green's functions and involving decoupling approximations that are guided by conservation of moments of the dissipation spectrum, has been developed by Toigo and Woodruff.[65] We shall here follow the derivation given by Kugler[66] which is somewhat simpler. We consider the response of the Wigner distribution operator to an external potential,

$$\chi_{\mathbf{p}\sigma}(\mathbf{k}, t) = -(i/\hbar)\vartheta(t)\langle[\hat{f}_{\mathbf{p}\sigma}(\mathbf{k}, t), \rho(-\mathbf{k}, 0)]\rangle, \qquad (5.27)$$

whose equation of motion is

$$\left[i\hbar\frac{\partial}{\partial t} - \frac{\hbar}{m}\mathbf{p}\cdot\mathbf{k} - \frac{\hbar^2k^2}{2m}\right]\chi_{\mathbf{p}\sigma}(\mathbf{k}, t)$$

$$= -\delta(t)[n_{\mathbf{p}+\mathbf{k}\sigma} - n_{\mathbf{k}\sigma}] + \sum_{\mathbf{q}} v(q)\chi^{(3)}_{\mathbf{p}\sigma}(\mathbf{q}, \mathbf{k}; t), \qquad (5.28)$$

[63] D. J. W. Geldart and R. Taylor, *Can. J. Phys.* **48**, 155, 167 (1970).

[64] D. J. W. Geldart, *Can. J. Phys.* **45**, 3139 (1967).

[65] F. Toigo and T. O. Woodruff, *Phys. Rev. B: Solid State* [3] **2**, 3958 (1970); **4**, 371 (1971).

[66] A. A. Kugler, *J. Stat. Phys.* **12**, 35 (1975).

where the three-particle response function $\chi^{(3)}$ is

$$\chi^{(3)}_{\mathbf{p}\sigma}(\mathbf{q}, \mathbf{k}; t - t') = -(i/\hbar)\vartheta(t - t')\langle[a^{+}_{\mathbf{p}\sigma}\rho(\mathbf{q})a_{\mathbf{p+k-q},\sigma}(t) - a^{+}_{\mathbf{p+q},\sigma}\rho(\mathbf{q})a_{\mathbf{p+k},\sigma}(t), \rho(-\mathbf{k}', t')]\rangle. \quad (5.29)$$

The decoupling scheme of Toigo and Woodruff is equivalent to replacing the last term in Eq. (5.28) by

$$\sum_{\mathbf{q}} v(q)\chi^{(3)}_{\mathbf{p}\sigma}(\mathbf{q}, \mathbf{k}; t) = A_{\mathbf{p}\sigma}(\mathbf{k})\chi(k, t), \quad (5.30)$$

with the requirement that $A_{\mathbf{p}\sigma}(\mathbf{k})$ be determined by the condition

$$\sum_{\mathbf{q}} v(q) \int \frac{d\omega}{2\pi}\,\omega\,\mathrm{Im}\,\chi^{(3)}_{\mathbf{p}\sigma}(\mathbf{q}, \mathbf{k}; \omega) = A_{\mathbf{p}\sigma}(\mathbf{k}) \int \frac{d\omega}{2\pi}\,\omega\,\mathrm{Im}\,\chi(k, \omega) \quad (5.31)$$

where the integral on the right-hand side is just $nk^2/m$ [see Eq. (2.16)]. This condition implies that the third moment sum rule for Im $\chi(k, \omega)$, Eq. (2.17), is satisfied. However, since the evaluation of the left-hand side of Eq. (5.31) involves two-particle density matrices at equilibrium, which are *a priori* unknown, additional approximations of the Hartree–Fock type are needed for a full determination of $A_{\mathbf{p}\sigma}(\mathbf{k})$. This, of course, would lead to a violation of the third moment.

The final result for the response function can be cast in the form (5.10) with a local field factor $G$ which depends on $\omega$. Tabulations and graphs of this function can be found in the original articles.[65,66] The long wavelength limit of the static response function of Toigo and Woodruff yields a value of the compressibility that is the same as one obtains from the Hartree–Fock energy, Eq. (5.12). As noted earlier, this is in fact quite close to the "true" value. As regards the pair correlation function $g(r)$, it is somewhat better behaved than what one obtains in the Hubbard approximation (see Figs. 1 and 2).

In subsequent work Pathak and Vashishta,[67] assuming the mean-field form (5.10) for the response, have determined $G(k)$ by requiring that the potential part of the third moment (2.17) be satisfied with self-consistency in $S(k)$ via the fluctuation-dissipation theorem. This approach does not give any better results for $g(r)$ or $K_0/K$.

## 6. Static Dielectric Function

It is well known that the dielectric function of an electron gas plays an important role in determining many of the properties of simple metals.

[67] K. N. Pathak and P. Vashishta, *Phys. Rev. B: Solid State* [3] 7, 3649 (1973).

Since the ions in a metal move much more slowly than the electrons, it is the static dielectric function that is of primary interest in evaluating, for example, the phonon dispersion curves and the properties of lattice defects.

The two static dielectric functions discussed in Section 1,b and given in Eqs. (1.31) and (1.32) determine, respectively, the screened ion–ion potential and the screened electron–ion potential (assuming, of course, that the bare electron–ion potential is weak). Comparing Eq. (5.10) with Eq. (1.31), we see that the exchange-correlation factor is related to the local field factor by

$$f_{xc}(k) = -(4\pi e^2/k^2)G(k). \tag{6.1}$$

From Eq. (2.3), we have the important relation

$$\lim_{k\to 0} G(k) = (1 - K_0/K)(k^2/k_{TF}^2) \tag{6.2}$$

which is verified by approximate schemes satisfying the compressibility sum rule. In Table III we have given the values of the coefficients of the low-$k$ expansion for $G(k)$,

$$G(k) = \alpha(k/k_F)^2 + \beta(k/k_F)^4 + \gamma(k/k_F)^6 + \cdots, \tag{6.3}$$

for different values of $r_s$. As we shall see later, the coefficients $\beta$ and $\gamma$ are related to the coefficients in the gradient expansion of the exchange-correlation energy of an inhomogeneous electron gas. The values of Gupta and Singwi[55] were obtained by the Vashishta–Singwi scheme with an exact determination of the parameter $a$ in Eq. (5.17) at each $r_s$, since the values of the higher coefficients in (6.3) are very sensitive to the precise value of this parameter. The values due to Rasolt and Geldart[68] were calculated by a diagrammatic technique that includes the high-density limiting result of Ma and Brueckner[69] and lowest order RPA effects exactly, and higher-order contributions approximately. Values of $\beta$ calculated by Kleinman[70] and by Rajagopal and Ray[71] are smaller by a factor of 2–3 than the results given in Table III.

In Fig. 3 the detailed shape of $G(k)$ as obtained in various theories of the dielectric function is shown at a value of $r_s$ corresponding to the density of Al ($r_s \simeq 2$), and in Fig. 4 the function $G(k)/k^2$ is reported at the density of Na ($r_s \simeq 4$) to highlight the differences between the various theoretical

[68] M. Rasolt and D. J. W. Geldart, *Phys. Rev. Lett.* **35,** 1234 (1975); D. J. W. Geldart and M. Rasolt, *Phys. Rev. B: Solid State* [3] **13,** 1477 (1976).

[69] S. Ma and K. A. Brueckner, *Phys. Rev.* [2] **165,** 18 (1968).

[70] L. Kleinman, *Phys. Rev. B: Solid State* [3] **10,** 2221 (1974).

[71] A. K. Rajagopal and S. Ray, *Phys. Rev. B: Solid State* [3] **12,** 3129 (1975).

TABLE III. COEFFICIENTS OF THE LOW-$k$ EXPANSION OF $G(k)$

| | $r_s$ | 0.985 | 1.975 | 2.975 | 3.95 | 4.93 | 5.90 |
|---|---|---|---|---|---|---|---|
| $\alpha$: | Vashishta–Singwi[a] | 0.24537 | 0.26389 | 0.27689 | 0.28658 | 0.29431 | 0.30059 |
| | Gupta–Singwi[b] | 0.25968 | 0.26890 | 0.27773 | 0.28492 | 0.29038 | 0.29395 |
| | Toigo–Woodruff[c] | 1/4 | 1/4 | 1/4 | 1/4 | 1/4 | 1/4 |
| $\beta$: | Gupta–Singwi[b] | −0.03641 | −0.02639 | −0.02231 | −0.02002 | −0.01823 | −0.01653 |
| | Rasolt–Geldart[d] | −0.037 | −0.035 | −0.032 | −0.029 | −0.027 | −0.025 |
| $\gamma$: | Gupta–Singwi[b] | −0.00265 | −0.00200 | −0.00150 | −0.00114 | −0.00090 | −0.00074 |

[a] P. Vashishta and K. S. Singwi, *Phys. Rev. B: Solid State* [3] **6,** 875, 4883 (1972).
[b] A. K. Gupta and K. S. Singwi, *Phys. Rev. B: Solid State* [3] **15,** 1801 (1977).
[c] F. Toigo and T. O. Woodruff, *Phys. Rev. B: Solid State* [3] **2,** 3958 (1970).
[d] M. Rasolt and D. J. W. Geldart, *Phys. Rev. Lett.* **35,** 1234 (1975). The values given in this table have been read from a graph and are therefore approximate.

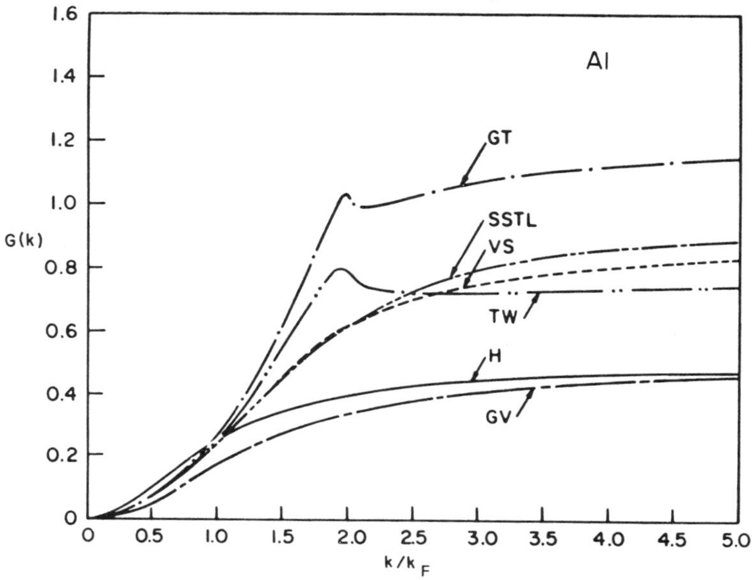


FIG. 3. The function $G(k)$ versus $k/k_F$ for the electron density of Al. The abbreviations refer to the following articles: H: J. Hubbard, *Proc. R. Soc. London, Ser. A* **240,** 539 (1957); **243,** 336 (1957); GV: D. J. W. Geldart and S. H. Vosko, *Can. J. Phys.* **44,** 2137 (1966); SSTL: K. S. Singwi, A. Sjölander, M. P. Tosi, and R. H. Land, *Phys. Rev. B: Solid State* [3] **1,** 1044 (1970); VS: P. Vashishta and K. S. Singwi, *Phys. Rev. B: Solid State* [3] **6,** 875 (1972); GT: D. J. W. Geldart and R. Taylor, *Can. J. Phys.* **48,** 155, 167 (1970); TW: F. Toigo and T. O. Woodruff, *Phys. Rev. B: Solid State* [3] **2,** 3958 (1970).

results for small values of $k$. There are two significant points to note: ($a$) the initial curvature of $G(k)/k^2$ should be negative according to the recent calculations of $\beta$ reported in Table III, which does not seem to be the case for the Geldart–Taylor and Toigo–Woodruff forms of $G(k)$; and ($b$) both these theories exhibit a rather sharp drop of $G(k)/k^2$ around $k = 2k_F$ which is not present in the other theories. This behavior of $G(k)$ around $k = 2k_F$ is a remnant of its behavior in the Hartree–Fock approximation. Just how much of the Hartree–Fock behavior is washed out by correlation is an important and as yet unsettled theoretical point.

For $k \to \infty$, it is obvious from Eq. (5.25) that $G(k)$ must satisfy the inequality

$$\tfrac{1}{3} \leq G(k) \leq \tfrac{2}{3}, \qquad (6.4)$$

with the lower bound providing the correct value for $r_s \to 0$. As seen from Fig. 3, various theories do not conform to this inequality, but the precise values of $G(k)$ for $k \gg 2k_F$ are not too important for practical purposes.

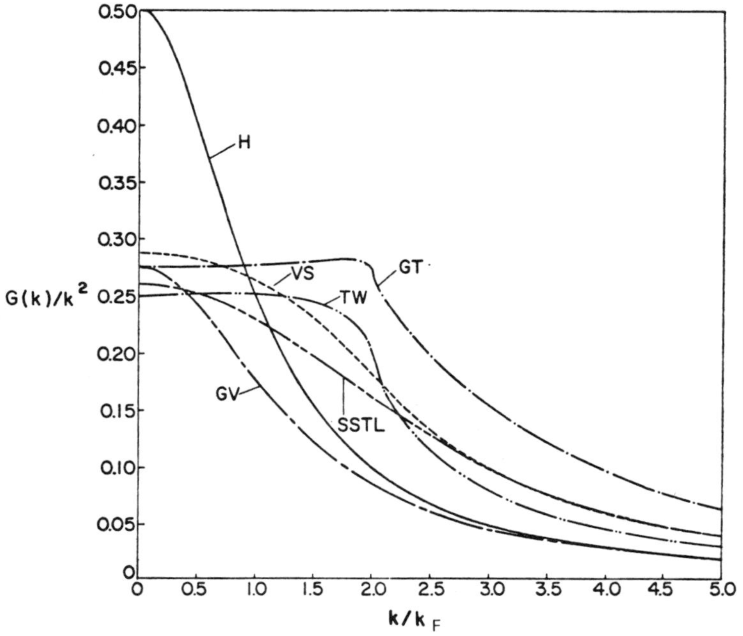


FIG. 4. The function $G(k)/k^2$ versus $k/k_F$ for the electron density of Na. The abbreviations are the same as in Fig. 3.

### *a. Interionic Potentials and Phonon Dispersion in Metals*

In the case where the electron–ion interaction is weak, one can write the interionic potential as[72]

$$V(r) = V_i(r) - \frac{2Z^2e^2}{\pi}\int_0^\infty dk\,\frac{\sin(kr)}{kr}\left|\frac{w(k)}{-4\pi Ze^2/k^2}\right|^2\left[1-\frac{1}{\varepsilon(k)}\right], \quad (6.5)$$

where $V_i(r)$ is the direct ion–ion potential, $Z$ is the number of conduction electrons per ion, and $w(k)$ is the bare-ion pseudopotential form factor. Figures 5 and 6, taken from the work of Rao,[73] report $V(r)$ for Na and Al

[72] W. A. Harrison, "Pseudopotentials in the Theory of Metals." Benjamin, New York, 1966; S. K. Joshi and A. K. Rajagopal, *Solid State Phys.* **22,** 160 (1968); V. Heine, *ibid.* **24,** 1 (1970).

[73] P. V. S. Rao, Ph.D. Thesis, Northwestern University, Chicago, Illinois, 1974 (unpublished); *Phys. Status Solidi B* **55,** 629 (1973); *J. Phys. Chem. Solids* **35,** 669 (1974); *J. Phys. F.* [3] **5,** 611 (1975); see also I. Ebbsjo, T. Kinell, and I. Waller, *J. Phys. C* [3] **11,** L501 (1978).

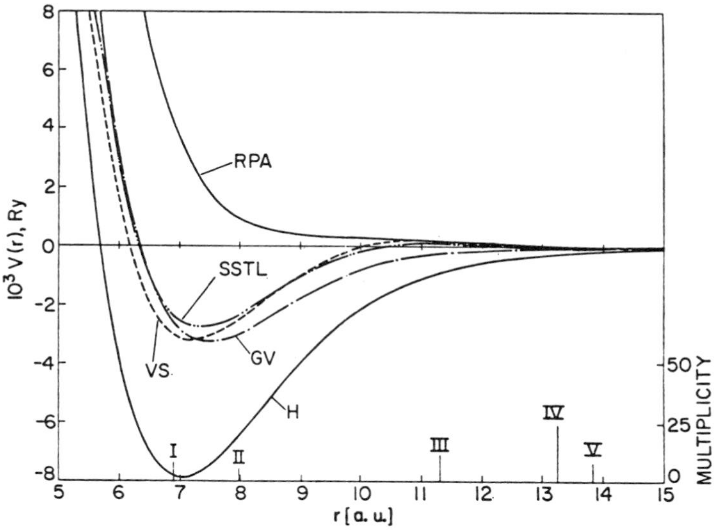


FIG. 5. Interionic potential $V(r)$ as a function of separation $r$ in Na, calculated by using a nonlocal optimized model potential screened with five different dielectric functions (abbreviations as in Fig. 3). The positions of the first few shells of neighbors are indicated by vertical bars of height proportional to the number of atoms in each shell. (From P. V. S. Rao, Ph.D. Thesis, Northwestern University, Chicago, Illinois, 1974.)

as calculated with different screening functions using the nonlocal optimized model potential of Shaw.[74] An interesting point to note for Al is the presence of a shoulder in $V(r)$ at distances between first and second neighbors and the occurrence of the main minimum just beyond second neighbors. These features, which have important consequences in defect calculations, are present in all the forms of $V(r)$ yielded by the more recent dielectric functions, and are quantitatively but no qualitatively affected by changing the bare pseudopotential as shown by Rao.[73]

There have been a vast number of calculations of phonon dispersion curves in metals with interionic potentials determined from the dielectric approach since the pioneering work of Toya.[75] To illustrate the degree of agreement with experiment achieved by the theory for simple metals as well as the sensitivity of the results to different choices of the screening function, we report in Fig. 7 from Rao[73] the phonon dispersion curves in Al. These are based on the potentials reported in Fig. 6. An exhaustive review of the field of lattice dynamics of nontransition metals has been

[74] R. W. Shaw, Jr., *Phys. Rev.* [2] **174,** 769 (1968).
[75] T. Toya, *J. Res. Inst. Catal. Hokkaido Univ.* **6,** 161, 183 (1958).

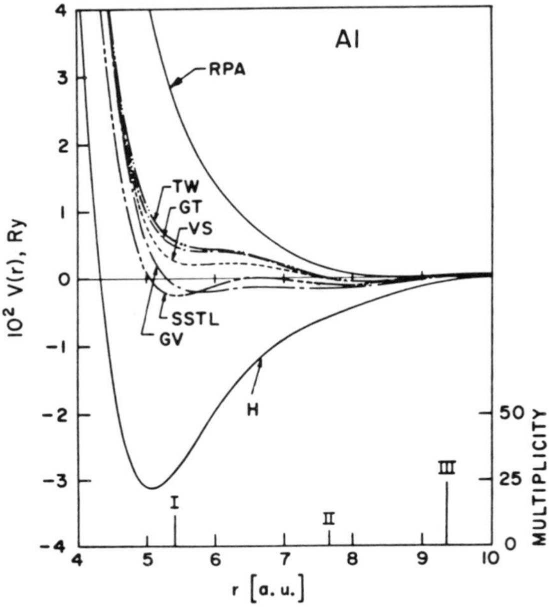


FIG. 6. Interionic potential $V(r)$ as a function of separation $r$ in Al, calculated by using a nonlocal optimized model potential screened with seven different dielectric functions (abbreviations as in Fig. 3). The positions of the first few shells of neighbors are indicated by vertical bars of height proportional to the number of atoms in each shell. (From P. V. S. Rao, Ph.D. Thesis, Northwestern University, Chicago, Illinois, 1974.)

given recently by Brovman and Kagan.[76] Among the variety of applications of interionic potentials to metal physics we may recall the *a priori* calculations of the properties of lattice defects[77] and the computer simulation of the structure and dynamics of liquid metals.[78]

A necessary criterion for the choice of the pseudopotential is that it should lead to a correct distribution of electronic charge density around each ion. This criterion has recently been adopted by Rasolt and Taylor[79] who adjust the parameters in a nonlocal pseudopotential such that the

[76] E. G. Brovman and Yu. M. Kagan, *in* "Dynamical Properties of Solids" (G. K. Horton and A. A. Maradudin, eds.), Vol. I. North-Holland Publ., Amsterdam, 1974.

[77] See, e.g., P. C. Gehlen, J. R. Beeler, Jr., and R. I. Jaffee, eds., "Interatomic Potentials and Simulation of Lattice Defects." Plenum, New York, 1972.

[78] See, e.g., R. Evans and D. A. Greenwood, eds., "Liquid Metals." Inst. Phys. Bristol, 1976.

[79] M. Rasolt and R. Taylor, *Phys. Rev. B: Solid State* [3] **11**, 2717 (1975); L. Dagens, M. Rasolt, and R. Taylor, *ibid.* p. 2726 (1975).

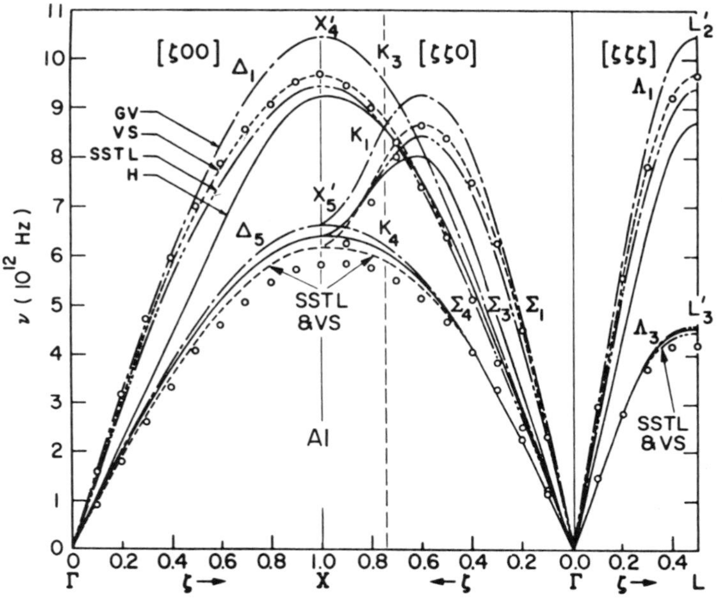


FIG. 7. Comparison of calculated and experimental phonon dispersion relations for Al. Results for the nonlocal optimized model potential are shown with four different types of screening (abbreviations as in Fig. 3). Open points represent the experimental data of R. Stedman, L. Almqvist, and G. Nilsson, *Phys. Rev.* [2] **162,** 549 (1967). (From P. V. S. Rao, Ph.D. Thesis, Northwestern University, Chicago, Illinois, 1974.)

charge density outside the ionic core as calculated in a linear response theory agrees with the results of an HKS self-consistent calculation. These authors show that the interionic potential thus obtained gives good results for phonon dispersion curves in various metals.

The effects of lattice periodicity on the screening function entering the interionic potential (6.5) in a metal have been considered by several authors[80] with special attention to the evaluation of the phonon dispersion curves. An external perburbation $V_e(\mathbf{k}, \omega)$ applied to an electronic system in a lattice generates not just a component $n(\mathbf{k}, \omega)$ of the polarization but an infinite set of components $n(\mathbf{k} + \mathbf{G}, \omega)$, where $\mathbf{G}$ is any reciprocal lattice vector. As a consequence, the dielectric function becomes a function of two wave vectors differing by a reciprocal lattice vector and is

[80] D. C. Wallace, *Phys. Rev.* [2] **182,** 778 (1969); E. G. Brovman, Yu. Kagan, and A. Holas, *Sov. Phys.—JETP* (*Engl. Transl.*) **30,** 883 (1970); R. Johnson, *Mat.-Fys. Medd.—K. Dan. Vidensk. Selsk.* **37,** (9) and (10)(1970); R. Johnson and A. Westin, Rept. AE-365. AB Atomenergi, Stockholm, 1969; C. J. Pethick, *Phys. Rev. B: Solid State* [3] **2,** 1789 (1970).

best represented by a matrix $\varepsilon_{GG'}(\mathbf{q}, \omega)$, where $\mathbf{q}$ is a wave vector inside the Brillouin zone. Therefore the effective interaction between two ions depends not only on their separation but also on their positions relative to the other ions in the metals. In simple metals one can take advantage of the weakness of the electron–ion interaction to treat these lattice effects by perturbation techniques. A full discussion of the long-wavelength limit has been given by Pethick[80] with special regard to the compressibility sum rule for the metal. Calculations of the phonon dispersion curves have been carried out by Johnson and Westin.[80] A brief discussion of these effects in semiconductors will be taken up in Section 11.

A characteristic lattice effect is the appearance, in certain circumstances, of new structure in the calculated phonon dispersion curves, whose origin is analogous to the Kohn effect, and can be qualitatively understood by simply considering the diagonal part $\chi^{\circ}_{GG}(\mathbf{q}, 0)$ of the noninteracting electron polarizability. A peak in this function may arise because of a feature of the Fermi surface known as "nesting," where two regions of the Fermi surface are separated by a fairly constant wave vector $\mathbf{k}_0$. Transitions with this wave vector, from states just inside to states just outside the Fermi surface, may thus yield a large contribution to $\chi_0$ for $\mathbf{q} + \mathbf{G} = \mathbf{k}_0$, and the result[81] is a dip in the phonon dispersion curve at $\mathbf{q} = \mathbf{k}_0 - \mathbf{G}$. This could also lead to the formation of spin density waves,[82] since $\chi_0$ enters in the expression for the paramagnetic susceptibility. Thus in systems like the transition and rare earth metals one can attempt[83] to correlate the positions of phonon anomalies, the wave vector of magnetic ordering, and features of the Fermi surface.

*b. Weakly Inhomogeneous Electron Gas*

So far we have mainly discussed a homogeneous interacting electron gas. However, in nature one is forced to deal with an inhomogeneous electron gas, an extreme example of which is the electron density profile at a metal surface. Although the latter is a very complicated problem to treat, it is possible to make some progress by treating the case of a weakly inhomogeneous electron gas, which can be done in terms of the properties of a homogeneous electron gas. That this is possible is based on an elegant theorem due to Hohenberg and Kohn[8] according to which the full many-particle ground state in the presence of an external potential is a unique functional of the density $n(\mathbf{r})$.

[81] See, e.g., S. K. Sinha and B. N. Harmon, *in* "Superconductivity in d- and f-band Metals" (D. H. Douglass, ed.), p. 269. Plenum, New York, 1976.

[82] See, e.g., C. Herring, *Magnetism* **4** (1966).

[83] R. P. Gupta and S. K. Sinha, *Phys. Rev. B: Solid State* [3] **3**, 2401 (1971); S. H. Liu, R. P. Gupta, and S. K. Sinha, *ibid.* **4**, 1100 (1971).

In the density functional formalism of Hohenberg, Kohn, and Sham,[8] the ground-state energy of a paramagnetic electron gas in an external potential $V_e(\mathbf{r})$ is a functional of the density $n(\mathbf{r})$ and is

$$E_{gs}[n(\mathbf{r})] = \int V_e(\mathbf{r})n(\mathbf{r})\, d\mathbf{r} + E_0[n(\mathbf{r})] + \frac{1}{2}\int \frac{n(\mathbf{r})n(\mathbf{r}')}{|\mathbf{r}-\mathbf{r}'|}\, d\mathbf{r}\, d\mathbf{r}' + E_{xc}[n(\mathbf{r})], \tag{6.6}$$

where $E_0[n(\mathbf{r})]$ is the kinetic energy functional of a system of noninteracting electrons of density $n(\mathbf{r})$, the third term is the classical Coulomb energy, and the last term is the exchange-correlation contribution given in Eq. (3.31). The functional $E_{gs}[n(\mathbf{r})]$ has the property that its minimum value is the exact ground-state energy attained when $n(\mathbf{r})$ is the correct ground-state density. The variational principle then leads to the HKS equation (3.30).

For a gas with a weak distortion from homogeneity,

$$n(\mathbf{r}) = n + \bar{n}(\mathbf{r}) \tag{6.7}$$

with $\bar{n}(\mathbf{r})/n \ll 1$ and $\int \bar{n}(\mathbf{r})\, d\mathbf{r} = 0$, one can formally write:

$$E_0[n(\mathbf{r})] = E_0(n) + \frac{1}{2}\iint d\mathbf{r}\, d\mathbf{r}' K_0(\mathbf{r}-\mathbf{r}';\, n)\bar{n}(\mathbf{r})\bar{n}(\mathbf{r}') + O(\bar{n}^3), \tag{6.8}$$

$$E_{xc}[n(\mathbf{r})] = E_{xc}(n) + \frac{1}{2}\iint d\mathbf{r}\, d\mathbf{r}' K_{xc}(\mathbf{r}-\mathbf{r}';\, n)\bar{n}(\mathbf{r})\bar{n}(\mathbf{r}') + O(\bar{n}^3). \tag{6.9}$$

Using these equations, Eq. (3.30) becomes

$$\int d\mathbf{r}'[K_0(\mathbf{r}-\mathbf{r}';\, n) + K_{xc}(\mathbf{r}-\mathbf{r}';\, n)]\bar{n}(\mathbf{r}') + V_H(\mathbf{r}) = 0. \tag{6.10}$$

In the mean-field approach the equivalent equation is

$$\left[-\frac{1}{\chi_0(k)} - \frac{4\pi e^2}{k^2}\, G(k)\right]\bar{n}(\mathbf{k}) + V_H(\mathbf{k}) = 0, \tag{6.11}$$

which by comparison with the Fourier transform of Eq. (6.10) yields

$$K_0(k) = -1/\chi_0(k), \tag{6.12}$$

$$K_{xc}(k) = -\frac{4\pi e^2}{k^2}\, G(k). \tag{6.13}$$

Kohn and Sham suggested expanding $K_{xc}(k)$ in powers of $k$, but they noted that such an expansion is not appropriate for $K_0(k)$. Therefore, they reformulated their theory in terms of a Schrödinger equation, where $K_{xc}(r)$ enters as an effective exchange-correlation potential in addition to the Hartree potential, and $K_0(r)$ goes over to $-\hbar^2\nabla^2/2m$.

In the limit of long wavelength,

$$\lim_{k\to 0} K_{xc}(k) \equiv K_{xc}(n) = \frac{1}{V}\frac{\partial^2 E_{xc}(n)}{\partial n^2} = \frac{\partial \mu_{xc}(n)}{\partial n} \tag{6.14}$$

where $\mu_{xc}(n)$ is the exchange-correlation chemical potential. This function is completely determined by the compressibility of the homogeneous electron gas, according to Eqs. (6.13) and (6.2). In particular the exchange part of $\mu_{xc}$ is, from Eq. (5.12),

$$\mu_x(n) = \int_\infty^n dn\, K_x(n) = -\frac{e^2}{\pi}(3\pi^2 n)^{1/3} \tag{6.15}$$

which is two-thirds of Slater's value. Our knowledge of $\mu_{xc}(n)$ in the homogeneous case, enables us to solve the HKS single-particle equation in a self-consistent manner for the inhomogeneous case for which the density variation is slow. Using the values of the correlation energy of Singwi *et al.*,[43] Hedin and Lundqvist[84] have given the exchange-correlation potential in a parametrized form. This local potential has been used in many recent band structure calculations.[85] More recently, Moruzzi *et al.*[86] have used it in self-consistent *ab initio* calculations of the cohesive energy for 26 metals in the third and fourth rows of the periodic table, obtaining remarkable agreement with experiment. An important implication of these results is that the local density approximation for the exchange-correlation potential in the HKS scheme is a good one. From a rigorous theoretical point of view, it is not clear as to why this approximation works as well as it does. Overhauser,[87] in particular, has drawn attention to the fact that the exchange-correlation potential is nonlocal and that this nonlocality should be taken into account in band structure calculations.

We would now like to go beyond the local density approximation (LDA), i.e., estimate the gradient corrections to the exchange and correlation energy. Equation (6.9) can be written as[55]

$$\begin{aligned} E_{xc}[n(\mathbf{R})] = &\int d\mathbf{R}\, n(\mathbf{R}) e_{xc}(n(\mathbf{R})) \\ &- \tfrac{1}{3}\pi \int d\mathbf{R}\, |\nabla n(\mathbf{R})|^2 \int_0^\infty r^4\, dr\, K_{xc}(r; n(\mathbf{R})) \\ &- \tfrac{1}{60}\pi \int d\mathbf{R}\, \nabla n(\mathbf{R}) \cdot \nabla[\nabla^2 n(\mathbf{R})] \int_0^\infty r^6\, dr\, K_{xc}(r; n(\mathbf{R})) + \cdots . \end{aligned} \tag{6.16}$$

[84] L. Hedin and B. I. Lundqvist, *J. Phys. C* [3] **4,** 2064 (1971).

[85] See, e.g., J. F. Janak, A. R. Williams, and V. L. Moruzzi, *Phys. Rev. B: Solid State* [3] **11,** 1522 (1975); J. F. Janak, V. L. Moruzzi, and A. R. Williams, *ibid.* **12,** 1257 (1975).

[86] V. L. Moruzzi, A. R. Williams, and J. F. Janak, *Phys. Rev. B: Solid State* [3] **15,** 2854 (1977).

[87] A. W. Overhauser, *Phys. Rev. B: Solid State* [3] **2,** 874 (1970); R. J. Duff and A. W. Overhauser, *ibid.* **5,** 2799 (1972); A. W. Overhauser, *ibid.* **10,** 4918 (1974).

Introducing the Fourier transform

$$K_{xc}(k;\ n) = \frac{4\pi}{k}\int_0^\infty r\,dr\,\sin(kr)K_{xc}(r;\ n), \tag{6.17}$$

it follows that

$$E_{xc}[n(\mathbf{R})] = \int d\mathbf{R}\ n(\mathbf{R})e_{xc}(n(\mathbf{R})) + \frac{1}{2}\int d\mathbf{R}\ g_{xc}^{(2)}(n(\mathbf{R}))\,|\nabla n(\mathbf{R})|^2$$

$$+ \frac{1}{6}\int d\mathbf{R}\ g_{xc}^{(3)}(n(\mathbf{R}))\nabla n(\mathbf{R})\cdot\nabla[\nabla^2 n(\mathbf{R})] + \cdots, \tag{6.18}$$

where

$$g_{xc}^{(2)}(n) = \frac{1}{2}\left.\frac{\partial^2 K_{xc}(k;\ n)}{\partial k^2}\right|_{k=0}, \tag{6.19}$$

$$g_{xc}^{(3)}(n) = -\frac{1}{8}\left.\frac{\partial^4 K_{xc}(k;\ n)}{\partial k^4}\right|_{k=0}. \tag{6.20}$$

Thus the coefficients $g_{xc}$ in the gradient expansion of the exchange-correlation energy are related to the coefficients of the $k^4$ and $k^6$ terms in the small-$k$ expansion of the local field factor $G(k)$. The latter have been discussed earlier in this section, and we have given their numerical values in Table III. In terms of these coefficients,

$$g_{xc}^{(2)}(n) = -4\pi e^2\beta(r_s)/(3\pi^2 n)^{4/3} \tag{6.21}$$

$$g_{xc}^{(3)}(n) = 12\pi e^2\gamma(r_s)/(3\pi^2 n)^2 \tag{6.22}$$

It should be borne in mind that $g_{xc}^{(3)}$ given in Eq. (6.22) represents the contribution from linear polarizability only and does not include contributions from nonlinear terms.[8]

As an application of the foregoing considerations, we shall here outline the calculation of surface energy of simple metals in a "jellium" model. The electron profile at the metal surface is calculated using the HKS scheme in which the external potential is provided by an abrupt termination of the positive background. The surface charge density is given by

$$n(x) = \frac{1}{\pi^2}\int_0^{k_F} dk\,(k_F^2 - k^2)\,|\psi_k(x)|^2, \tag{6.23}$$

where $\psi_k(x)$ is the solution of the equation

$$\left[-\frac{1}{2}\frac{d^2}{dx^2} + V_{eff}(x;\ n)\right]\psi_k(x) = \frac{1}{2}(k^2 - k_F^2)\psi_k(x) \tag{6.24}$$

with

$$V_{eff}(x;\ n) = V_{es}(x;\ n) + \frac{\delta E_{xc}[n]}{\delta n(x)}. \tag{6.25}$$

The first term is the electrostatic potential determined by Poisson's equation,

$$\nabla^2 V_{\text{es}}(x) = -4\pi[n(x) - n\vartheta(-x)], \tag{6.26}$$

and the second term is the exchange-correlation potential. $E_{\text{xc}}[n]$ is

$$E_{\text{xc}}[n] = \int_{-\infty}^{\infty} dx\, n(x)e_{\text{xc}}(n(x)) + \frac{1}{2}\int_{-\infty}^{\infty} dx\, g_{\text{xc}}^{(2)}(n(x)) \left[\frac{dn(x)}{dx}\right]^2, \tag{6.27}$$

retaining only the first gradient correction term. Therefore,

$$V_{\text{xc}}(x;\, n) = \frac{d}{dn}[ne_{\text{xc}}(n)] - \left[\frac{1}{2}\frac{dg_{\text{xc}}^{(2)}(n)}{dn}\left(\frac{dn}{dx}\right)^2 + g_{\text{xc}}^{(2)}(n)\frac{d^2n}{dx^2}\right]. \tag{6.28}$$

The self-consistent calculations of Lang and Kohn[88] included only the first term in Eq. (6.28), i.e., within the LDA.

In Fig. 8 we have given the results[88] of such a calculation for $r_s = 4$ (jellium model of Na). One is immediately struck by the fact that the electron density profile varies rather rapidly near the metal surface. The condition that the density should vary slowly, i.e., $|\nabla n|/n \ll k_F$ so that one is justified in using the LDA, is violated rather seriously near the surface. We also see that the surface barrier arises mainly from $V_{\text{xc}}$ and is nearly twice the Fermi energy. An accurate knowledge of $V_{\text{xc}}$ is thus important in a surface calculation. The calculated work function, defined by $V_{\text{eff}}(\infty) - \varepsilon_F$, is 3.06 eV, which agrees within 10% with the measured work function for Na. The calculated surface energy $\sigma$ is also in reasonable agreement with the measured value for Na. On the one hand, this remarkable agreement between theory and experiment is gratifying, but on the other hand, it is somewhat disturbing for two reasons: (*a*) no effect of the ionic lattice has been incorporated in the calculation; and (*b*) gradient corrections to $V_{\text{xc}}$ have been neglected. As regards the former, it is understandable since the pseudopotential for Na is very weak. As regards the latter, all that one can do at present is to calculate the effect of the first gradient correction to $\sigma$.

Using first-order perturbation theory, the change in surface energy is

$$\delta\sigma = \frac{1}{2}\int g_{\text{xc}}^{(2)}(n)\left(\frac{dn}{dx}\right)^2 dx \tag{6.29}$$

where for $n(x)$ one uses the values obtained in LDA. Such a calculation[55] yields that the change in $\sigma$ due to the first gradient correction is an increase by 20–25%. Inclusion of the second term in the gradient series

[88] N. D. Lang and W. Kohn, *Phys. Rev. B: Solid State* [3] **1,** 4555 (1970); **3,** 1215 (1971); N. D. Lang, *Solid State Phys.* **28,** 225 (1973).

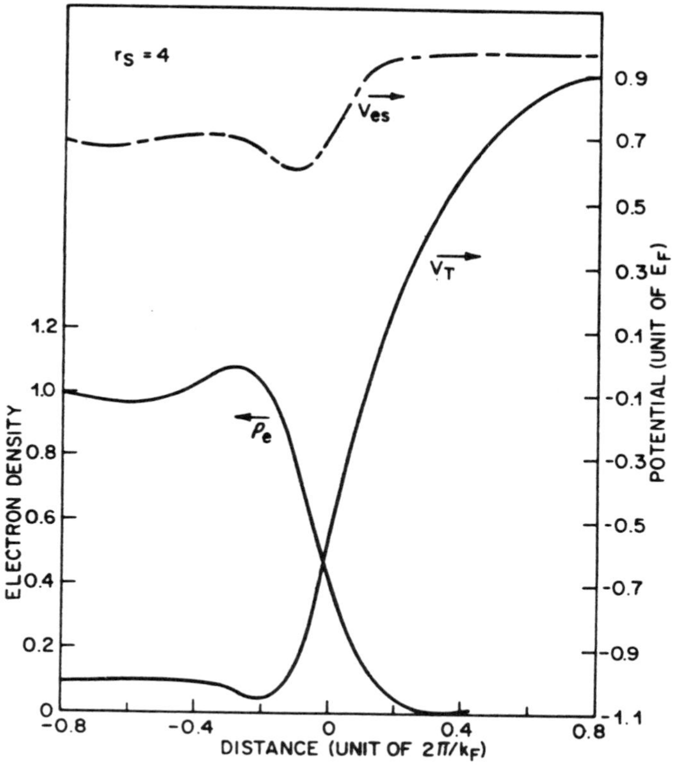


FIG. 8. Electron charge density, electrostatic potential, and total potential in the surface region for the jellium model of Na versus distance normal to the surface. [From J. A. Appelbaum and D. R. Hamann, *Rev. Mod. Phys.* **68,** 479 (1976), based on work by N. D. Lang and W. Kohn, *Phys. Rev. B: Solid State* [3] **1,** 4555 (1970).]

yields a correction of the opposite sign of no more than a few percent. The conclusion then is that the inclusion of nonlocal gradient corrections within a linear theory gives a contribution of no more than +20% to the surface energy calculated by LDA.

The use of LDA for exchange and correlation has given rise to some controversy during recent years. This question has been analyzed in detail by Langreth and Perdew,[89] who have considered the relative contributions of long versus short wavelength fluctuations to the exchange-correlation energy at the surface of a jellium metal. Their analysis shows that at small wave vectors the LDA is very bad when compared with their

[89] D. C. Langreth and J. P. Perdew, *Phys. Rev. B: Solid State* [3] **15,** 2884 (1977).

exact form, whereas at large wave vectors LDA becomes exact. They have given a recipe for an interpolation scheme between the two limits, which they have tested against the soluble infinite barrier model in RPA. Their final conclusion is that the LDA is accurate to better than 10% for the exchange-correlation energy.

For a metal like Al, with $r_s = 2$, the calculated jellium surface energy is negative. As shown by Lang and Kohn,[88] this unphysical result is due to the neglect of the discreteness of the lattice and can be rectified by a first-order perturbation calculation. The jellium model obviously has its limitations and one has to include the presence of the ionic lattice in all the self-consistent calculations.[90] This makes the calculations very tedious.

More recently Sahni and Gruenebaum[90a] have attempted to circumvent this difficulty for real metals by employing a variational scheme. These authors use a one-parameter variational density, obtained from single-particle wavefunctions generated by a linear potential model, to minimize the surface energy functional. The latter contains the ionic contribution via a local pseudopotential and also gradient corrections to the exchange and correlation energy functional. In this scheme the ionic contribution is not treated in a perturbative way. Sahni and Gruenebaum have used this variational scheme with success to calculate the surface energy, dipole barrier, and work function of a large number of simple metals.

A testing ground for some of the approximations that go into the calculation of the surface energy is provided by the electron–hole liquid (EHL) in semiconductors such as Ge. This system[91] can be schematized, through the use of experimentally known parameters such as effective masses, as a many-component Fermi liquid on a uniform dielectric background. Detailed numerical calculations of the surface energy of the EHL in both normal and stressed Ge and Si have recently been made by Vashishta *et al.*[92] The calculations involve the knowledge of the exchange-correlation energy of the liquid, which is independently tested against measured cohesive properties of the liquid, as well as the LDA in the HKS scheme. The results are in reasonably good agreement with experimental values wherever available.

As a final comment it is worth pointing out the existence of the empirical relation[92a] $K\sigma \sim L$ between the bulk compressibility, the

[90] See J. A. Appelbaum and D. R. Hamann, *Rev. Mod. Phys.* **48,** 479 (1976).

[90a] V. Sahni and J. Gruenebaum, *Phys. Rev. B: Condens. Matter* [3] **19,** 1840 (1979).

[91] T. M. Rice, *Solid State Phys.* **32,** 1 (1977).

[92] P. Vashishta, R. K. Kalia, and K. S. Singwi, *in* "Physics of Highly Excited States in Solids" (M. Ueta and Y. Nishina, eds.), p. 186. Springer-Verlag, Berlin and New York, 1976; R. K. Kalia and P. Vashishta, *Phys. Rev. B: Condens. Matter* [3] **17,** 2655 (1978); for other references see the article of Rice.[91]

[92a] P. A. Egelstaff and B. Widom, *J. Chem. Phys.* **53,** 2667 (1970); R. C. Brown and N. H. March, *J. Phys. C* [3] **6,** L363 (1973); K. S. Singwi and M. P. Tosi, *Solid State Commun.* **34,** 209 (1980); *Phys. Rev.* **B23,** 1640 (1981).

surface tension, and the "surface thickness" $L$, both for liquid metals and electron–hole liquids. Although such a relation can be derived on the basis of a gradient expansion for the kinetic energy functional, a more general proof is still lacking.

## 7. Dynamical Behavior of the Electron Liquid

Complete information concerning the elementary excitations of density fluctuations in the electron fluid is contained in the function $S(k, \omega)$ which is related to the dielectric function through Eq. (1.20). In the Born approximation, the differential cross section for scattering against density fluctuations is given by $S(k, \omega)$ except for some trivial multiplying factors. Inelastic electron scattering techniques have been used for many years to study the plasmon excitation in metals. Reviews of the early experimental and theoretical work can be found in the articles by Pines,[93] Raether,[94] and Platzman and Wolff.[95]

During recent years both inelastic electron and X-ray scattering experiments have been performed to measure the energy loss spectra over a wide range of momentum transfer. These experiments are analogous to the cold neutron inelastic scattering experiments to study the excitation spectra of condensed systems such as quantum liquids and solids. Such experiments have indeed been responsible for furthering our understanding of the microscopic atomic dynamics of the condensed systems. The more important results of recent experiments in simple metals can be summarized as follows:

1. The plasmon dispersion relation in the small wave vector region $k \ll k_c$ ($k_c$ being the critical wave vector at which the plasmon joins the particle–hole continuum) is well represented by

$$\omega_p(k) = \omega_p(0) + \alpha\hbar k^2/m \tag{7.1}$$

with the coefficient $\alpha$ for Al ranging from $0.38 \pm 0.02$ (Batson *et al.*[96]) to 0.42 (Gibbons *et al.*[97]). Höhberger *et al.*[98] give $\alpha = 0.401$, whereas Zacharias's value[99] is 0.40.

[93] D. Pines, *Physica (Amsterdam)* **26,** S103 (1960).

[94] H. Raether, *Springer Tracts Mod. Phys.* **38,** 84 (1965).

[95] P. M. Platzman and P. A. Wolff, "Waves and Interactions in Solid State Plasmas," Solid State Phys. Suppl. 13. Academic Press, New York, 1973.

[96] P. E. Batson, C. H. Chen, and J. Silcox, *Phys. Rev. Lett.* **37,** 937 (1976); see also P. E. Batson, Thesis, Mater. Sci. Cent. Rep. 2673. Cornell University, Ithaca, New York, 1976.

[97] P. C. Gibbons, S. E. Schantterly, J. J. Ritsko, and J. R. Fields, *Phys. Rev. B: Solid State* [3] **13,** 2451 (1976).

[98] H. J. Höhberger, A. Otto, and E. Petri, *Solid State Commun.* **16,** 175 (1975).

[99] P. Zacharias, *J. Phys. F* [3] **5,** 645 (1975).

2. The above law seems to hold approximately for $k$ up to $k_c$ ($k_c \simeq 0.8\,k_F$ in Al), and in this region the width of the plasmon peak increases slowly from its value at $k = 0$ in a manner proportional to $k^2$. A similar behavior of the dispersion curve up to $k_c$ is observed[100] by X-ray scattering from Be, with anisotropic $\alpha$ and damping. At $k \gtrsim k_c$, the line shape is strongly broadened and is asymmetric. The more recent electron scattering results of Batson *et al.*[96] for the dispersion curve, which extend up to $k \simeq 1.5k_F$, are reported in Fig. 9. In this region, Zacharias[99] sees an almost dispersionless peak, which is believed to be due to multiple scattering.[96] The observed[96] loss function for Al at $k = 1.14k_F$ is shown in Fig. 10.

3. In the region $k \gtrsim 1.5k_F$ the electron scattering experiments are affected by poor energy resolution. In this region, up to $k \simeq 2k_F$, X-ray scattering experiments[101] yield unexpected results. The excitation spectrum has a double peak structure or a shoulder on the high-frequency side of the main peak, as shown in Fig. 11. For $k < 2k_F$ the latter has larger strength than the former, and at $k \simeq 2k_F$ there occurs a switching over of the two strengths. This peak is reported to show almost no dispersion in the entire region $k_c < k < 2k_F$. These features have been observed to be common to such diverse systems as Al, Be, and graphite, thus suggesting that they are not one-electron band structure effects. There seems to be some discrepancy between the electron scattering results of Batson *et al.*[96] and the X-ray scattering results of Platzman and Eisenberger[101] in the region of $k$ where they overlap.

4. For $k > 2k_F$, the X-ray scattering experiment[101] shows a single broad peak corresponding to free particle excitations.

The question now is how much of the aforementioned spectral features can be understood on the basis of the approximate theories of the dielectric function that we have discussed earlier. First of all, the mean-field theory expression (5.10) for the response function yields the following limiting law for the plasmon dispersion relation,

$$\omega_p^2(k) = \omega_p^2 + k^2 \left\{ \frac{6}{5} \frac{\varepsilon_F}{m} - \omega_p^2 \lim_{k \to 0} \left[ \frac{G(k)}{k^2} \right] \right\} + \cdots \tag{7.2}$$

Clearly if $G(k)$ is independent of $\omega$ as is the case for Eq. (5.10), one obtains the result that the interaction contribution to the coefficient $\alpha$ in Eq. (7.1) is determined by the interaction contribution to the static screening length. According to the discussion given in Section 2, we should instead expect a relation between $\alpha$ and the third moment at least if the damping is very small [see Eq. (2.20)], and a shift from such a value for $\alpha$ to a value

[100] P. Eisenberger, P. M. Platzman, and K. C. Pandy, *Phys. Rev. Lett.* **31**, 311 (1973).

[101] P. M. Platzman and P. Eisenberger, *Phys. Rev. Lett.* **33**, 152 (1974).

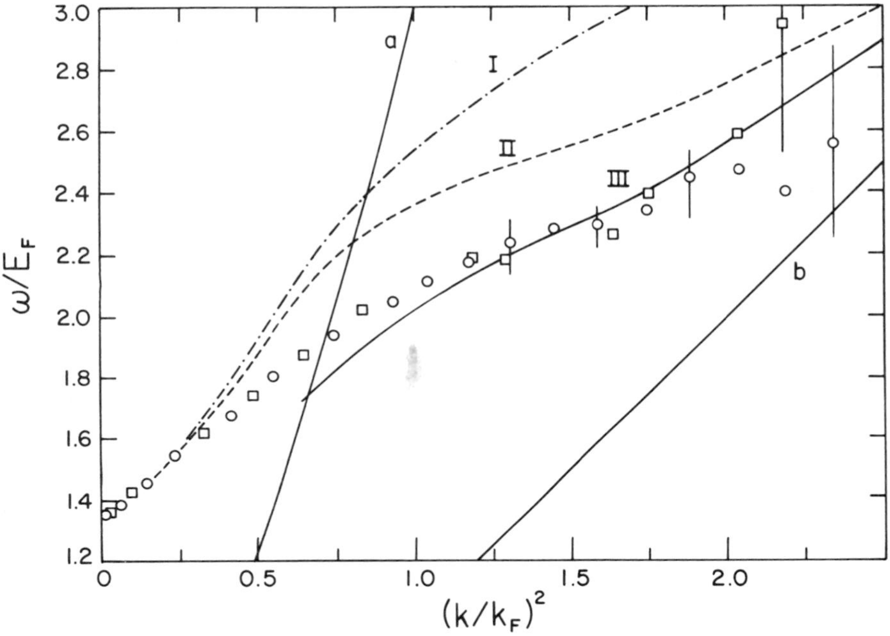


FIG. 9. Position of the peak in $S(k, \omega)$ of Al versus the square of the wave number. The experimental points are due to P. E. Batson, C. H. Chen, and J. Silcox, *Phys. Rev. Lett.* **37**, 937 (1976). The curves marked with Roman numerals represent various theoretical results: RPA (I), VS (II), and first-order perturbation theory (III). For small $k$ ($k < k_c$) curve III (not shown) coincides with curve I. For a discussion of curve III in the transition region around $k_c$ see A. Holas, P. K. Aravind, and K. S. Singwi [*Phys. Rev* **B20**, 4912 (1979)]. Curve $a$ is the edge of the particle–hole continuum, and curve $b$ is the free-particle dispersion curve $\omega = \hbar q^2/2m$.

yielded by the full compressibility as damping increases [see Eq. (2.22)]. For $r_s = 2$ corresponding to the density of Al, Eq. (7.2) yields $\alpha_{\text{RPA}} = 0.451$ in the RPA and $\alpha_{\text{VS}} = 0.363$ in the Vashishta–Singwi scheme, against a value 0.429 from the third moment sum rule and a value 0.163 from the full compressibility. The experimental values fall in the range 0.38–0.42, as reported above. We conclude that one is indeed in an almost collisionless regime and that inclusion of a frequency dependence of the local field factor is needed for quantitative accuracy.

Such a frequency dependence is in fact already needed at order $k^2$ to account for the damping of the plasmon via its decay into multipair excitations. To see this, let us write the response function in the form

$$\chi(k, \omega) = \frac{\chi_0(k, \omega)}{1 - (4\pi e^2/k^2)[1 - G(k, \omega)]\chi_0(k, \omega)} \tag{7.3}$$

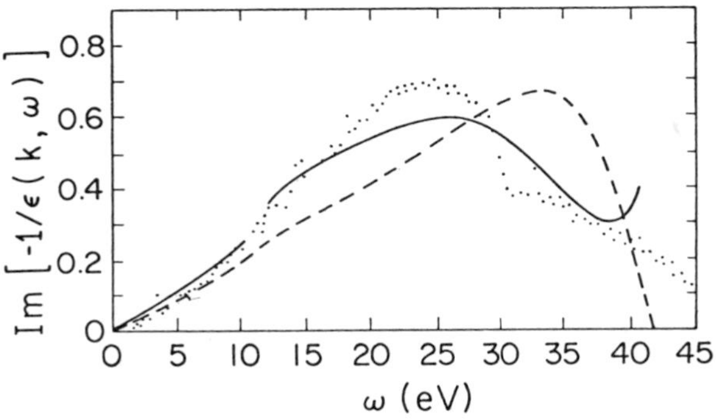


FIG. 10. Energy loss function Im[$-1/\varepsilon(k, \omega)$] for Al at $k = 1.14k_F$; the solid curve is calculated from first-order perturbation theory, and the broken curve is the RPA result. The dots are experimental points from P. E. Batson, C. H. Chen, and J. Silcox, *Phys. Rev. Lett.* **37,** 937 (1976). [From A. Holas, P. K. Aravind, and K. S. Singwi, *Phys. Rev.* **B20,** 4912 (1979).

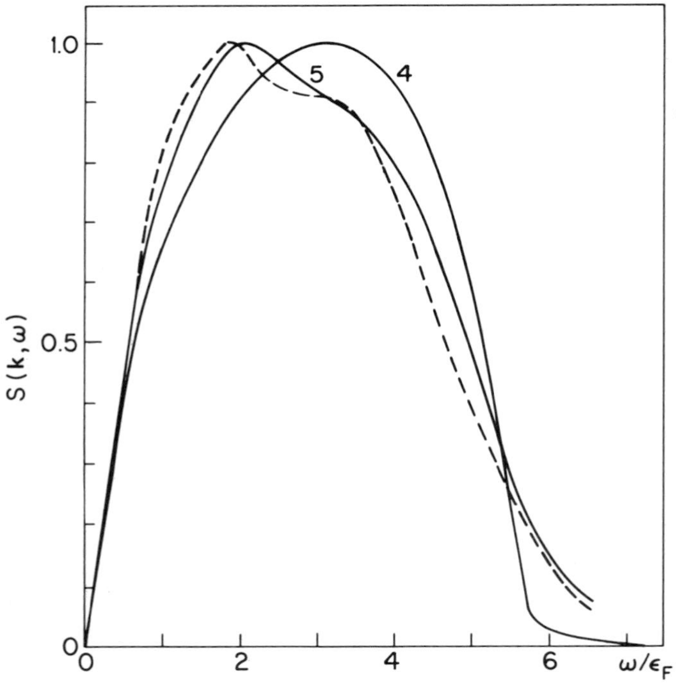


FIG. 11. $S(k, \omega)$ of Al versus frequency at $k = 1.6k_F$. The dashed curve reports the X-ray scattering results of P. M. Platzman and P. Eisenberger, *Phys. Rev. Lett.* **33,** 152 (1974). Curve 5 gives the theoretical results of G. Mukhopadhyay, R. K. Kalia, and K. S. Singwi, *Phys. Rev. Lett.* **34,** 950 (1975); curve 4 is a typical Mori-formalism result. [From G. Mukhopadhyay and A. Sjölander, *Phys. Rev. B: Condens. Matter* [3] **17,** 3589 (1978).]

where $G(k, \omega)$ must satisfy on account of causality the Kramers–Kronig relation

$$\text{Re } G(k, \omega) - G(k, \infty) = P \int_{-\infty}^{+\infty} \frac{d\omega'}{\pi} \frac{\text{Im } G(k, \omega')}{\omega' - \omega}. \tag{7.4}$$

The plasmon dispersion and damping are determined by the zero of the denominator in (7.3). In particular for long wavelengths, the plasmon width is given by

$$\Delta\omega_p(k) = \frac{\omega_p(k)}{2} \left[ \frac{m\omega_p^2(k)}{nk^2} \text{Im } \chi_0(k, \omega_p(k)) - \text{Im } G(k, \omega_p(k)) \right]. \tag{7.5}$$

Since Im $\chi_0$ vanishes outside the single-pair continuum, the plasmon width is determined in this region of wavelengths entirely (apart from Umklapp processes) by the imaginary part of $G$.

Detailed calculations of the shape of the loss function for wavenumbers above $k_c$ in Al[96,97] and Na[97] show that it cannot be accounted for by taking a purely real local field. Dispersion curves in Al as computed in the RPA (curve I) and in the Vashishta–Singwi static theory (curve II) are compared in Fig. 9 with the experimental data. It is seen that the static local field correction leads to a downward shift of the dispersion curve above $k_c$ by several electron volts but is still far away from the experimental data.

### *a. Dynamic Correlations in First-Order Perturbation Theory*

To formulate a microscopic approach to the problem of dynamical correlations that is physically transparent and at the same time tractable enough to yield results that can be compared with experiment is a difficult task. A lowest-order perturbation calculation for the proper polarizability $\tilde{\chi}(k, \omega)$ is still feasible. The latter was first evaluated in the static limit by Geldart and Taylor,[63] as discussed in Section 5,c. Here we shall discuss the results of a dynamic calculation.[102]

In first-order perturbation theory the problem is to evaluate $\chi_1(k, \omega)$ of Eq. (3.58). As we have already noted, this function has two contributions, one coming from self-energy corrections and the other from the exchange local field. The imaginary part of the former can be evaluated analytically, whereas the evaluation of the imaginary part of the latter can be reduced to a one-dimensional numerical integration. It is thus easy to see that Im $\chi_1$ has jump singularities at the characteristic frequencies $\omega_s = (k^2/2) + k$ and $|(k^2/2) - k|$, arising from the self-energy part. Re $\chi_1$ is determined

[102] A. Holas, P. K. Aravind, and K. S. Singwi, *Phys. Rev.* **B20**, 4912 (1979).

from Im $\chi_1$ by Hilbert transform, and it is easily shown that the jump discontinuities at $\omega_s$ give rise to logarithmic singularities in Re $\chi_1$ at the same frequencies. If the self-energy contribution is evaluated to all orders (in an effective mass approximation), the above singularities are rounded off, but the results are not much affected away from the singularity frequencies.

In the present theory the expression for the dynamic local field is

$$G(k, \omega) = -\frac{k^2}{4\pi e^2} \frac{\chi_1(k, \omega)}{\chi_0(k, \omega)[\chi_0(k, \omega) + \chi_1(k, \omega)]}. \tag{7.6}$$

The results at $k = k_F$ are reported in Fig. 12 for two values of $r_s$, together with the results for $G_1(k, \omega)$.

$$G_1(k, \omega) = -\frac{k^2}{4\pi e^2} \frac{\chi_1(k, \omega)}{\chi_0^2(k, \omega)}. \tag{7.7}$$

which is actually independent of $r_s$. The arrows show the positions of the characteristic frequencies $\omega_s$ where, as noted above, the curves cannot be trusted. We see that both Re $G$ and Re $G_1$ are initially flat but then rise to a maximum located slightly below the upper edge of the single-pair continuum. Thereafter they drop sharply and become negative before finally approaching a positive asymptotic value which satisfies the third frequency moment sum rule to the appropriate degree of approximation. The noteworthy feature in the imaginary part of the local field is its large rise near the upper characteristic frequency, whereas the absence of a tail at higher frequencies leads to the plasmon being undamped outside the single-pair continuum [see Eq. (7.5)]. One should therefore be careful while approximating the local field correction by a frequency-independent function if one is interested in the high-frequency behavior of the electron gas.

Although the above perturbative theory applies for small $r_s$, it seems to yield reasonable results up to values of $r_s$ as large as 2. This confidence is gained[102] by examining the detailed shape of the pair correlation function $g(r)$ and by a direct comparison with the experimental data[96] for the electron energy loss in Al. This comparison is reported in Fig. 9 for the dispersion curve (curve III) and in Fig. 10 for the energy loss function at $k = 1.14k_F$. In particular the coefficient $\alpha$ in Eq. (7.1) is

$$\alpha = \alpha_{RPA}[1 - (\omega_p^2/4)] \tag{7.8}$$

which yields a value of 0.39 for Al. This result was derived earlier by DuBois[15] and by Nozières and Pines.[103]

[103] P. Nozières and D. Pines, *Phys. Rev.* [2] **111**, 442 (1958).

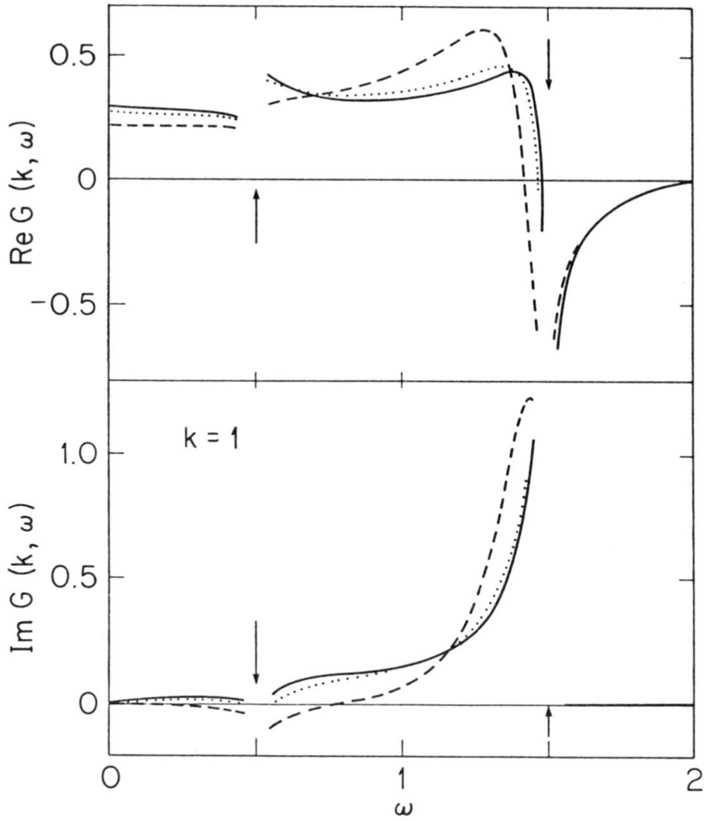


FIG. 12. Real and imaginary parts of the local field correction $G(k, \omega)$ at $k = k_F$ as calculated from first-order perturbation theory. The dotted curve is for $r_s = 0.5$, and the broken curve is for $r_s = 2$. The solid curve reports the function $G_1(k, \omega)$ given by Eq. (7.7). [From A. Holas, P. K. Aravind, and K. S. Singwi, *Phys. Rev.* **B20**, 4912 (1979)].

It is worth mentioning here similar recent work of Brosens *et al.*[104] These authors start with the Hartree–Fock equation of motion for the one-particle Wigner distribution function, Eq. (3.52), and solve it approximately by a variational method. Their result for $G(k, \omega)$ is the same as Eq. (7.7), which they proceed to evaluate for $r_s = 3$. Although the presence of singularities in $\chi_1$ was noted by these authors, the role of singularities in $G(k, 0)$, $S(k, \omega)$, and the plasmon dispersion is considered

[104] F. Brosens, L. F. Lemmens, and J. T. Devreese, *Phys. Status Solidi B* **74,** 45 (1976); F. Brosens, J. T. Devreese, and L. F. Lemmens, *ibid.* **80,** 99 (1977); J. T. Devreese, F. Brosens, and L. F. Lemmens, *Phys. Rev.* **B21,** 1349 (1980).

significant rather than unphysical. Other authors[105] have been led to the result (7.7) but have failed to recognize the presence of singularities, and their numerical results appear[102] to be seriously in error. A perturbation series for $G$ has been written down by Dharma-wardana,[105a] but no attempt has been made to evaluate it.

*b. Kinetic Equation Approach*

In the preceding section we have seen that the first-order perturbation theory is restricted to $r_s \lesssim 2$ and misses qualitative features such as the damping of the plasmon outside the single-pair continuum. A tractable approach to transcend the first-order theory has been based on a truncation scheme[106] of the equation of motion (3.46) for the two-particle Wigner distribution function. The scheme consists of including only the Hartree potential term on the right-hand side of Eq. (3.46). This procedure is equivalent to making an RPA-like approximation on the two-particle Wigner function. The approximation subsumes both the low-order theory just discussed and the exact limiting behaviors at large $k$ or large $\omega$ discussed by Niklasson[30] (see Section 3,c). It also keeps some account of simultaneous excitations of two particle–hole pairs.

Within the above approximation it is straightforward to write the expression for the proper polarizability. However, its numerical evaluation requires a knowledge of the equilibrium two-particle Wigner function. In the absence of such a knowledge two different ansatzes were explored[106] subject to the exact requirement

$$\frac{1}{N}\sum_{\mathbf{p}\sigma}\sum_{\mathbf{p}'\sigma'} f^{(2)}_{\mathbf{p}\sigma;\mathbf{p}'\sigma'}(k) = S(k) - 1. \tag{7.9}$$

The results are similar for the two ansatzes and are best summarized by comparison with the first-order theory as follows:

1. The local field $G(k, \omega)$ still has a resonance structure at the characteristic frequencies $\omega_s$ as was illustrated in Fig. 12 for the first-order theory. However, Im $G(k, \omega)$ has now acquired a high-frequency tail, and Re $G(k, \omega)$ at low frequencies leads to a good value for the compressibility at least for $r_s \leq 2$.

[105] D. N. Tripathy and S. S. Mandal, *Phys. Rev. B: Solid State* [3] **16,** 231 (1977); B. K. Rao, S. S. Mandal, and D. N. Tripathy, *J. Phys. F* **9,** L51 (1979). In other publications [S. S. Mandal, B. K. Rao, and D. N. Tripathy, *Phys. Rev. B: Condens. Matter* [3] **18,** 2524 (1978); D. N. Tripathy, B. K. Rao, and S. S. Mandal, *Solid State Commun.* **22,** 83 (1977)] these authors have proceeded to evaluate $g(r)$ and $S(k)$ for the electron gas by replacing $G_1(k, \omega)$ by $G_1(k, 0)$. This function has an unphysical large peak at $k = 2k_F$, which is strongly reduced upon inclusion of higher-order effects.[63] As a consequence these authors report an unphysical structure of the electron gas. Even within low order perturbation

2. The shape of the energy loss function in Al is similar to that reported for the first-order theory in Fig. 10, but there is a quantitative improvement in the absolute value of the loss at its peak.

3. The peak positions for $k > k_c$ in Al are practically unaltered relative to the results of the first-order theory that were compared with experiment in Fig. 9.

4. The width of the plasmon at long wavelength is proportional to $k^2$, and its magnitude for both Al and Na is in fair agreement with the experimental data. On the other hand, the dispersion coefficient $\alpha$ is sizably underestimated.

Although the approximation we have been discussing leads on the whole to improvement, there is obviously a need for a better theory, especially for increasing values of $r_s$. Higher-order terms should also lead to rounding off of the singularities in $G(k, \omega)$, leaving only broad resonances in its structure. A close examination of the terms that have been neglected in Eq. (3.46) shows that these include additional self-energy corrections and that strong cancellation between these and the remaining terms are to be expected.

### *c. X-Ray Inelastic Scattering*

We now turn to the large $k$-region where the X-ray scattering experiments[101] have revealed a "double-peak" structure in $S(k, \omega)$. In comparing theory with experiment, it should be borne in mind that the present experimental situation is perhaps not completely settled. Assuming that the reported spectrum is a genuine jellium property, none of the previously discussed theories yields such a structure. Although the possibility of obtaining such a structure on inclusion of lifetime effects of the single-particle states has been pointed out by Mukhopadhyay *et al.*[107] the shoulder of the high-energy side of the "peak," as seen in Fig. 11, is so weak that it could easily be washed out when convoluted with the experimental resolution function. Besides, this theory violates the continuity equation. A calculation of the lineshape resembling the experimental spectrum has also been reported by Barnea[108] by including a loss

theory this structure in $S(k)$ is washed out[102] upon inclusion of the frequency dependence of $G_1(k, \omega)$.

[105a] M. W. C. Dharma-wardana, *J. Phys. C* [3] **9**, 1919 (1976).

[106] P. K. Aravind, A. Holas, and K. S. Singwi, *Phys. Rev. B: Condens. Matter* [3] (to be published); see also K. Utsumi and S. Ichimaru; *Phys. Rev.* **B22**, 1522 (1980); *ibid.* 5203 (1980) and *Phys. Rev.* **B23**, 3291 (1981).

[107] G. Mukhopadhyay, R. K. Kalia, and K. S. Singwi, *Phys. Rev. Lett.* **34**, 950 (1975); see also K. Awa, H. Yasuhara, and T. Asahi, *Solid State Commun.* **38**, 1285 (1981).

[108] G. Barnea, *J. Phys. C* [3] **12**, L263 (1979).

process due to excitation of a long-wavelength plasmon and a particle–hole pair.

This problem has been reexamined by several authors[109] in the framework of Mori's formalism. Mukhopadhyay and Sjölander write the response function in the form

$$\chi(k, \omega) = \frac{\bar{\chi}_0(k, \omega)}{1 - (4\pi e^2/k^2)\{1 - G_0(k) + [G_0(k) - G_\infty(k)][-i\omega\Gamma(k, \omega)]\}\bar{\chi}_0(k, \omega)}, \tag{7.10}$$

where $\bar{\chi}_0(k, \omega)$ is the ideal gas response calculated with the true momentum distribution, $G_0$ and $G_\infty$ are local field factors appropriate for low and high frequencies, and $\Gamma(k, \omega)$ is a memory function for which these authors have tried various simple forms. Typical results obtained from this theory are shown in Fig. 11, and clearly they are in disagreement with experiment. Introduction of a relaxation time into the theory by Jindal *et al.*[109] can shift the broad peak to lower energies, without showing a double-peak structure.

In the foregoing discussion we have seen how the introduction of a frequency-dependent local field $G(k, \omega)$ gives rise to a sizable lowering of the peak position of $S(k, \omega)$ from its RPA value and to damping of the plasmon. One should expect a similar situation to occur in the case of liquid $He^3$. Indeed the observed dispersion[110] of zero sound in $He^3$ in the momentum transfer region $0.8 \text{ Å}^{-1} < k < 1.5 \text{ Å}^{-1}$ is almost flat. In contrast to the electron gas case, the mass renormalization in $He^3$ is very large ($m^* = 3m$). Such a microscopic theory of $He^3$ is lacking.

## IV. Some Applications

### 8. Two-Component Quantum Plasma

An interesting and useful generalization of the theory of electron correlations of Singwi *et al.*[43] to a two-compoent quantum plasma was made by Sjölander and Stott[111]—a generalization that provides a new

[109] G. Mukhopadhyay and A. Sjölander, *Phys. Rev. B: Condens. Matter* [3] **17,** 3589 (1978); V. K. Jindal, H. B. Singh, and K. N. Pathak, *Phys. Rev. B: Solid State* [3] **15,** 252 (1977). For a formal discussion of the memory function see M. Nusair and B. Goodman, *Phys. Lett. A* **70A,** 275 (1979).

[110] K. Sköld, C. A. Pelizzari, R. Kleb, and G. E. Ostrowski, *Phys. Rev. Lett.* **37,** 842 (1976); see also C. H. Aldrich, III, C. J. Pethick, and D. Pines, *ibid.* p. 845.

[111] A. Sjölander and M. J. Stott, *Solid State Commun.* **8,** 1881 (1970); *Phys. Rev. B: Solid State* [3] **5,** 2109 (1972).

method to treat a nonlinear problem such as the calculation of the screening charge around an impurity like a proton or a positron in an electron gas. In fact these authors calculated positron annihilation rates in simple metals that were in excellent agreement with experiment (for $r_s \le 4$), thus demonstrating that the theory was capable of handling multiple-scattering processes[112] between the positron and the electron in an electron fluid. The Sjölander–Stott generalization was later extended by Vahishta *et al.*[113] to a multicomponent plasma and was successfully used to calculate the ground state properties of the electron–hole liquid in semiconductors such as germanium and silicon. The same was also extended by Tosi *et al.*[114] to develop a unified theory of liquid metals. In this section we discuss very briefly these developments and refer the reader for details to the original papers.

Consider a neutral two-component degenerate plasma consisting of electrons and positively charged particles, say holes. As an artifice, let us apply a weak external perturbation such that $V_e^i(\mathbf{r}, t)$ is the external potential that acts on the $i$th component of the plasma. The perturbation Hamiltonian is

$$H'(t) = \sum_i \int d\mathbf{r}\ \rho_i(\mathbf{r}) V_e^i(\mathbf{r}, t), \tag{8.1}$$

and the induced density change of the $i$th component can be written in linear response theory as

$$n_i(\mathbf{k}, \omega) = \sum_j \chi_{ij}(k, \omega) V_e^j(\mathbf{k}, \omega). \tag{8.2}$$

This defines a matrix of partial density response functions $\chi_{ij}(k, \omega)$ from which the dielectric function can be calculated as

$$1/\varepsilon(k, \omega) = 1 + \sum_{ij} \varphi_{ij}(k)\chi_{ij}(k, \omega), \tag{8.3}$$

where $\varphi_{ij}(k) = 4\pi e_i e_j/k^2$. In the spirit of a generalized polarization-potential theory we write

$$n_1(\mathbf{k}, \omega) = \chi_{01}(k, \omega)[V_e^1(\mathbf{k}, \omega) + f_{11}(k) n_1(\mathbf{k}, \omega) + f_{12}(k) n_2(\mathbf{k}, \omega)] \tag{8.4}$$

[112] S. Kahana, *Phys. Rev.* [2] **129,** 1622 (1963); J. P. Carbotte, *ibid.* **155,** 197 (1967); J. Crowell, V. E. Anderson, and R. H. Ritche, *ibid.* **150,** 243 (1966).

[113] P. Vashishta, P. Bhattacharyya, and K. S. Singwi, *Nuovo Cimento Soc. Ital. Fis. B* [11] **23B,** 172 (1974).

[114] M. P. Tosi, M. Parrinello, and N. H. March, *Nuovo Cimento Soc. Ital. Fis. B* [11] **23B,** 195 (1974).

and a similar expression for $n_2(\mathbf{k}, \omega)$, where $\chi_{0i}(k, \omega)$ are the free-particle polarizabilities, and $f_{ij}(k)$ are effective potentials. Solving explicitly for $n_i(\mathbf{k}, \omega)$ and comparing these with Eqs. (8.2), we have the following expressions for the partial polarizabilities:

$$\chi_{11}(k, \omega) = \chi_{01}(k, \omega)[1 - f_{22}(k)\chi_{02}(k, \omega)]/\Delta(k, \omega), \tag{8.5a}$$

$$\chi_{12}(k, \omega) = f_{21}(k)\chi_{01}(k, \omega)\chi_{02}(k, \omega)/\Delta(k, \omega), \tag{8.5b}$$

and

$$\chi_{22}(k, \omega) = \chi_{02}(k, \omega)[1 - f_{11}(k)\chi_{01}(k, \omega)]/\Delta(k, \omega), \tag{8.5c}$$

where

$$\Delta(k,\omega) = [1 - f_{11}(k)\chi_{01}(k, \omega)][1 - f_{22}(k)\chi_{02}(k, \omega)] - f_{12}(k)f_{21}(k)\chi_{01}(k, \omega)\chi_{02}(k, \omega). \tag{8.6}$$

From symmetry, $f_{21} = f_{12}$ and $\chi_{21} = \chi_{12}$.

The effective potentials $f_{ij}(k)$ entering the above mean-field approximation can be written as

$$f_{ij}(k) = \varphi_{ij}(k)[1 - G_{ij}(k)], \tag{8.7}$$

where the $G_{ij}(k)$ are local field factors that take different values in different approximations. In RPA, $G_{ij}(k) = 0$ and it follows from the above equations that

$$\varepsilon_{\text{RPA}}(k, \omega) = 1 - \frac{4\pi e^2}{k^2}\chi_{01}(k, \omega) - \frac{4\pi e^2}{k^2}\chi_{02}(k, \omega). \tag{8.8}$$

In the Hubbard approximation,

$$G_{ij}(k) = \tfrac{1}{2}\delta_{ij}k^2/(k^2 + k_{\text{F}i}^2) \tag{8.9}$$

where $k_{\text{F}i}$ is the Fermi wave number of the $i$th component. The corresponding dielectric function still is the sum of independent polarizabilities of the two components, corrected for exchange. This is no longer the case for the STLS approximation, where correlations (in particular, electron–hole correlations) are introduced by writing

$$G_{ij}(k) = -\frac{1}{n}\int \frac{d\mathbf{q}}{(2\pi)^3}\frac{\mathbf{k}\cdot\mathbf{q}}{q^2}\gamma_{ij}(|\mathbf{k} - \mathbf{q}|). \tag{8.10}$$

Here, $n$ is the number density of electrons and the structure factors $\gamma_{ij}(k)$ are related to the partial pair correlation functions $g_{ij}(r)$ by

$$g_{ij}(r) = 1 + \frac{1}{n}\int \frac{d\mathbf{q}}{(2\pi)^3}\exp(i\mathbf{q}\cdot\mathbf{r})\gamma_{ij}(q). \tag{8.11}$$

The fluctuation-dissipation theorem in a multicomponent case yields

$$\langle \rho_i(\mathbf{k})\rho_j(-\mathbf{k})\rangle = n_i[\delta_{ij} + (n_i/n)\gamma_{ij}(k)] = -\frac{\hbar}{\pi}\int_0^\infty d\omega \,\mathrm{Im}\,\chi_{ij}(k,\omega). \quad (8.12)$$

The set of equations (8.5), (8.10), and (8.12) have to be solved self-consistently. For example, in a two-component case we have three coupled linear integral equations, one for each of the $\gamma_{ij}(k)$. A knowledge of these functions enables one to calculate the correlation energy.

*a. Positron Annihilation in Metals*

Considerable simplifications occur in applying the above results to the positron annihilation problem owing to the fact that the density $n_+$ of positrons is negligible compared to that of electrons. If we let $n_+ \to 0$ and remember that $\chi_{0+}$ is proportional to $n_+$, it can be verified[111] that we have just the following integral equation to solve:

$$\gamma_{+-}(k) = f(k)[1 + \frac{1}{n}\int \frac{d\mathbf{q}}{(2\pi)^3}\frac{\mathbf{k}\cdot\mathbf{q}}{q^2}\gamma_{+-}(|\mathbf{k}-\mathbf{q}|), \quad (8.13)$$

where

$$f(k) = (\hbar/\pi n_+)\int_0^\infty d\omega\,\mathrm{Im}[\chi_{0+}(k,\omega)/\varepsilon(k,\omega)]. \quad (8.14)$$

Here, $\varepsilon(k, \omega)$ is the dielectric function of the electron gas, and in the limit $n_+ \to 0$

$$\chi_{0+}(k,\omega) = \frac{n_+}{\hbar}\left(\frac{1}{\omega - \hbar k^2/2m_+ + i\eta} - \frac{1}{\omega_+\hbar k^2/2m_+ + i\eta}\right)\cdot \quad (8.15)$$

If the solution $\gamma_{+-}(k)$ is known, its integral determines the annihilation rate through the formula[115]

$$\lambda = \frac{12.0}{r_s^3}\,g_{+-}(0)\times 10^9 \quad \mathrm{sec}^{-1}. \quad (8.16)$$

The results of Sjölander and Stott for $\lambda$ are in very good agreement with experiment for densities $r_s \lesssim 4$, but at lower densities they seem to diverge as do the calculations of Kahana and other authors[112] based on a ladder approximation. This divergence is attributed to the occurrence of an incipient bound state. The same difficulty is met already at values of $r_s \gtrsim 2$ when one tries to calculate the nonlinear screening charge around a fixed proton.[111] Therefore the multicomponent generalization of the STLS theory cannot be trusted when one is approaching a situation where bound

[115] R. A. Ferrell, *Rev. Mod. Phys.* **28**, 308 (1956).

states start appearing. Later on, Bhattacharyya and Singwi[116] tried to rectify this situation, but in an ad hoc fashion by introducing a parameter in the theory. The annihilation rate as a function of electron density is shown in Fig. 13.

*b. Electron–Hole Liquid (EHL)*

Under intense illumination such as that of laser light, a dense gas of excitons can be formed in a semiconductor such as germanium. This gas at very low temperatures undergoes a phase transition[117] into a metallic "liquid"—a transition quite analogous to the condensation of water vapor into water droplets. The electron–hole fluid should be viewed as two interpenetrating Fermi fluids of electrons and holes. This condensate represents a state of matter quite unique in nature because of its extreme quantum character. Fortunately, the condensation energy can be measured directly by either optical or thermodynamic measurements, thus providing a crucial test of the many-body approximations that go into its first-principles calculation. We shall not go into the details of the actual calculations for which we refer the reader to the original articles[118–120] and to a recent review by Rice.[121] We shall merely point out the salient features and the results of the calculatons.

To illustrate our main point, we shall consider an idealized model with one parabolic conduction band and one parabolic hole band. The Hartree–Fock energy per electron–hole pair is

$$E_{\mathrm{HF}} = \left(\frac{2.21}{r_s^2} - \frac{1.832}{r_s}\right) E_x \ \mathrm{Ry}. \tag{8.17}$$

The ground-state energy is the sum of the Hartree–Fock energy and the correlation energy, the latter can be calculated only approximately. The

[116] P. Bhattacharyya and K. S. Singwi, *Phys. Rev. Lett.* **29,** 22 (1972); see also *Phys. Lett. A* **41A,** 457 (1972).

[117] L. V. Keldysh, *Proc. Int. Conf. Phys. Semicond. 9th, 1968* p. 1303 (1968); Ya E. Pokrovskii, *Phys. Status Solidi A* **11,** 385 (1972).

[118] W. F. Brinkman, T. M. Rice, P. W. Anderson, and S. T. Chui, *Phys. Rev. Lett.* **28,** 961 (1972); W. F. Brinkman and T. M. Rice, *Phys. Rev. B: Solid State* [3] **7,** 1508 (1973).

[119] M. Combescot and P. Nozières, *J. Phys. C* [3] **5,** 2369 (1972); M. Combescot, *Phys. Rev. Lett.* **32,** 15 (1974).

[120] P. Vashishta, P. Bhattacharyya, and K. S. Singwi, *Phys. Rev. Lett.* **30,** 1248 (1973); *Phys. Rev. B: Solid State* [3] **10,** 5108 (1974); P. Bhattacharyya, V. Massida, K. S. Singwi, and P. Vashishta, *ibid.* p. 5127 (1974); P. Vashishta, S. G. Das, and K. S. Singwi, *Phys. Rev. Lett.* **33,** 911 (1974).

[121] T. M. Rice, *Solid State Phys.* **32,** 1 (1977). For a recent review of the experimental situation, see J. C. Hensel, T. G. Phillips, and G. A. Thomas, *ibid.* p. 88.

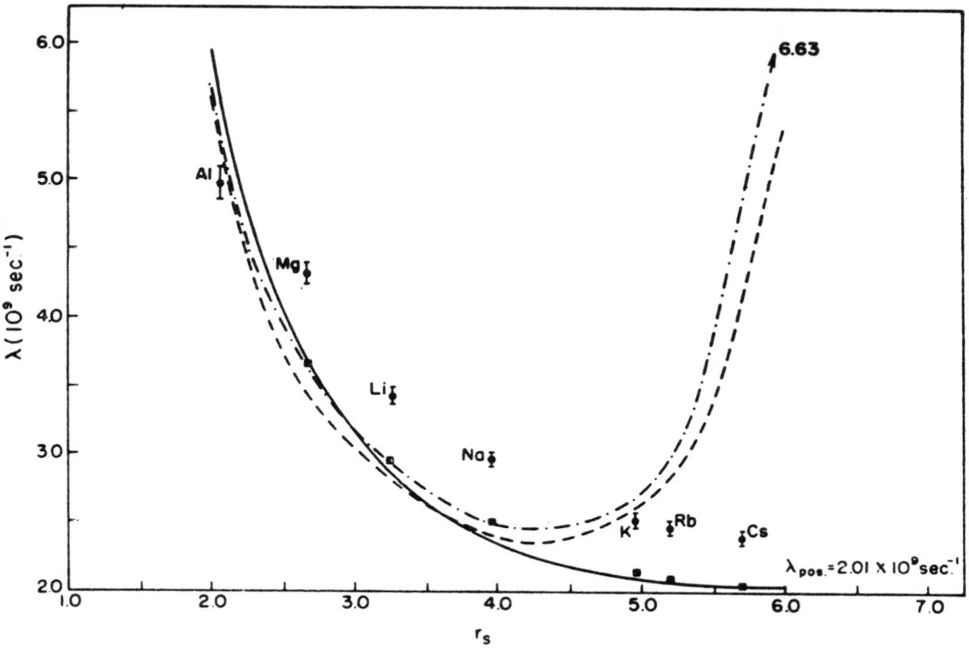


FIG. 13. Positron annihilation rate λ versus $r_s$. Dashed curve, from A. Sjölander and M. J. Stott, *Solid State Commun.* **8,** 1881 (1970); solid curve, from P. Bhattacharyya and K. S. Singwi, *Phys. Rev. Lett.* **29,** 22 (1972); dash–dot curve is based on the ladder approximation [from J. Crowell, V. E. Anderson, and R. H. Ritchie, *Phys. Rev.* [2] **150,** 243 (1966)]. Experimental data are from H. Weisberg and S. Berko, *Phys. Rev.* [2] **154,** 249 (1967); the squares show the estimated contribution due to conduction electrons only (15% core removed).

simplest approximation is the RPA, which is good in the high-density limit $r_s \ll 1$. An improved version of the RPA is the Hubbard approximation, which includes exchange between particles of the same species with parallel spins. This approximation was used by Brinkman *et al.*[118] Then there is the interpolation scheme of Nozières and Pines,[2] which was used extensively by Combescot and Nozières[119] in their study of the ground-state energy of the EHL. The correlations between electrons and holes, which are responsible for the exciton bound state, become of increasing importance as $r_s$ increases. The STLS approximation, which treats correlations between electrons and holes on an equal footing, has been used by Vashishta *et al.*[120] in their ground-state energy calculations for the EHL. Since all these schemes are approximate, the only way to assess their relative merits is to compare their results with experiment. Of course, for $r_s \lesssim 1$ all these schemes give very similar results.

In the model case under consideration ($m_e = m_h$), the calculated ground-state energy by Brinkman *et al.* is $-0.86E_x$ at $r_s = 1.7$, whereas the corresponding values of Vashishta *et al.* are $-0.99E_x$ at $r_s = 2$. This is still higher than $-E_x$, the binding energy of a single exciton, and thus the condensate should not be formed. In the actual case, it is the multiplicity of bands and their anisotropy that leads to the formation of the condensate. A significant difference between the above two calculations is in the value of the electron–hole pair correlation function for zero separation, i.e., $g_{eh}(0)$. Whereas Brinkman and Rice give $g_{eh}(0) \simeq 2$, Vashishta *et al.* find $g_{eh}(0) \simeq 8$ for $r_s = 1.7$. This large enhancement, as in the positron annihilation problem, is a nonlinear effect arising from multiple scatterings between the electron and the hole. Recent experiments of Chou and Wong[122] indicate that the $r_s$ dependence of $g_{eh}(0)$ is in accord with the STLS theory. In Fig. 14 the values of the ground-state energy as a function of $r_s$ are given for this model system. (See Note Added in Proof, p. 266.)

The model we have considered, although an idealized one, is not very far from reality. The band structure of normal Ge consists of four equivalent conduction valleys at the zone edge and two hole bands at the center of the zone. The energy–wavenumber relations for these bands are fairly complex. This complex band structure is considerably simplified under a very high uniaxial [111] stress. In this case, one has to deal with a single conduction valley and a single hole band—a situation quite analogous to the model considered above with the difference that the bands are ellipsoidal rather than parabolic. This complication can be incorporated in the calculations. The calculations of Brinkman *et al.*[118] using the Hubbard approximation and those of Combescot and Nozières[119] give such a small value for the binding energy ($\simeq 0.07E_x$) that there arises a doubt as to whether the metallic state exists. On the other hand, Vashishta *et al.*[120] find that the binding energy is $\simeq 0.17E_x$, and conclude that the metallic state is bound at a density $n = 1.1 \times 10^{16}$ cm$^{-3}$. Their value for $g_{eh}(0)$ is 7 at the equilibrium density. The experiments by Feldman *et al.*[123] show that a condensate still exists under a uniform stress as high as 13 kg/mm$^2$, where band degeneracies have been completely removed but the nonparabolic character of the valence band is still appreciable. Their value of the binding energy is $0.65 \pm 0.07$ meV and is in reasonable agreement with the theoretical estimates.

Similar calculations in Ge and Si based on the STLS theory, although numerically very involved, have been done by Vashishta *et al.*[120] Some of their results are given in Table IV and compared with other theories and with experiment.

[122] H.-h. Chou and G. K. Wong, *Phys. Rev. Lett.* **41,** 1677 (1978).

[123] B. J. Feldman, H. H. Chou, and G. K. Wong, *Solid State Commun.* **26,** 209 (1978).

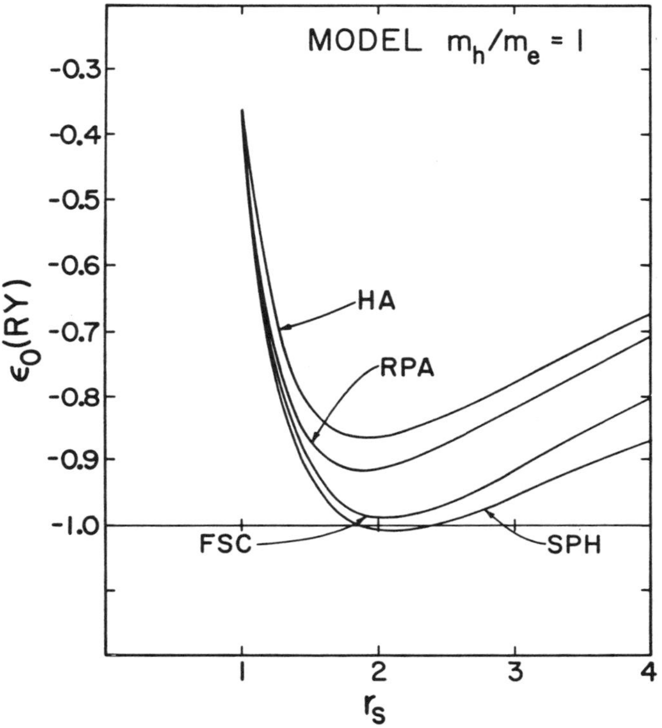


FIG. 14. Ground-state energy per electron–hole pair, in excitonic rydbergs, versus $r_s$ for a model semiconductor. The various curves report the results of the RPA, the Hubbard approximation (HA), and the STLS approximation (FSC). [From P. Vashishta, P. Bhattacharyya, and K. S. Singwi, *Phys. Rev. B: Solid State* [3] **10,** 5108 (1974).]

### *c. Electronic Structure of Hydrogen in Metals*

Hydrogen in metals forms the simplest and yet the most interesting impurity. Since the proton provides a very strong perturbing potential on the electron gas, the problem of calculating the screening charge around it is a nonlinear one. As mentioned in Section 8,a, the Sjölander–Stott theory fails in this case because of its inability to treat incipient bound states. The HKS formalism is, however, well suited to treat this problem in the case of a heavy impurity.[124] In this scheme, one solves self-

[124] Z. D. Popović, M. J. Stott, J. P. Carbotte, and G. R. Piercy, *Phys. Rev. B: Solid State* [3] **13,** 590 (1976); C. O. Almbladh, U. von Barth, Z. D. Popović and M. J. Stott, *ibid.* **14,** 2250 (1976); E. Zaremba, L. M. Sander, H. B. Shore, and J. H. Rose, *J. Phys. F* [3] **7,** 1763 (1977); P. Jena and K. S. Singwi, *Phys. Rev. B: Condens. Matter* [3] **17,** 3518 (1978).

TABLE IV. PROPERTIES OF THE ELECTRON–HOLE LIQUID[a]

| | Ge | Si | Ge[111] | Si[100] |
|---|---|---|---|---|
| Ground state energy (meV) | | | | |
| BRAC | −5.3 | −20.4 | −2.8 | — |
| CN | −5.6 | −19.4 | −2.6 | — |
| VDS | −5.9 | −22.0 | −3.1 | −14.7 |
| Binding energy (meV) | | | | |
| BRAC | 1.15 | 5.7 | 0.2 | — |
| CN | 1.45 | 4.7 | 0 | — |
| VDS | 1.75 | 7.3 | 0.5 | 1.9 |
| Experiment | 1.8 ± 0.2[b] | 8.2 ± 0.1[c] | 0.65 ± 0.07[d] | — |
| Equilibrium density ($10^{17}$ cm$^{-3}$) | | | | |
| BRAC | 1.8 | 34 | 0.12 | — |
| CN | 2.0 | 36 | 0.13 | — |
| VDS | 2.2 | 32 | 0.11 | 4.5 |
| Experiment | 2.38 ± 0.05[b] | 33.3 ± 0.5[c] | 0.27 ± 0.03[d] | — |
| Critical temperature (K) | | | | |
| CN | 8 | 28 | — | — |
| VDS | 5.9 | 20.8 | 2.6 | 14.3 |
| Experiment | 6.5 ± 0.1[e] | 27 ± 1[f] | 3.5 ± 0.5[d] | — |

[a] BRAC: W. F. Brinkman, T. M. Rice, P. W. Anderson, and S. T. Chui, *Phys. Rev. Lett.* **28,** 961 (1972); W. F. Brinkman and T. M. Rice, *Phys. Rev. B: Solid State* [3] **7,** 1508 (1973). CN: M. Combescot and P. Nozières, *J. Phys. C* [3] **5,** 2369 (1972); M. Combescot, *Phys. Rev. Lett.* **32,** 15 (1974). VDS: P. Vashishta, S. G. Das, and K. S. Singwi, *Phys. Rev. Lett.* **33,** 911 (1974).

[b] G. A. Thomas, A. Frova, J. C. Hensel, R. E. Miller, and P. A. Lee, *Phys. Rev. B: Solid State* [3] **13,** 1692 (1976).

[c] R. B. Hammond, T. C. McGill, and J. W. Mayer, *Phys. Rev. B: Solid State* [3] **13,** 3566 (1976).

[d] B. J. Feldman, H. H. Chou, and G. K. Wong, *Solid State Commun.* **26,** 209 (1978).

[e] G. A. Thomas, T. M. Rice, and J. C. Hensel, *Phys. Rev. Lett.* **33,** 219 (1974).

[f] J. Shah, M. Combescot, and A. H. Dayem, *Phys. Rev. Lett.* **38,** 1497 (1977).

consistently the following set of equations:

$$\left[-\frac{\hbar^2}{2m}\nabla^2 + V_{\text{eff}}(\mathbf{r};\, n(\mathbf{r})\right]\psi_i(\mathbf{r}) = \varepsilon_i\psi_i(\mathbf{r}), \tag{8.18}$$

where

$$n(\mathbf{r}) = \sum_i |\psi_i(\mathbf{r})|^2, \tag{8.19}$$

and

$$V_{\text{eff}}(\mathbf{r};\, n(\mathbf{r})) = -\frac{e^2}{r} + e^2\int d\mathbf{r}'\,\frac{n(\mathbf{r}')}{|\mathbf{r}-\mathbf{r}'|} + V_{\text{xc}}[n(\mathbf{r})]. \tag{8.20}$$

In the local density approximation,

$$V_{\text{xc}}[n(\mathbf{r})] = \frac{d}{dn}[n\varepsilon_{\text{xc}}(n)] - \mu_{\text{xc}}(n_o), \tag{8.21}$$

where $\varepsilon_{xc}(n)$ is the exchange-correlation energy per particle and $V_{xc}[n]$ has been defined with respect to the exchange-correlation potential $\mu_{xc}$ at the average density $n_0$ of the host metal. As $r \to \infty$, $V_{eff}[n]$ tends to zero. One can also include gradient corrections to $V_{xc}[n]$ without any difficulty. The numerical procedure for solving the above equations self-consistently for both the scattering states and the bound states is a standard one. In each iteration, one has to ensure that the Friedel sum rule is satisfied to a high degree of accuracy.

The results of such a calculation are shown in Fig. 15 for various values of $r_s$. Inclusion of the first gradient correction enhances the electron density in the vicinity of the proton by about 15% (for $r_s = 5$) over the corresponding value in the local density approximation, whereas the differences are no more than a couple of percents for large $r$. Shallow bound states are found to exist for values of $r_s$ larger than 2. The screening

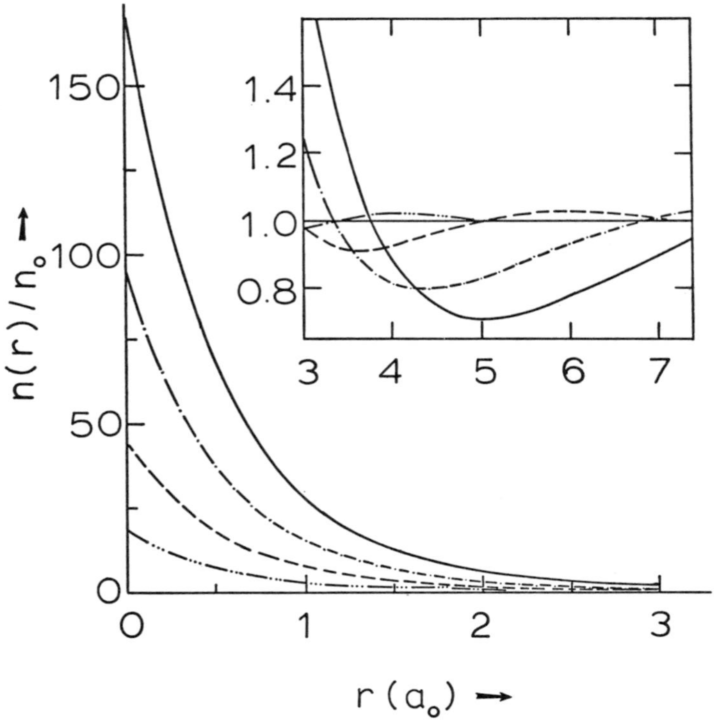


FIG. 15. Normalized electron density distribution $n(r)/n_0$ around a proton in the electron gas, at various vlaues of $r_s$ (in ascending order, $r_s$ = 2.07, 3, 4, and 5). The inset shows the Friedel oscillations on a magnified scale. [From P. Jena and K. S. Singwi, *Phys. Rev. B: Condens. Matter* [3] **17**, 3518 (1978).]

charge profiles around the proton have been used in *a priori* calculations of the heat of solution of hydrogen in Al and Mg, with results in good agreement with experiment.

## 9. Spin Correlations

It is well known that the enhancement of the paramagnetic susceptibility of an interacting electron gas over its Pauli value is due to short-range Coulomb and exchange effects.[125] Early calculations of this enhancement were carried our[126] within the framework of the Landau theory of Fermi liquids. Here, we wish to outline a self-consistent theory, although approximate, for the frequency and wavenumber–dependent paramagnetic response $\chi_M(k, \omega)$. The development of this theory proceeds on lines quite analogous to the extension of the STLS theory to a two-component plasma, discussed in Section 8. The two components now correspond to electrons with spin up and spin down, and the external perburbation is a magnetic field. The magnetic response of the electron gas in the paramagnetic state ($n_\uparrow = n_\downarrow$) is determined by the difference in the induced polarizations of spin-up and spin-down electrons. One arrives at the following expression[127] for the generalized magnetic susceptibility,

$$\chi_M(k, \omega) = -g_0^2\mu_B^2\chi_0(k, \omega)/[1 - I(k)\chi_0(k, \omega)], \tag{9.1}$$

where $g_0\mu_B$ is the magnetic moment of an electron, $\chi_0(k, \omega)$ is the Lindhard polarizability, and $I(k)$ is given by

$$I(k) = -\frac{2\pi e^2}{k^2}[G_{\uparrow\uparrow}(k) - G_{\uparrow\downarrow}(k)]. \tag{9.2}$$

The local field factor $G_{\sigma\sigma'}(k)$ is

$$G_{\sigma\sigma'}(k) = -\frac{1}{n}\int \frac{d\mathbf{q}}{(2\pi)^3}\frac{\mathbf{q}\cdot\mathbf{k}}{q^2}\gamma_{\sigma\sigma'}(|\mathbf{k} - \mathbf{q}|), \tag{9.3}$$

and the partial structure factors $\gamma_{\sigma\sigma'}(k)$ are related to the corresponding pair correlation functions by

$$g_{\sigma\sigma'}(r) = 1 + \frac{1}{n}\int \frac{d\mathbf{q}}{(2\pi)^3}\exp(-i\mathbf{q}\cdot\mathbf{r})\gamma_{\sigma\sigma'}(q). \tag{9.4}$$

[125] P. A. Wolff, *Phys. Rev.* [2] **120,** 814 (1960); M. Bailyn, *ibid.* **136,** A1321 (1964).

[126] D. Silverstein, *Phys. Rev.* [2] **128,** 631 (1963); **130,** 912 (1963); T. M. Rice, *Ann. Phys.* (*N.Y.*) **31,** 100 (1965); *Phys. Rev.* [2] **175,** 858 (1968); L. Hedin, *ibid.* **139,** A796 (1965); L. Hedin and S. Lundqvist, *Solid State Phys.* **23,**1 (1969).

[127] R. Lobo, K. S. Singwi, and M. P. Tosi, *Phys. Rev.* [2] **186,** 470 (1969).

The expression (9.1) for the dynamic susceptibility is the same as given earlier by Izuyama *et al.*[128] with the difference that the local field factor, instead of being a constant parameter of the theory, is a function of the wave number to be determined microscopically in a self-consistent manner through the use of the appropriate fluctuation-dissipation theorem. This allows one to determine $g_{\uparrow\uparrow}(r) - g_{\uparrow\downarrow}(r)$, which when combined with the knowledge of $g(r) = \frac{1}{2}[g_{\uparrow\uparrow}(r) + g_{\uparrow\downarrow}(r)]$ obtained from the dielectric problem, yields separately $g_{\uparrow\uparrow}(r)$ *and* $g_{\uparrow\downarrow}(r)$. The results[127] show that $g_{\uparrow\uparrow}(r)$ in the theory does not vanish exactly for $r \to 0$, implying a violation of the Pauli principle, The function $I(k)$ is found to drop rather gently with $k$ and becomes quite small for $k > 2k_F$—a result that is in reasonable agreement with experiment[129] for paramagnetic Ni.

The static paramagnetic susceptibility is given by

$$\chi_M = \lim_{k\to 0} \chi_M(k, 0) = \chi_F/(1 - \gamma k_{TF}^2/k_F^2), \tag{9.5}$$

where $\chi_F$ is the Pauli susceptibility and the enhancement factor $\gamma$ is

$$\gamma = -\frac{k_F^2}{4\pi e^2} \lim_{k\to 0} I(k). \tag{9.6}$$

As was the case for the dielectric problem the expressions (9.2) and (9.3) for $I(k)$ yield a poor value for the enhancement factor, in the sense that this value does not agree with the one calculated from the dependence of the ground state energy on magnetization. A considerable improvement is again obtained[50] by screening the interaction in the integrand of Eq. (9.3). These results were applied[130] to calculate the nuclear spin-lattice relaxation rates in alkali metals, with satisfactory results.

The theory of Lobo *et al.* was later extended[131] to account for the dependence of the pair correlation functions on magnetization, in a manner analogous to the extension of the STLS theory for the dielectric problem discussed in Section 5,b. Here too the value of the parameter $a$ was taken to be $\frac{2}{3}$. The results for $\chi_M/\chi_F$ together with the theoretical results of other authors are compared with experiment in Fig. 16. It is gratifying that all recent theoretical calculations of the susceptibility[132]

[128] T. Izuyama, D. J. Kim, and R. Kubo, *J. Phys. Soc. Jpn.* **18,** 1025 (1963).

[129] R. D. Lowde and C. G. Windsor, *Adv. Phys.* **19,** 813 (1970).

[130] P. Bhattacharyya, K. N. Pathak, and K. S. Singwi, *Phys. Rev. B: Solid State* [3] **3,** 1568 (1971).

[131] P. Vashishta and K. S. Singwi, *Solid State Commun.* **13,** 901 (1973).

[132] D. Hamann and A. W. Overhauser, *Phys. Rev.* [2] **143,** 183 (1966); R. Dupree and D. J. W. Geldart. *Solid State Commun.* **9,** 145 (1971); G. Pizzimenti, M. P. Tosi, and A. Villari, *Nuovo Cimento Lett.* **2,** 81 (1971); U. von Barth and L. Hedin, *J. Phys. C* [3] **5,** 1629 (1972).

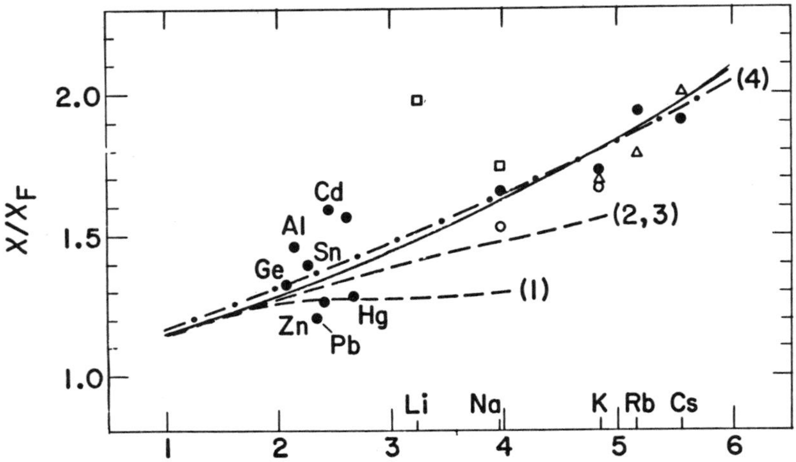


FIG. 16. Paramagnetic susceptibility of the electron gas, in units of the Pauli susceptibility $\chi_F$ versus $r_s$. Curves 1–3 give older results based on the Landau theory, cited in the text under Ref. 126; curve 4 is typical of recent theoretical results, cited in the text under Ref. 132; the solid curve gives the limiting values of the dynamic susceptibility obtained in the self-consistent theory. Experimental values: squares, from conduction electron spin resonance data, corrected for band mass [R. T. Schumacher and C. O. Slichter, *Phys. Rev.* [2] **101,** 58 (1956)]; empty circles, from spin wave data [L. Hedin, *Phys. Rev.* [2] **139,** A796 (1965)]; triangles, from Knight shift data in alloys, corrected for band mass [J. A. Kaeck, *Phys. Rev.* [2] **175,** 897 (1968)]; full circles, from total susceptibility [R. Dupree and D. J. W. Geldart, *Solid State Commun.* **9,** 145 (1971)]. [From P. Vashishta and K. S. Singwi, *Solid State Commun.* **13,** 901 (1973).]

agree to within a few percent and are in good agreement with experiment. It will be seen from Fig. 16, however, that there is a large discrepancy between theory and experiment in the case of Li. This has been accounted for by Vosko *et al.*[133] in a calculation that includes lattice effects by the HKS density-functional formalism. The same calculation shows that these effects are quite small in the other alkali metals.

There is, of course, a wider interest[134] in the study of an inhomogeneous spin-polarized electron gas by the HKS formalism in connection, for instance, with surface properties of magnetic systems[135] and with the

[133] S. H. Vosko, J. P. Perdew, and A. H. MacDonald, *Phys. Rev. Lett.* **35,** 1725 (1975).

[134] J. C. Stoddart and N. H. March, *Ann. Phys.* (*N.Y.*) **64,** 174 (1970); U. von Barth and L. Hedin, *J. Phys. C* [3] **5,** 1629 (1972); A. K. Rajagopal and J. Callaway, *Phys. Rev. B: Solid State* [3] **7,** 1912 (1973); D. Gunnarson, B. I. Lundqvist, and J. W. Wilkins, *ibid.* **10,** 1319 (1974).

[135] R. L. Kautz and B. B. Schwartz, *Phys. Rev. B: Solid State* [3] **14,** 1560 (1977).

calculation of the net magnetic field seen by a muon in a magnetic metal.[136] At present these calculations are done in the local density approximation for which a knowledge of the exchange-correlation energy as a function of local density and magnetization is sufficient. A refined knowledge of the wavenumber–dependent local fields in a spin-polarized system will permit the inclusion of the first gradient correction terms.

## 10. Two-Dimensional Electron Fluid

Interest in electron gases in spaces with dimensionality different from three stems partly from the realization that much insight can be gained by studying the physical properties of a system as a function of space dimensionality. More importantly, there are physical situations where a schematization in terms of a quasi-two-dimensional or even a quasi-one-dimensional electron gas is applicable. For instance, electrons trapped at the surface of liquid helium, or in an inversion layer in a metal–insulator–semiconductor system form a quasi-two-dimensional electron gas. The latter system in particular has been extensively studied[137] both experimentally and theoretically.

The inversion layer is realized by the application of a voltage to the metal, which bends the bands near the surface of the semiconductor inducing a surface layer where the majority carrier is the minority carrier in the bulk. An important advantage of this device is that the carrier density in the layer can be varied over a wide range by simply varying the voltage. If the field is sufficiently strong, the carriers are confined in a narrow, approximately triangular well leading to quantization[138] of motion in the $z$ direction. Corresponding to each of these quantized states, one has a band of energies varying continuously with $k$ parallel to the surface (surface subband). At low temperatures the carriers reside mostly in the lowest subband. There is ample evidence[139] to show that the dynamics of the carriers in this system is indeed two-dimensional.

[136] K. G. Petzinger and R. Munjal, *Phys. Rev. B: Solid State* [3] **3,** 1560 (1977); P. Jena, K. S. Singwi, and R. M. Nieminen, *Phys. Rev. B: Condens. Matter* [3] **17,** 301 (1978).

[137] G. Dorda, *Festkoerperprobleme* **13,** 215 (1973); F. Stern, *CRC Crit. Rev. Solid State Sci.* **4,** 499 (1974); F. Stern, *Surf. Sci.* **58,** 333 (1976), and other papers in the same journal issue and in *Surf. Sci.* **73** (1978); N. F. Mott, M. Pepper, S. Pollitt, R. H. Wallis, and C. J. Adkins, *Proc. R. Soc. London, Ser. A* **345,** 169 (1975). For a review of electronic trapping at the surface of liquid He see M. W. Cole, *Rev. Mod. Phys.* **46,** 451 (1974).

[138] J. R. Schrieffer, *in* "Semiconductor Surface Physics" (R. H. Kingston, ed.), p. 55. Univ. of Pennsylvania Press, Philadelphia, 1957.

[139] A. B. Fowler, F. F. Fang, W. E. Howard, and P. J. Stiles, *Phys. Rev. Lett.* **16,** 901 (1966); F. F. Fang and P. J. Stiles, *Phys. Rev.* [2] **174,** 823 (1968).

For definiteness we shall consider the case in which the carriers are electrons at a [001]Si–$SiO_2$ interface. The conduction band of Si consists of six equivalent valleys with longitudinal effective mass $m_l = 0.98m$ and transverse effective masses $m_t = 0.19m$. The electric field splits the sixfold degeneracy into a twofold state with $m_l$ perpendicular to the surface and a fourfold state with $m_t$ perpendicular to the surface. From each of these two states, a ladder of $z$-quantized states are obtained with energies

$$E_i(k) = E_i + \frac{\hbar^2 k_x^2}{2m_x} + \frac{\hbar^2 k_y^2}{2m_y} \tag{10.1}$$

where $E_i$ are the energies normal to the surface and (for the lowest-lying ladder) $m_x = m_y = m_t$. The corresponding wavefunction is

$$\psi_i(x, y, z) = \zeta_i(z)\exp(ik_x x + ik_y y), \tag{10.2}$$

where $\zeta_0(z)$ can be determined by solving[140] the Schrödinger equation self-consistently with Poisson's equation. An approximate variational form for $\zeta_0(z)$ in the extreme quantum limit is

$$\zeta_0(z) = (b^3/2)^{1/2} z \exp(-zb/2). \tag{10.3}$$

A number of calculations of the exchange and correlation effects have been carried out for this system in the extreme quantum limit in recent years, since the early RPA treatment of Stern.[141] One extreme schematization is to assume that the electrons move in an infinitely thin sheet, interacting via the Coulomb potential $e^2/r$. Effects of finite thickness can easily be incorporated in an approximation in which the electron dynamics is still confined to two dimensions but the wavefunction (10.3) enters to determine an effective two-dimensional potential by a suitable average over the third dimension. This effective potential[142] reduces to $e^2/r$ for $r \to \infty$ (where $r$ is the separation between two electrons in the plane) but tends to $b \ln(2/\gamma br)$, where $\gamma$ is Euler's constant for $r \to 0$, thus taking in the latter limit a form similar to that for an infinite string of charges.

In the two-dimensional case the exchange energy per electron is[143]

$$\varepsilon_x = -\frac{8}{3\pi r_s}\left(\frac{2}{n_v}\right)^{1/2} \text{Ry}^*, \tag{10.4}$$

where $n_v$ is the degeneracy of the subband, $r_s$ is related to the electron density by $n^{-1} = \pi r_s^2 a_B^{*2}$ with $a_B^* = \varepsilon\hbar^2/(m^* e^2)$, and $\text{Ry}^* = m^* e^4/(2\varepsilon^2\hbar^2)$.

[140] F. Stern and W. E. Howard, *Phys. Rev.* [2] **163,** 816 (1967).
[141] F. Stern, *Phys. Rev. Lett.* **18,** 546 (1967).
[142] M. Jonson, *J. Phys. C* [3] **9,** 3055 (1976).
[143] F. Stern, *Phys. Rev. Lett.* **30,** 278 (1973).

The correlation energy in the high-density limit is[144]

$$\varepsilon_c = -0.38 - 0.17\, r_s \ln r_s + \cdots. \tag{10.5}$$

At finite density (up to $r_s = 16$), Jonson[142] has calculated the pair correlation function and the correlation energy in various approximations, i.e., the RPA and the Hubbard and STLS approximations. The general result is that the inclusion of short-range correlations is much more important in the two-dimensional case than in the three-dimensional case, and that taking account of the finite thickness of the layer reduces somewhat their effects. Jonson's conclusion is that it is crucial for realistic many-body calculations to include the finite extent of the inversion layer and to go beyond the Hubbard approximation for $r_s \gtrsim 4$.

A remarkable property of the collective plasmon excitation in the two-dimensional electron gas is that its frequency does not tend to a constant value with $k \to 0$, but vanishes according to the law[141,145]

$$\omega_p^2(k) = (2\pi n e^2/m)k[1 + \tfrac{3}{4}(k/me^2)] + \cdots. \tag{10.6}$$

This behavior has been verified experimentally for electrons on liquid helium.[146] The coefficient of the $k^2$ term in Eq. (10.6) is the RPA results. Effects of exchange and correlation on the dispersion curves have been evaluated by Rajagopal[147] and by Jonson.[142] Figure 17 shows the plasmon dispersion curve as calculated in various approximation.

Landau theory parameters for a two-dimensional electron gas, such as the many-body enhancement of the effective mass and the susceptibility, have been calculated by several authors.[148] These effects on the effective mass are quite sizable as compared to the three-dimensional case, in agreement with experiment. The existence of subbands has been verified experimentally[149] by direct spectroscopic measurements, and there now exist accurate measurements of subband splittings. Calculations[150] of the

[144] A. K. Rajagopal and J. C. Kimball, *Phys. Rev. B: Solid State* [3] **15**, 2819 (1977).

[145] A. B. Chaplik, *Sov. Phys.—JETP (Engl. Transl.)* **33**, 997 (1971).

[146] C. C. Grimes and G. Adams, *Phys. Rev. Lett.* **36**, 145 (1976).

[147] A. K. Rajagopal, *Phys. Rev. B: Solid State* [3] **15**, 4264 (1977). Also, A. Czachor, A. Holas, S. Sharma and K. S. Singwi have recently studied in the first-order perturbation theory the dynamical correlations in a two-dimensional electron liquid (*Phys. Rev. B* to be published).

[148] C. S. Ting, T. K. Lee, and J. J. Quinn, *Phys. Rev. Lett.* **34**, 870 (1975); T. K. Lee, C. S. Ting, and J. J. Quinn, *Solid State Commun.* **16**, 1309 (1975); *Phys. Rev. Lett.* **35**, 1048 (1975); B. Vinter, *Phys. Rev. Lett.* **35**, 1044 (1975).

[149] See, for instance, F. Neppl, J. P. Kotthaus, and J. F. Koch, *Phys. Rev. B: Solid State* [3] **16**, 1519 (1977).

[150] See, for instance, B. Vinter, *Phys. Rev. B: Solid State* [3] **13**, 447 (1976); **15**, 3947 (1977); T. Ando, *ibid.* **13**, 3468 (1976).

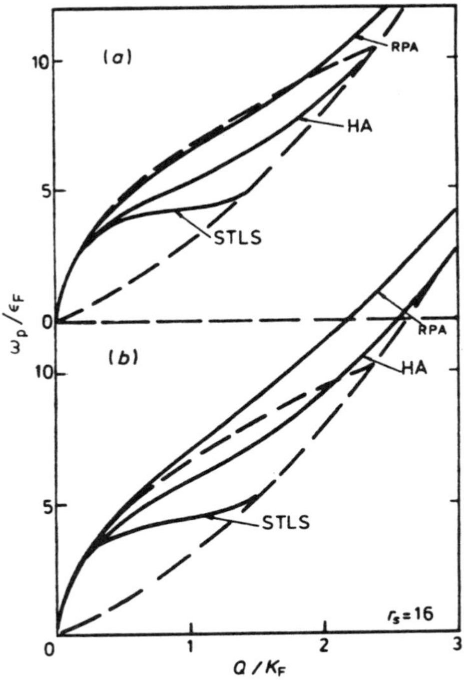


FIG. 17. Plasmon dispersion curves for a two-dimensional electron gas for $r_s = 16$ in various approximations. Part (b) shows the effect of the finite thickness of the layer. The broken curves give the onset of the particle–hole continuum and the limiting law $\omega_p^2(k) = (2\pi ne^2/m)k$. [From M. Jonson, *J. Phys. C* **9**, 3055 (1976).]

many-body effects on the electron energy levels are in good agreement with observations.

## 11. Dielectric Response of Semiconductors and Insulators

We have briefly mentioned in Section 6,a that a dielectric matrix formulation is needed in principle to deal with the response of an electron gas in a lattice. Important physical consequences, which have been the subject of a number of investigations in recent years, derive from the matrix nature of the response in a semiconductor or an insulator. To indicate one such consequence, we recall that the static (electronic) dielectric function of a strong insulator, in the extreme tight binding limit where the crystal can be broken up into distinct polarizable units, obeys the Lorentz–Lorenz (or Clausius–Mossotti) formula,

$$\lim_{k\to 0} \varepsilon(k, 0) = 1 + 4\pi\alpha/(1 - (4\pi/3)\alpha), \qquad (11.1)$$

where $\alpha$ is the polarizability of the lattice cell. The denominator in the above expression arises from the difference between the local field seen by a localized electron cloud, due to the polarization of all the *other* electron clouds, and the Hartree field, due to the *total* polarization. Obviously, the latter field includes the interaction of each localized cloud with itself. This introduces an error that (for $k \to 0$) becomes of order $1/V$ and is thus negligible only in the opposite limit where the electrons are completely delocalized.

The relation between local field corrections in an insulator and the matrix character of the response was clarified in early work of Adler and Wiser.[151] We introduce the density response matrix by writing the polarization created by a weak external field of wave vector $\mathbf{q} + \mathbf{G}$, where $\mathbf{G}$ is a vector of the reciprocal lattice, and $\mathbf{q}$ is confined to lie in the Brillouin zone, as

$$\delta n(\mathbf{r}, \omega) = \int d\mathbf{r}' \, \chi(\mathbf{r}, \mathbf{r}'; \omega) V_e(\mathbf{q} + \mathbf{G}, \omega) \exp[i(\mathbf{q} + \mathbf{G}) \cdot \mathbf{r}']$$
$$= \sum_{\mathbf{G}'} \chi_{\mathbf{G}'\mathbf{G}}(\mathbf{q}, \omega) V_e(\mathbf{q} + \mathbf{G}, \omega) \exp[i(\mathbf{q} + \mathbf{G}') \cdot \mathbf{r}], \tag{11.2}$$

having set

$$\chi(\mathbf{r}, \mathbf{r}'; \omega) = \sum_{\mathbf{q}} \sum_{\mathbf{G},\mathbf{G}'} \chi_{\mathbf{G}\mathbf{G}'}(\mathbf{q}, \omega) \exp[i(\mathbf{q} + \mathbf{G}) \cdot \mathbf{r} - i(\mathbf{q} + \mathbf{G}') \cdot \mathbf{r}']. \tag{11.3}$$

The microscopic Hartree potential is

$$V_{\mathrm{H}}(\mathbf{r}, \omega) = V_e(\mathbf{r}, \omega) + \int d\mathbf{r}' \, v(\mathbf{r} - \mathbf{r}') \delta n(\mathbf{r}', \omega), \tag{11.4}$$

and the inverse dielectric matrix $\varepsilon^{-1}_{\mathbf{G}\mathbf{G}'}(\mathbf{q}, \omega)$ is introduced by relating the microscopic Hartree field to the external field through

$$V_{\mathrm{H}}(\mathbf{q} + \mathbf{G}, \omega) = \sum_{\mathbf{G}'} \varepsilon^{-1}_{\mathbf{G}\mathbf{G}'}(\mathbf{q}, \omega) V_e(\mathbf{q} + \mathbf{G}', \omega), \tag{11.5}$$

yielding

$$\varepsilon^{-1}_{\mathbf{G}\mathbf{G}'}(\mathbf{q}, \omega) = \delta_{\mathbf{G}\mathbf{G}'} + \frac{4\pi e^2}{|\mathbf{q} + \mathbf{G}|^2} \chi_{\mathbf{G}\mathbf{G}'}(\mathbf{q}, \omega). \tag{11.6}$$

Finally, the proper microscopic polarizability is introduced by relating the polarization to the Hartree potential through

$$n(\mathbf{q} + \mathbf{G}, \omega) = \sum_{\mathbf{G}'} \tilde{\chi}_{\mathbf{G}\mathbf{G}'}(\mathbf{q}, \omega) V_{\mathrm{H}}(\mathbf{q} + \mathbf{G}', \omega), \tag{11.7}$$

[151] S. Adler, *Phys. Rev.* [2] **126,** 413 (1962); N. Wiser, *ibid.* **129,** 62 (1963).

which yields

$$\chi_{\mathbf{GG}'}(\mathbf{q}, \omega) = \sum_{\mathbf{G}''} \tilde{\chi}_{\mathbf{GG}''}(\mathbf{q}, \omega)\varepsilon^{-1}_{\mathbf{G}''\mathbf{G}'}(\mathbf{q}, \omega) \tag{11.8}$$

and

$$\varepsilon_{\mathbf{GG}'}(\mathbf{q}, \omega) = \delta_{\mathbf{GG}'} - \frac{4\pi e^2}{|\mathbf{q} + \mathbf{G}|^2} \tilde{\chi}_{\mathbf{GG}'}(\mathbf{q}, \omega). \tag{11.9}$$

All the above relations are obvious extensions of the definitions that hold for a homogeneous electron gas.

To make contact with the macroscopic dielectric function such as given in Eq. (11.1), we must suppose that the external field varies slowly in space ($\mathbf{G} = 0$, $\mathbf{q} \to 0$) and average the microscopic quantities over the unit cell. In particular, the macroscopic Hartree potential at $\mathbf{R}$, $\bar{V}_{\mathrm{H}}(\mathbf{R}, \omega)$, is evaluated as the average of $V_{\mathrm{H}}(\mathbf{r}, \omega)$ over the lattice cell $\Delta_{\mathbf{R}}$ centered at $\mathbf{R}$, with the result

$$\begin{aligned}
\bar{V}_{\mathrm{H}}(\mathbf{R}, \omega) &= \frac{1}{\Delta} \int_{\Delta_{\mathbf{R}}} d\mathbf{r}\ V_{\mathrm{H}}(\mathbf{r}, \omega) \\
&= \frac{1}{\Delta} \sum_{\mathbf{q},\mathbf{G}} V_{\mathrm{H}}(\mathbf{q} + \mathbf{G}, \omega) \int_{\Delta_{\mathbf{R}}} d\mathbf{r} \exp[-i(\mathbf{q} + \mathbf{G})\cdot\mathbf{r}] \\
&\to \sum_{\mathbf{q}} V_{\mathrm{H}}(\mathbf{q}, \omega) \exp(-i\mathbf{q}\cdot\mathbf{R})
\end{aligned} \tag{11.10}$$

namely,

$$\bar{V}_{\mathrm{H}}(\mathbf{q}, \omega) = V_{\mathrm{H}}(\mathbf{q} + \mathbf{G}, \omega)|_{\mathbf{G}=0} = \varepsilon^{-1}_{00}(\mathbf{q}, \omega)V_{\mathrm{e}}(\mathbf{q}, \omega). \tag{11.11}$$

The dielectric function of the system, defined as usual as the ratio between the external potential and the macroscopic Hartree potential, is thus given[150] by the reciprocal of the $\mathbf{G} = \mathbf{G}' = 0$ element of the inverse dielectric matrix,

$$\varepsilon(\mathbf{q}, \omega) = 1/\varepsilon^{-1}_{00}(\mathbf{q},\omega). \tag{11.12}$$

A similar calculation for the macroscopic polarization yields

$$\begin{aligned}
\bar{n}(\mathbf{q}, \omega) &= n(\mathbf{q} + \mathbf{G}, \omega)|_{\mathbf{G}=0} = \sum_{\mathbf{G}'} \tilde{\chi}_{0\mathbf{G}'}(\mathbf{q}, \omega)V_{\mathrm{H}}(\mathbf{q} + \mathbf{G}', \omega) \\
&= \frac{\sum_{\mathbf{G}'} \tilde{\chi}_{0\mathbf{G}'}(\mathbf{q}, \omega)\varepsilon^{-1}_{\mathbf{G}'0}(\mathbf{q}, \omega)}{\varepsilon^{-1}_{00}(\mathbf{q}, \omega)} \bar{V}_{\mathrm{H}}(\mathbf{q}, \omega).
\end{aligned} \tag{11.13}$$

This expression embodies the local field corrections arising from the microscopic components of the Hartree field and is obviously different from the result $\bar{n}(\mathbf{q}, \omega) = \tilde{\chi}_{00}(\mathbf{q}, \omega)\bar{V}_{\mathrm{H}}(\mathbf{q}, \omega)$ that one would obtain upon neglect of the matrix character of the response.

In the RPA, one would naturally start the evaluation of the dielectric response from Eq. (11.9), by replacing the proper polarizability by the independent electron polarizability $\chi^0_{GG'}(\mathbf{q}, \omega)$. One is then faced with the problem of evaluating $\chi^0_{GG'}$ from the single-particle wavefunctions and energies, as well as with the problem of matrix inversion implied by Eq. (11.12). The latter problem can be formally overcome[152] by expanding the Bloch states into Wannier functions (or into tight-binding functions, in practice), since this leads to an exact factorization of the matrix $\chi^0_{GG'}$. We refer the reader for a full discussion to the review of Sham[153], and report here for illustrative purposes the results for the simplest model of a strong insulator, possessing just one valence band and one conduction band, each described by one strongly localized Wannier function. One finds for such a model

$$\chi^0_{GG'}(\mathbf{q}, \omega) = F_{vc}(\mathbf{q} + \mathbf{G})\gamma(\mathbf{q}, \omega)F^*_{vc}(\mathbf{q} + \mathbf{G}'), \tag{11.14}$$

where

$$F_{vc}(\mathbf{k}) = \int d\mathbf{r} \exp(-i\mathbf{k}\cdot\mathbf{r})w_v(\mathbf{r})w_c(\mathbf{r}) \tag{11.15}$$

and

$$\gamma(\mathbf{q}, \omega) = \sum_{\mathbf{k}} \left[\frac{n^v_{\mathbf{k}} - n^c_{\mathbf{k}+\mathbf{q}}}{\hbar\omega - \varepsilon_c(\mathbf{k} + \mathbf{q}) + \varepsilon_v(\mathbf{k}) + i\eta} - \frac{n^v_{\mathbf{k}+\mathbf{q}} - n^c_{\mathbf{k}}}{\hbar\omega - \varepsilon_v(\mathbf{k} + \mathbf{q}) + \varepsilon_c(\mathbf{k}) + i\eta}\right] \tag{11.16}$$

Here, $w(\mathbf{r})$, $n_{\mathbf{k}}$, and $\varepsilon(\mathbf{k})$ are the Wannier function, the occupation number, and the single-particle energy, respectively, with indices that refer to the two bands. Matrix inversion is immediate and yields

$$\chi^{RPA}_{GG'}(\mathbf{q}, \omega) = \chi^0_{GG'}(\mathbf{q}, \omega)\Big/\left\{1 - \frac{4\pi e^2}{q^2}[1 - G_{RPA}(\mathbf{q})\chi^0_{00}(\mathbf{q}, \omega)]\right\} \tag{11.17}$$

and

$$\varepsilon_{RPA}(\mathbf{q}, \omega) = 1 - \frac{4\pi e^2}{q^2}\chi^0_{00}(\mathbf{q}, \omega)\Big/\left[1 + \frac{4\pi e^2}{q^2}G_{RPA}(\mathbf{q})\chi^0_{00}(\mathbf{q}, \omega)\right], \tag{11.18}$$

[152] S. K. Sinha, *Phys. Rev.* [2] **177,** 1256 (1969); E. Hayashi and M. Shimizu, *J. Phys. Soc. Jpn.* **26,** 1396 (1969); R. Pick, *in* "Phonons" (M. A. Nusimovici, ed.), p. 20. Flammarion Press, Paris, 1971. L. J. Sham, *Phys. Rev. B: Solid State* [3] **6,** 3584 (1972); W. Hanke, *ibid.* **8,** 4585 (1973); W. Hanke and L. J. Sham, *Phys. Rev. Lett.* **33,** 582 (1974).

[153] L. J. Sham, *in* "Elementary Excitations in Solids, Molecules and Atoms" (J. T. Devreese, A. B. Kunz, and T. C. Collings, eds.). Plenum, New York, 1974.

where

$$G_{\mathrm{RPA}}(\mathbf{q}) = -\sum_{\mathbf{G}\neq 0} \frac{q^2}{|\mathbf{q} + \mathbf{G}|^2} |F_{\mathrm{vc}}(\mathbf{q} + \mathbf{G})|^2 / |F_{\mathrm{vc}}(\mathbf{q})^2. \tag{11.19}$$

For the same model one has

$$\lim_{k\to 0} \left[ -\frac{e^2}{q^2} \chi^0_{00}(\mathbf{q}, 0) \right] = \alpha, \tag{11.20}$$

a constant because of the orthogonality of the Wannier functions and of the gap in the single-particle spectrum. Comparison with the Lorentz–Lorenz expression (11.1) then shows that, since $G_{\mathrm{RPA}} \leq 0$, the local field correction has the wrong sign in this approximation. The inclusion of exchange and correlation effects is thus crucial in a mean-field theory of the dielectric response of an insulator, both in the static case[154] and in the dynamic case,[155] where these terms are responsible for the electron–hole interaction leading to excitonic states.

The dielectric matrix formalism has found important applications in the *a priori* theory of lattice dynamics in insulators and semiconductors. For a good insulator, the microscopic polarization of the valence electrons localized at the atomic sites is in essence equivalent to a set of localzed atomic dipoles, so that the dielectric theory can be used[152] to provide a derivation and a generalization of the empirical shell model for lattice dynamics. The problem is a more subtle one in the case of a semiconductor. Let us first note that, if one starts with bare ions of charge $Ze$ and $Z$ valence electrons per atom, and if one were to neglect the matrix character of the response, one would find a longitudinal mode that has optic rather than acoustic character, since the dielectric function goes to a constant value for $q \to 0$ instead of diverging as $q^{-2}$. Empirically, one could get around this difficulty by inserting additional charges in the cell so as to neutralize the residual charge $Ze/\sqrt{\varepsilon}$ on each ion. The location of these extra charges is also crucial, however, since noncentral forces are needed for stability of the transverse acoustic modes. The "bond charge" model,[156] reasonably enough, locates the extra charges at the centers of the interatomic bonds.

[154] S. K. Sinha, R. P. Gupta, and D. L. Price, *Phys. Rev. Lett.* **26,** 1324 (1971); *Phys. Rev. B: Solid State* [3] **4,** 2564, 2573 (1974).

[155] P. V. Giaquinta, M. Parrinello, E. Tosatti, and M. P. Tosi, *J. Phys. C* [3] **9,** 2031 (1976); P. V. Giaquinta, E. Tosatti, and M. P. Tosi, *Solid State Commun.* **19,** 123 (1976).

[156] J. C. Phillips, *Phys. Rev.* [2] **166,** 832 (1968); **168,** 917 (1968); R. M. Martin, *ibid.* **186,** 871 (1969).

The above model has found a basic justification in the dielectric matrix formalism. A detailed analysis[157] of static screening for $q \to 0$ shows that the elements $\varepsilon_{0G}^{-1}(\mathbf{q})$ of the inverse dielectric matrix diverge as $q^{-1}$ in this limit, and that these singular terms contribute a Coulomb-like term to the dynamical matrix which must cancel the residual Coulomb term from diagonal screening. Explicit calculations of the dielectric matrix for Si yield phonon dispersion curves[154,158] which have the correct physical behavior and are in reasonable agreement with experiment. The same dielectric matrix has been used[159] to evaluate the distribution of valence electrons in a linear-response pseudopotential scheme, showing that the theory leads to a pile-up of bonding charge between the ions.

Finally, we would also like to point out an implication of the dielectric matrix formalism on the behavior of the plasmon in crystal lattices, in addition to the obvious lattice effects which are a frequency shift and a damping (even at $\mathbf{q} = 0$, because of local field terms and of Umklapp decays into electron–hole paris), and anisotropy effects. The microscopic resonance condition can be obtained from Eq. (11.5) by requesting that the microscopic Hartree field be different from zero at the plasmon frequency for $V_e = 0$, and is

$$\det[\varepsilon_{GG'}(\mathbf{q}, \omega)] = 0. \tag{11.21}$$

This condition implies that the plasmon becomes a Bloch excitation,[160] which, compatibly with damping, presents gaps at the zone boundary. Explicit calculations[161] indicate that these effects are negligible in most simple metals, but may be observable in semiconductors.

The calculation of dielectric matrices from realistic band structures has been attempted so far for a few materials only.[162] Applications include the

[157] L. J. Sham, *Phys. Rev.* [2] **188,** 1431 (1969); R. Pick, M. H. Cohen, and R. M. Martin, *Phys. Rev. B: Solid State* [3] **1,** 910 (1970).

[158] C. M. Bertoni, V. Bortolani, C. Calandra, and E. Tosatti, *Phys. Rev. Lett.* **28,** 1578 (1972); *Phys. Rev. B: Solid State* [3] **9,** 1710 (1974); see also A. Baldereschi and K. Maschke, *in* "Lattice Dynamics" (M. Balkanski, ed.), p. 36. Flammarion Sciences, Paris, 1978.

[159] E. Tosatti, C. Calandra, V. Bortolani, and C. M. Bertoni, *J. Phys. C* [3] **5,** L299 (1972).

[160] N. H. March and M. P. Tosi, *Proc. R. Soc. London, Ser. A* **330,** 373 (1972); W. M. Saslow and G. F. Reiter, *Phys. Rev. B: Solid State* [3] **7,** 2995 (1973); K. C. Pandy, P. M. Platzman, P. Eisenberger, and E. M. Foo, *ibid.* **9,** 5046 (1974).

[161] R. Girlanda, M. Parrinello, and E. Tosatti, *Phys. Rev. Lett.* **36,** 1386 (1976).

[162] J. A. Van Vechten and R. M. Martin, *Phys. Rev. Lett.* **28,** 446 (1972); S. G. Louie, J. R. Chelikovsky, and M. L. Cohen, *ibid.* **34,** 155 (1975); M. E. Brener, *Phys. Rev. B: Solid State* [3] **12,** 1487 (1975); S. P. Singhal, *ibid.* p. 564; A. Baldereschi and E. Tosatti, *Phys. Rev. B: Condens. Matter* [3] **17,** 4710 (1978); see also A. Baldereschi and E. Tosatti, *Solid State Commun.* **29,** 131 (1979).

calculation of optical properties,[163] of microscopic local fields and screening of charged impurities,[164] and of collective excitations in layered structures[165] and in quasi-one-dimensional conductors.[166]

ACKNOWLEDGMENTS

The authors wish to express their sincere thanks to Alf Sjölander for a critical reading of the manuscript and for many suggestions.

## Note Added in Proof

Vashishta *et al.*[120] have also calculated the ground state energy as a function of $r_s$ for different values of the mass ratio $m_h/m_e$ for this model system. Some interesting results emerge from this calculation: (1) In the STLS scheme one does not obtain a metallic binding, at least up to $m_h/m_e = 6$, and the value of the energy minimum almost remains the same ($-0.99E_x$) with increasing value of this ratio. (2) In the Hubbard approximation one does obtain a metallic binding for $m_h/m_e > 10$. (3) The $r_s$ value at which the minimum in $\varepsilon_0$ occurs shifts very slightly to lower values with increasing mass ratio. From the trend of these results, Vashishta *et al.* were led to conjecture that metallic hydrogen ($m_H/m_e \sim 2000$) in the liquid phase might exist for $r_s$ around 1.8. This conjecture seems to be borne out by a very recent calculation of Chakravarty and Ashcroft[167] who have obtained a variational upper bound to the ground state energies of metallic hydrogen and concluded that the possibility of a $T = 0$ liquid metallic phase can not be ruled out for $r_s \simeq 1.64$.

[163] W. Hanke, *Festkoerperprobleme* (*Adv. Solid State Phys.* **19** (1979).
[164] A. Baldereschi, R. Car, and E. Tosatti, *Conf. Ser—Inst. Phys.* **43,** 1207 (1979); R. Car and A. Selloni, *Phys. Rev. Lett.* **42,** 1365 (1979).
[165] A. J. Holden and J. C. Inkson, *J. Phys. C* [3] **10,** 49; L251 (1978).
[166] P. F. Williams and A. N. Bloch, *Phys. Rev. B: Solid State* [3] **9,** 1246 (1974); G. Guitliani, E. Tosatti, and M. P. Tosi, *Nuovo Cimento Lett.* **16,** 385 (1976).
[167] S. Chakravarty and N. W. Ashcroft, *Phys. Rev.* **B18,** 4588 (1978).

# Author Index

Numbers in parentheses are reference numbers and indicate that an author's work is referred to although his name is not cited in the text.

## C

**H**

**Q**



**R**

## S

# Subject Index

**F**